AF595458

Article

Comparison of Methods to Select Candidates for High-Density Genotyping; Practical Observations in a Cattle Breeding Program

Rudi A. McEwin [1,*], Michelle L. Hebart [1], Helena Oakey [2], Rick Tearle [1], Joe Grose [3], Greg Popplewell [4] and Wayne S. Pitchford [1]

1 Davies Livestock Research Centre, School of Animal and Veterinary Sciences, University of Adelaide, Roseworthy, SA 5371, Australia; michelle.hebart@adelaide.edu.au (M.L.H.); rick.tearle@adelaide.edu.au (R.T.); wayne.pitchford@adelaide.edu.au (W.S.P.)
2 Robinson Research Institute, Adelaide Medical School, University of Adelaide, North Adelaide, SA 5006, Australia; helena.oakey@adelaide.edu.au
3 3D Genetics Pty LTD, 939 Pukawidgi Rd, Bukkulla, NSW 2360, Australia; joe.grose@3dgenetics.com.au
4 Popplewell Genetics, 33 Tom Schmidt Court, Mount Samson, QLD 4520, Australia; greg@popplewell.com.au
* Correspondence: rudi.mcewin@adelaide.edu.au

Abstract: Imputation can be used to obtain a large number of high-density genotypes at the cost of procuring low-density panels. Accurate imputation requires a well-formed reference population of high-density genotypes to enable statistical inference. Five methods were compared using commercial Wagyu genotype data to identify individuals to produce a "well-formed" reference population. Two methods utilised a relationship matrix (MCG and MCA), two of which utilised a haplotype block library (AHAP2 and IWS), and the last selected high influential sires with greater than 10 progeny (PROG). The efficacy of the methods was assessed based on the total proportion of genetic variance accounted for and the number of haplotypes captured, as well as practical considerations in implementing these methods. Concordance was high between the MCG and MCA and between AHAP2 and IWS but was low between these groupings. PROG-selected animals were most similar to MCA. MCG accounted for the greatest proportion of genetic variance in the population (35%, while the other methods accounted for approximately 30%) and the greatest number of unique haplotypes when a frequency threshold was applied. MCG was also relatively simple to implement, although modifications need to be made to account for DNA availability when running over a whole population. Of the methods compared, MCG is the recommended starting point for an ongoing sequencing project.

Keywords: high density genotyping; imputation; sequencing; reference population

Citation: McEwin, R.A.; Hebart, M.L.; Oakey, H.; Tearle, R.; Grose, J.; Popplewell, G.; Pitchford, W.S. Comparison of Methods to Select Candidates for High-Density Genotyping; Practical Observations in a Cattle Breeding Program. *Agriculture* **2022**, *12*, 276. https://doi.org/10.3390/agriculture12020276

Academic Editor: Dongxiao Sun

Received: 19 January 2022
Accepted: 10 February 2022
Published: 15 February 2022

1. Introduction

Genomic selection [1] has been rapidly adopted by many breeding sectors following its successful introduction to the dairy industry. This is due to realised gains in prediction accuracy of genomic estimated breeding values that have increased the response to selection for key economic traits as greater proportions of genetic variation are explained and generation intervals are decreased [2–4].

In genomic selection, a sufficiently dense single nucleotide polymorphism (SNP) panel that covers the entire genome is utilised with the expectation that all quantitative trait loci (QTL) are in linkage disequilibrium with at least one SNP. This allows the prediction of QTL effects across the population over generations. For traits with few underlying QTL, lower density SNP panels may be sufficient to capture these effects, assuming close proximity of at least one SNP. However, where there are many underlying QTL, denser SNP panels may be required [2]. This is often the requirement for many traits in cattle breeding, such as fertility, where no QTL of major effect has been found, unlike milk fat percentage in Dairy [5]. Denser SNP panels have been shown to increase breeding value accuracy [6,7]. If there are many QTL of minor effect contributing to variation in a desired trait, a large number of phenotypic records will be required to achieve reasonable estimation accuracies relative to trait heritability [8].

With the size of the reference population clearly having an impact on the accuracy of genomic prediction in the target population, there is a clear need to identify cost-effective methods to procure more phenotypes. One solution would be to capitalise on the large numbers of phenotypes available in commercial herds using genotyping to replace often incomplete/missing pedigree data. However, this solution would be accompanied by high genotyping costs, which usually only nucleus herds have means for.

In 2010, the Illumina BovineHD chip became available with 777,962 SNPs, and now whole-genome sequencing is the new frontier [9,10]. However, the high price of sequencing and HD chips is a barrier to their application across large numbers of animals. Imputation can add value here. By investing in a good reference population of dense genotypes, imputation can then utilise cheaper, less dense SNP panels, which reduces the overall cost of genotyping while capitalising on high-density results. Given this, the key question is which animals should be densely genotyped to form the best reference set for imputation of sparsely genotyped animals? An ideal approach would be to select founder animals of the population, but the availability of this option is limited depending on population age (are the founders still alive/have DNA stored, i.e., semen). A second approach would be to select influential animals with large numbers of effective progeny. However, this may bias certain high-performing family groups by selecting relatives from a few family lines.

This work details the results and observations of an actual field trial to select candidates that represent the Australian Wagyu population for whole-genome sequencing to facilitate imputation to sequence for BLUP prediction. Four methods have been compared to just selecting highly influential sires herein that were expected to achieve high imputation accuracy to higher density/sequence arrays [11,12]. Strategies were compared that fall under two categories; (1) strategies that utilise relationship matrix data already available in routine BLUP and GBLUP analyses, and (2) Strategies that take a more bioinformatics approach based on population haplotype frequency. Measures of how efficiently animals were selected and similarities between animals selected are discussed as well as practical implications.

2. Materials and Methods

In total, five methods were trialed and compared to identify candidates for whole-genome sequencing in an Australian Wagyu population. Assuming a sequencing budget of $100,000 and sequencing per animal costing $1000, 100 animals were selected from each method. The first two methods are denoted the MCA and MCG method, respectively [11]. Candidates are selected for whole-genome sequencing through these methods by minimising the genetic variation of the target population relative to the selected candidates, improving imputation accuracy. The MCA method utilises (Wrights) numerator relationship matrix (**A**) such that;

$$\mathbf{A}_{11}{}^{*} = \mathbf{A}_{11} - \mathbf{A}_{12}\mathbf{A}_{22}{}^{-1}\mathbf{A}_{21} \quad (1)$$

where the 1 subscript denotes the set of target animals and 2 subscript denotes the set of animals selected to be sequenced. The diagonal elements of $\mathbf{A}_{11}{}^{*}$ are the residual variances that are expected to remain if sequence data were to be obtained from the selected individuals and used to predict/impute genotypes of the target set. Animals were selected using an iterative process which is described in full by [11]. Briefly, all animals start in the target set. The aim is to minimise *trace* ($\mathbf{A}_{11}{}^{*}$), each animal is tested to see who reduces *trace* ($\mathbf{A}_{11}{}^{*}$) the most, resulting in selected candidate *i*. After selecting animal *i*, the entire relationship matrix is made conditional on the genotype of that animal, then the process starts again, looking for the next animal that most reduces *trace* ($\mathbf{A}_{11}{}^{*}$) until the desired number of candidates are selected. An Australian Full-Blood Wagyu pedigree comprised of 10,549 individuals with a depth of up to 9 generations from the current generation was utilised to construct A through the R package pedigreemm [13]. For the animals in the pedigree that were genotyped (see below), the average number of great-grandparents recorded in the pedigree was 7.3 with a median value of 8, indicating a high level of pedigree completeness.

The second method utilises a genomic relationship matrix (**G**) in place of **A**, denoted the MCG method. Utilising genotype information on 5334 individuals genotyped with 30 K GGP-LD (Neogen: GeneSeek Operations) or Bovine VersaSNP 50 K (Weatherbys Scientific) chips, **G** was constructed as per VanRaden's first method [14]. Animals that were genotyped on the 50 K platform were imputed to 30 K using the 11,484 SNPs that overlapped between the chips. This decision was made due to the significantly larger reference population available on the 30 K chip (4940 vs. 394). Fimpute 2.2 [15] was used to perform the imputation. After, SNPs were kept that had a minor allele frequency greater than or equal to 0.05, retaining 21,094 SNPs for GRM construction. All genotyped animals were present in the pedigree resulting in an overlap of 5334 animals between the numerator (**A**) and genomic (**G**) relationship matrices.

The third and fourth methods were described by [12] and referred to as AHAP2 ([12] modified the AHAP method presented by [16]) and the inverse weight selection method (IWS). Both methods require the construction of a haplotype "block" library. This library was constructed utilising the 5334 post imputation genotypes to construct **G**, using FindHap v3 (http://aipl.arsusda.gov/software/findhap/; accessed on 21 May 2020). Program settings included 4 iterations at 3 haplotype block widths (50, 75 and 100 SNPs). Only the 100 SNP wide blocks were retained for analysis. Haplotype blocks, which by definition are non-overlapping, were assigned a unique ID and their frequency in the dataset was calculated. It was assumed that haplotype frequencies in this population are reflective of the Australian Wagyu industry. In total, 339,824 unique haplotypes were identified with a mean haplotype frequency of 0.07% and a minimum and maximum haplotype frequency of 0.005 and 0.28%, respectively.

The distribution of haplotype frequencies on the log scale (Figure 1) clearly indicated a skewed distribution towards lower frequencies. Due to exponential increases in haplotype counts at lower frequencies, haplotypes with a frequency lower than 0.1% were excluded from consideration. This brought the total number of haplotypes under consideration for sampling down to 20,854 of which 588 had a haplotype frequency $\leq$5% (Common), 3666 had a frequency $\geq$1% but <5% (Uncommon) and 16,600 had a haplotype frequency $\geq$0.1% but <1% (Rare). A haplotype frequency threshold of 0.1% was chosen to allow for 1 in 1000 error in genotype calls.

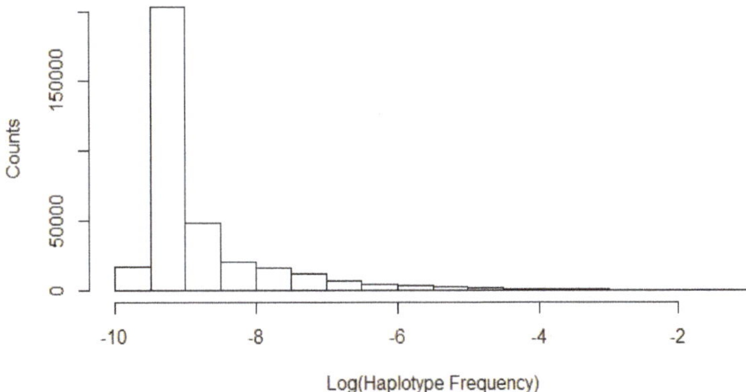

Figure 1. Distribution of haplotype block frequency (log scale) of 339,824 blocks, 100 SNPs in width, estimated from a population of 5334 genotyped Australian Wagyu.

Both the AHAP2 and IWS methods were designed to maximise the haplotype coverage from the population while minimising the redundancy of haplotype sampling. Both methods choose candidates to maximise the number of haplotypes sampled per dollar invested in sequencing, achieved through a weighting system, however, two separate approaches are used to achieve this.

The AHAP2 method, which is an iterative modification on the AHAP method described by [16], utilises the following equation;

$$\text{Sample weight} = \sum_{i=1}^{NHAP} f_i \quad \text{if } i = \text{homozygous}.$$

The frequency of the haplotype in the population is defined by f_i as determined by FindHap, and *NHAP* is the total number of haplotypes under consideration. Only haplotypes that are homozygous within a potential candidate are counted towards the weighting for selection. All individuals in the imputed genotype set were considered as potential candidates. After calculating the weight for all individuals, the individual with the highest weighting is selected as the sequencing priority. Once a candidate is chosen, all homozygous haplotypes that this candidate contained are removed from consideration for all remaining samples. Sample weights are then recalculated and the next sequencing candidate is selected until the desired number of candidates (n = 100) are sampled.

In reverse to the AHAP2 method, the IWS method preferentially selects candidates that carry rare frequency haplotypes. Ref. [12] developed an inverted parabolic function that calculated sequencing priority (weighting) under the following equation;

$$\text{Sample weight} = \sum_{i=1}^{NHAP} f_i^2 - 2f_i + 1 \quad \text{if } i = \text{homozygous}.$$

As f_i approaches 0, the haplotypes score approaches 1, increasing the weighting. More frequent haplotypes give an increasingly smaller weighting to the sample.

The final method is a more traditional approach that selects animals based on influence in pedigree. This was to assess a previous attempt to genotype animals that "describe" the population. Previously, 166 Full-Blood Wagyu animals were genotyped on the Illumina 770 K platform. These animals were selected as influential due to having greater than 10 progeny nationwide (PROG), with effective progeny numbers of 1 to 437, mean = 47, in the pedigreed population described herein. One hundred of the 166 animals were randomly chosen for comparison against the other methods were appropriate.

3. Results

3.1. Overlap between Chosen Candidates

The MCA and MCG methods had a high degree of similarity between them, with MCA selecting 70/100 individuals (Table 1) that were selected by MCG. A strong positive rank correlation of 0.82 (Figure 2) demonstrates high-rank concordance between the animals selected in common between the two methods. As MCA contains animals that are not available in MCG, a modified version of the MCA method was run (data not shown) where only the 5334 genotyped animals could be chosen but still relative to the whole pedigreed population, i.e., genotyped animals were selected based on their relationship to all animals in the pedigree. This produced similar results, with 73 animals being selected in common between MCA modified and MCG.

There was little overlap between the relationship matrices' methods and the haplotype methods AHAP2 and IWS (Table 1). For example, the specific animals themselves selected by IWS are all progeny or grand-progeny of those selected by MCG and/or MCA. There was a moderate similarity between animals selected by IWS and AHAP2. Differences are due to different emphasis weights on rare versus common haplotypes. It is important to reiterate that all methods used the same starting population of 5334 genotyped animals where appropriate (i.e., MCA utilised a much bigger pedigreed population). Additionally, all genotyped animals were in the pedigree.

Table 1. The degree of overlap, i.e., the number of animals selected in common, between the MCA, MCG, IWS, AHAP2, and PROG [A] methods. The number of animals sampled by each method is displayed on the diagonal.

	MCA	MCG	IWS	AHAP2	PROG
MCA	100				
MCG	70	100			
IWS	5	7	100		
AHAP2	2	4	61	100	
PROG	80	78	7	4	

[A] PROG in this instance refers to the full list of 166 Full-Blood Wagyu animals genotyped on the Illumina 770 K platform and the overlap between these 166 animals and selected candidates from other methods.

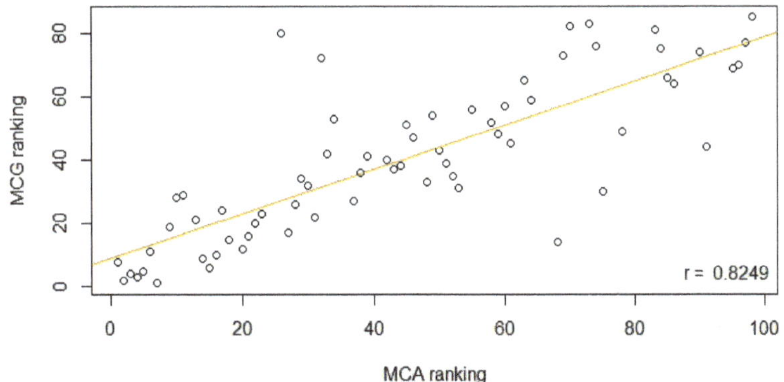

Figure 2. A plot of ranks of candidates selected for whole-genome sequencing using the MCA or MCG methods, respectively.

The animals selected by these four methods were then compared to the full list of 166 animals genotyped on 770 K due to being identified as influential sires (PROG). The MCA and MCG methods are most similar, in regards to animals selected, to this influential sire methodology, with an overlap of 80 and 78 animals, respectively (Table 1). As expected, this resulted in a lower overlap with IWS and AHAP2 methods.

3.2. *Percentage of Genetic Variance Explained*

The MCG method accounted for more genetic variance (34.6%) when 100 animals were selected compared to 30% when the MCA method was used. The first 20 selected candidates accounted for 19 and 21% of the genetic variance for the MCG and MCA method, respectively, with each additional animal thereafter contributing less information (Figure 3). Where the number of selected candidates was low, MCA outperformed the MCG method until approximately 30 candidates where MCG became superior. IWS was superior to AHAP2, accounting for 23.3% of the genetic variance compared to 22.9% when selecting 100 candidates, although both methods accounted for significantly less genetic variance compared to methods utilising a relationship matrix. For PROG, the mean percentage of genetic variance accounted for when randomly sampling 100 of the most influential sires for 5 replicates was 29.6% (SD = 0.40, data not shown), equivalent to the MCA method. MCA modified, where only genotyped animals are available for selection relative to the whole pedigree, account for 29.3% of the genetic variance, giving very similar results to MCA.

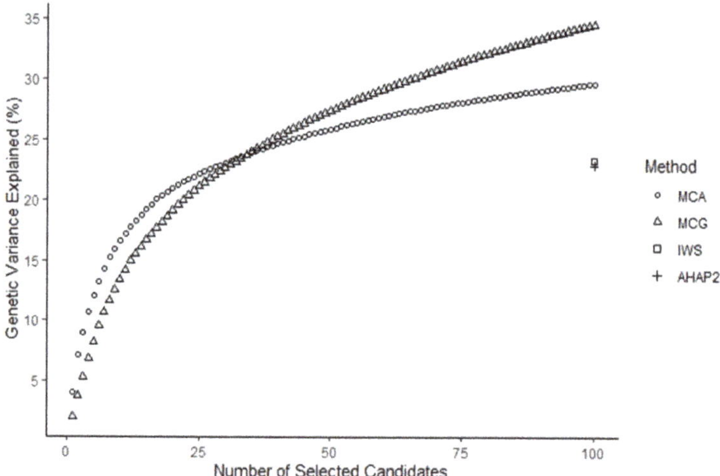

Figure 3. Diagonal values of **A*** representing the percentage of genetic variance explained for each additional selected candidate for whole-genome sequencing using the MCG method (**top**) or MCA method (**bottom**). The IWS and AHAP2 methods are presented as singular dots where 100 animals have been sampled.

3.3. Number of Unique Haplotypes Accounted for

Haplotype blocks were categorised into common, uncommon and rare classifications based on frequency in the population. The number of haplotypes accounted for within each group was then assessed for three methods (Table 2). All methods were able to account for the 588 unique common haplotypes in the population and a similar number of uncommon haplotypes; approximately 3500 haplotypes out of the 3666 in the population. The three methods begin to clearly separate where rare haplotypes are considered. MCG accounted for 8175 rare haplotypes followed by IWS and AHAP2 with 6492 and 5137, respectively. This resulted in MCG accounting for the highest total number of haplotypes (12,320) compared to IWS and AHAP2.

Table 2. Number of unique haplotypes accounted for when 100 animals are selected as whole-genome sequencing candidates using varying methods that utilise a relationship matrix (MCA/MCG) or haplotype library (IWS/AHAP2), respectively.

Method	Common ≥5%	Uncommon 1%–<5%	Rare 0.1%–<1%	Total
MCA [A]	-	-	-	-
MCG	588	3557	8175	12,320
IWS	588	3507	6492	10,587
AHAP2	588	3524	5137	9249
Max # [B]	588	3666	16,600	20,854

[A] As not all MCA selected animals were genotyped, the number of unique haplotypes accounted for cannot be estimated. [B] Max # denotes the maximum number of haplotypes in each category that can be sampled.

4. Discussion

4.1. Comparison of Relationship Matrix Methods

The methods which utilised a relationship matrix, MCA and MCG, had very high concordance between them in regard to specific candidates selected (Table 1). The rank correlation reported of 0.82 (Figure 2) is a stronger relationship than previously reported [11]. One explanation is that Wagyu are known to already have a very small effective population size; 43.4 in Australia [17], with only a small number of animals serving as the founder population for Australia's herd today. Given this, the MCG and MCA method are more likely to select identical candidates than the population in the original study, which was a Norwegian pig population pedigree with simulated genotype data [11].

MCA performed better where the number of selected candidates was low (Figure 3). This is likely due to the MCA method having access to the full pedigree of 10,549 individuals with a depth of up to 9 generations, whereas only 5334 of these animals were available for selection under MCG. There are some population structure implications in the data behind this. The pedigree includes deeper information on original "imported" founder animals in the population and a larger number of descendants, whereas MCG only includes genotypes on these founders and a subset of their descendants. The additional depth and breadth of pedigree appears advantageous to better inform selection decisions of early selected candidates. MCG appeared robust as the genomic relationships were able to compensate for the lack of pedigree depth after a certain number of selected candidates due to more detailed relationship information regarding Mendelian sampling. When only the genotyped animals could be selected as candidates (MCA modified), it performed extremely similarly to MCA on a whole. This supports the conclusion that the pedigree used in constructing A is not adding any information above and beyond what G captures. MCG also demonstrates a steady increase in genetic variance accounted for as the number of candidates approaches 100 whereas MCA begins to level off. This can again be attributed to more variation being able to be discerned through genomic relationships, which can better describe animals, particularly where relationships would be traditionally low (zero) in A and between full-sibs. MCG is also advantageous to MCA in that it can be run without concern for completeness of pedigree, assuming individuals in the population under consideration can be genotyped.

4.2. Comparison of Haplotype Block Methods

The methods which utilised 100 SNP wide haplotype blocks, IWS and AHAP2, had moderate concordance between the animals selected with 61/100 animals in common. In contrast, concordance between these methods and candidates selected by MCA and MCG was poor (Table 1). An analysis of the pedigree reveals the specific animals themselves selected by IWS, in particular, are all progeny or grand-progeny of those selected by MCG/MCA. This makes sense as only homozygous haplotypes are considered in the calculation of the weighting. Influential haplotypes being targeted (those accounted for by MCA/MCG) must be passed on across generations through paternal and maternal lines to be selected by IWS, and to a lesser degree, the AHAP2 method.

While MCG accounted for the greatest number of haplotypes with a frequency of 0.1% or greater (12,320, Table 2), it did not account for the greatest number of haplotypes overall when counting haplotypes below this frequency. Candidates selected using the cut-off restrictions were compared to the unrestricted raw data to get a view of the incidental rare haplotypes that were sampled in passing. IWS, AHAP2 and MCG accounted for an additional 9842, 7221 and 2631 haplotypes respectively below a frequency of 0.1% resulting in grand-totals of 20,429, 16,470 and 14,951 haplotypes sampled out of 339,824 respectively. Given this metric, IWS was the best where total number of haplotypes are concerned. Results from [12] are consistent with those above, with IWS demonstrating it accounted for the greatest number of haplotypes while selecting the least number of candidates compared to AHAP2. Additionally, given a set number of candidates, IWS accounted for more haplotypes than AHAP2, which is a more comparable metric to the study herein.

A study on simulated dairy data performed by [18] demonstrated similar findings to the study herein, with IWS accounting for a greater proportion of unique haplotypes (when all incidental haplotypes are included) than a method analogous to MCG. In addition, the overlap of selected candidates was very low between these methods across varying selection densities (50 to 1200 individuals). However, IWS did not outperform MCG in terms of genetic variance accounted for (Figure 3). Initial thoughts in this study were that the more haplotypes accounted for, the greater the degree of genetic variance explained, but

Figure 3 demonstrates that is clearly not the case. There could be a couple of explanations for this.

The IWS method is intentionally selecting animals that are more distantly related to others by preferentially selecting rare haplotypes. Animals that are homozygous for a rare haplotype had to receive one copy from each of the paternal and maternal lines, which to occur suggests the paternal and maternal lines were already likely related, i.e., IWS selects animals from the ends of different family branches rather than the bulk of the whole family tree. Additionally, given the haplotype blocks used aren't representative of "actual" haplotypes segregating in the population, they are merely chunks of SNPs in 100 SNP wide blocks; selection of individuals where these true haplotypes are essentially broken up could explain a loss in genetic variance accounted for. In contrast, the GRM utilises all SNPs and it can capture the similarity of true haplotypes between individuals in its estimation of relationships. Implementing the IWS and AHAP2 methods utilising a true haplotype library warrants further investigation.

Another point for consideration is that, while it could be expected that more haplotypes in the reference would yield higher imputation accuracies, IWS preferentially selected haplotypes with a low frequency. Ref. [19] demonstrated using initial data from the 1000 bulls genome project that the accuracy of imputed calls was high for SNPs with a MAF > 0.1 while it decreased rapidly for rarer variant sites. Ref. [18] demonstrated this nicely, showing imputation accuracy of specific variants increases with MAF bin. Additionally, [18] showed that reference populations selected by IWS were more effective at achieving high imputation accuracies for low MAF SNPs than other methods compared, but this advantage lessened with increasing reference population size.

As high-density genotyping and sequencing costs decrease, it would be more feasible to target lower frequency haplotypes by sequencing additional candidates to improve their accuracy of imputation. Methods, such as those proposed by [20,21] that allocate sequencing resources to specific haplotypes rather than individuals, would be suitable for this purpose, in fact, they propose an adjustment to IWS to allow for this. The benefit of the method proposed by [21] is that it assembles high-coverage sequence data through the accumulation of low coverage information over genome segments that are shared with many other individuals. This prevents these "census" haplotypes from being "over-sequenced" so that sequencing resources can then be allocated towards key-rare variants, for example. This method has been shown to achieve high imputation accuracies through hybrid peeling in deep pedigreed populations [22].

4.3. Practical Considerations

While the haplotype block methods appeared promising, their performance was inferior to relationship matrix-based methods given the metrics measured herein. One-hundred animals selected under MCG accounted for the most genetic variance and accounted for the greatest number of haplotypes (above a frequency of 0.1%). One key assumption was made here; both the MCA and MCG methods assumed that all potential candidates had DNA readily available. This is not always the case in a commercial pedigree. Both methods could easily be modified to account for DNA availability, i.e., the animal is still alive or has blood/semen/hair in storage. The animal that is selected, within an iteration, is logically that which reduces the residual genetic variance of the target population, i.e., Diag(A_{11}^{*}), the most. Multiplying each candidate's impact on the residual by a vector of 0 (no DNA available) or 1 (DNA available) ensures only candidates with DNA are chosen. This also prevents bias when selecting sequence candidates to form the reference if you were just to remove animals with no DNA from the analysis altogether. MCA clearly outperformed MCG where the number of samples selected was low and this could reflect a scenario where the sequencing budget is low. A strong depth of pedigree proved advantageous to the GRM, where the number of selected candidates is low. To capitalise on the depth of pedigree while utilising the detail of genomic relationships, an H matrix could be constructed, as is done for single-step GBLUP [23,24] with parameters set around DNA availability. Similar modifications could be made to the haplotype methods to account for DNA availability, though it is more likely that only potential candidates are included when running these methods. An important consideration is that MCA assumes a relatively high pedigree completeness to be effective. Naturally, animals not included in the pedigree cannot their genetic contribution to the population identified. Where pedigree is widely incomplete, the MCG method would be best using genomic relationships to replace those estimated from pedigree.

The relationship matrix methods also have one key advantage over haplotype methods when being applied within a breeding program. That is, they utilise data that is routinely constructed within a genetic evaluation program and are therefore simple and relatively quick to implement. This is compared to constructing haplotype libraries where cut-off decisions around haplotype inclusion must be made. This decision can impact the final animals that are selected for HD genotyping or sequencing. For example, the cut-off used for IWS by [12] was 4%, whereas it was 0.1% herein. In addition, the examples provided in this discussion assume selection within one population of animals and do not deeply discuss implications of across breed or crossbred populations.

The common method of selecting highly influential sires (PROG) performed equally as well as the MCA method and is clearly still a useful, cheap and easy method to select animals for high density genotyping. However, clear pitfalls of the method are that highly influential bulls tend to have highly influential sons and so on. While not explicitly outlined herein, immediate family links (siblings, progeny) do exist between the 166 influential animals, and this is only partially captured in Table 1 due to a lack of complete overlap between the methods. It is, therefore, necessary to adjust for kinship, genetic contribution, etc. Whether this is done ad hoc or using more scientific methods as in [11], this added complexity detracts from its usefulness, especially as the other four methods compared herein actively remove this laborious activity.

5. Conclusions

Selection using the MCG is highly recommended as a starting point for an ongoing sequencing project. Then the best method depends on the use case for the future set of sequences. If the aim is to select sequence candidates to allow for the overall imputation of the population, then it is better to select animals carrying common haplotypes in the first instance. If the resulting sequences from the selected animals are to be used for variant discovery or annotation of deleterious variants, animals carrying novel information should be selected.

Author Contributions: Conceptualisation, R.A.M., G.P., J.G. and W.S.P.; methodology, R.A.M., G.P. and R.T.; software, R.A.M. and R.T.; validation, R.A.M.; formal analysis, R.A.M.; investigation, R.A.M.; resources, W.S.P.; data curation, R.T.; writing—original draft preparation, R.A.M.; writing—review and editing, R.A.M., M.L.H., H.O. and W.S.P.; visualisation, R.A.M.; supervision, M.L.H., H.O. and W.S.P.; project administration, W.S.P.; funding acquisition, R.A.M. and W.S.P. All authors have read and agreed to the published version of the manuscript work reported.

Funding: R.A.M. was supported by scholarships from the Australian Government Research Training Program Scholarship and 3D Genetics.

Institutional Review Board Statement: Not Applicable.

Informed Consent Statement: Not Applicable.

Data Availability Statement: Restrictions apply to the availability of these data. Data were obtained from 3D Genetics Pty Ltd. and are available with the permission of J.G.

Conflicts of Interest: The authors declare no conflict of interest.

References

1. Meuwissen, T.H.E.; Hayes, B.J.; Goddard, M.E. Prediction of total genetic value using genome-wide dense marker maps. *Genetics* **2001**, *157*, 1819–1829. [CrossRef]
2. Hayes, B.J.; Bowman, P.J.; Chamberlain, A.; Goddard, M. Invited review: Genomic selection in dairy cattle: Progress and challenges. *J. Dairy Sci.* **2009**, *92*, 433–443. [CrossRef]
3. García-Ruiz, A.; Cole, J.B.; VanRaden, P.M.; Wiggans, G.R.; Ruiz-López, F.J.; Van Tassell, C.P. Changes in genetic selection differentials and generation intervals in US Holstein dairy cattle as a result of genomic selection. *Proc. Natl. Acad. Sci. USA* **2016**, *113*, E3995–E4004. [CrossRef] [PubMed]
4. Lee, Y.M.; Dang, C.G.; Alam, M.Z.; Kim, Y.S.; Cho, K.H.; Park, K.D.; Kim, J.J. The effectiveness of genomic selection for milk production traits of Holstein dairy cattle. *Asian-Australas. J. Anim. Sci.* **2020**, *33*, 382–389. [CrossRef] [PubMed]
5. Grisart, B.; Coppieters, W.; Farnir, F.; Karim, L.; Ford, C.; Berzi, P.; Cambisano, N.; Mni, M.; Reid, S.; Simon, P. Positional candidate cloning of a QTL in dairy cattle: Identification of a missense mutation in the bovine DGAT1 gene with major effect on milk yield and composition. *Genome Res.* **2002**, *12*, 222–231. [CrossRef] [PubMed]
6. Khatkar, M.S.; Moser, G.; Hayes, B.J.; Raadsma, H.W. Strategies and utility of imputed SNP genotypes for genomic analysis in dairy cattle. *BMC Genom.* **2012**, *13*, 538. [CrossRef] [PubMed]
7. Ogawa, S.; Matsuda, H.; Taniguchi, Y.; Watanabe, T.; Kitamura, Y.; Tabuchi, I.; Sugimoto, Y.; Iwaisaki, H. Genomic prediction for carcass traits in Japanese Black cattle using single nucleotide polymorphism markers of different densities. *Anim. Prod. Sci.* **2017**, *57*, 1631–1636. [CrossRef]
8. Goddard, M. Genomic selection: Prediction of accuracy and maximisation of long term response. *Genetica* **2009**, *136*, 245–257. [CrossRef]
9. Georges, M. Towards sequence-based genomic selection of cattle. *Nat. Genet.* **2009**, *46*, 807. [CrossRef] [PubMed]
10. VanRaden, P.M.; Tooker, M.E.; O'connell, J.R.; Cole, J.B.; Bickhart, D.M. Selecting sequence variants to improve genomic predictions for dairy cattle. *Genet. Sel. Evol.* **2017**, *49*, 32. [CrossRef]
11. Yu, X.; Woolliams, J.A.; Meuwissen, T.H. Prioritizing animals for dense genotyping in order to impute missing genotypes of sparsely genotyped animals. *Genet. Sel. Evol.* **2014**, *46*, 46. [CrossRef] [PubMed]
12. Bickhart, D.; Hutchison, J.; Null, D.; VanRaden, P.; Cole, J. Reducing animal sequencing redundancy by preferentially selecting animals with low-frequency haplotypes. *J. Dairy Sci.* **2016**, *99*, 5526–5534. [CrossRef] [PubMed]
13. Bates, D.; Vazquez, A.I. Pedigreemm: Pedigree-Based Mixed-Effects Models. R Package Version 0.0-3. Available online: https://CRAN.R-project.org/package=pedigreemm (accessed on 2 March 2021).
14. VanRaden, P.M. Efficient methods to compute genomic predictions. *J. Dairy Sci.* **2008**, *91*, 4414–4423. [CrossRef]
15. Sargolzaei, M.; Chesnais, J.; Schenkel, F. FImpute-An efficient imputation algorithm for dairy cattle populations. *J. Dairy Sci.* **2011**, *94*, 421.
16. Druet, T.; Macleod, I.; Hayes, B. Toward genomic prediction from whole-genome sequence data: Impact of sequencing design on genotype imputation and accuracy of predictions. *Heredity* **2014**, *112*, 39–47. [CrossRef] [PubMed]
17. Zhang, Y.D.; Banks, R. Genetic diversity and trends of Australian Japanese Black cattle. *Proc. Assoc. Adv. Anim. Breed. Genet.* **2021**, *24*, 451–454.
18. Butty, A.M.; Sargolzaei, M.; Miglior, F.; Stothard, P.; Schenkel, F.S.; Gredler-Grandl, B.; Baes, C.F. Optimizing selection of the reference population for genotype imputation from array to sequence variants. *Front. Genet.* **2019**, *10*, 510. [CrossRef]
19. Daetwyler, H.D.; Capitan, A.; Pausch, H.; Stothard, P.; Van Binsbergen, R.; Brøndum, R.F.; Liao, X.; Djari, A.; Rodriguez, S.C.; Grohs, C. Whole-genome sequencing of 234 bulls facilitates mapping of monogenic and complex traits in cattle. *Nat. Genet.* **2014**, *46*, 858. [CrossRef]
20. Gonen, S.; Ros-Freixedes, R.; Battagin, M.; Gorjanc, G.; Hickey, J.M. A method for the allocation of sequencing resources in genotyped livestock populations. *Genet. Sel. Evol.* **2017**, *49*, 47. [CrossRef]
21. Ros-Freixedes, R.; Gonen, S.; Gorjanc, G.; Hickey, J.M. A method for allocating low-coverage sequencing resources by targeting haplotypes rather than individuals. *Genet. Sel. Evol.* **2017**, *49*, 78. [CrossRef]
22. Ros-Freixedes, R.; Whalen, A.; Gorjanc, G.; Mileham, A.J.; Hickey, J.M. Evaluation of sequencing strategies for whole-genome imputation with hybrid peeling. *Genet. Sel. Evol.* **2020**, *52*, 18. [CrossRef] [PubMed]
23. Legarra, A.; Aguilar, I.; Misztal, I. A relationship matrix including full pedigree and genomic information. *J. Dairy Sci.* **2009**, *92*, 4656–4663. [CrossRef] [PubMed]
24. Christensen, O.F.; Lund, M.S. Genomic prediction when some animals are not genotyped. *Genet. Sel. Evol.* **2010**, *42*, 2. [CrossRef] [PubMed]

 agriculture

Article

Estimation of Pool Construction and Technical Error

John Keele [1,*], Tara McDaneld [1], Ty Lawrence [2], Jenny Jennings [3] and Larry Kuehn [1]

[1] USDA, ARS, U.S. Meat Animal Research Center, P.O. Box 166, Clay Center, NE 68933, USA; tara.mcdaneld@usda.gov (T.M.); larry.kuehn@usda.gov (L.K.)
[2] Department of Agricultural Sciences, West Texas A&M University, Canyon, TX 79016, USA; tlawrence@wtamu.edu
[3] Five Rivers Cattle Feeding, LLC, Amarillo, TX 79022, USA; jenny.s.jennings@gmail.com
* Correspondence: john.keele@usda.gov; Tel.: +1-402-762-4251

Abstract: Pooling animals with extreme phenotypes can improve the accuracy of genetic evaluation or provide genetic evaluation for novel traits at relatively low cost by exploiting large amounts of low-cost phenotypic data from animals in the commercial sector without pedigree (data from commercial ranches, feedlots, stocker grazing or processing plants). The average contribution of each animal to a pool is inversely proportional to the number of animals in the pool or pool size. We constructed pools with variable planned contributions from each animal to approximate errors with different numbers of animals per pool. We estimate pool construction error based on combining liver tissue, from pulverized frozen tissue mass from multiple animals, into eight sub-pools containing four animals with planned proportionality (1:2:3:4) by mass. Sub-pools were then extracted for DNA and genotyped using a commercial array. The extracted DNA from the sub-pools was used to form super pools based on DNA concentration as measured by spectrophotometry with planned contribution of sub-pools of 1:2:3:4. We estimate technical error by comparing estimated animal contribution using sub-samples of single nucleotide polymorphism (SNP). Overall, pool construction error increased with planned contribution of individual animals. Technical error in estimating animal contributions decreased with the number of SNP used.

Keywords: DNA pooling; genomic relationship; genomic prediction

Citation: Keele, J.; McDaneld, T.; Lawrence, T.; Jennings, J.; Kuehn, L. Estimation of Pool Construction and Technical Error. *Agriculture* **2021**, *11*, 1091. https://doi.org/10.3390/agriculture11111091

Academic Editors: Heather Burrow and Michael Goddard

Received: 20 September 2021
Accepted: 1 November 2021
Published: 4 November 2021

1. Introduction

Large scale genotyping using high-density arrays or next-generation sequencing techniques has revolutionized genetic prediction and identification of causal chromosomal regions through genome-wide association studies (GWAS). Furthermore, apportioning genetic variation to chromosomal regions (a goal of GWAS and an estimation of regional heritability using best linear unbiased prediction using a regional genomic relationship matrix and restricted maximum likelihood estimation of variance components) can improve genome prediction accuracy [1]. Genetic prediction accuracy is improved by increased variance in regional genomic relationships and higher, more consistent linkage disequilibrium between observed SNP and unknown or unobserved quantitative trait loci compared to the whole genome. Hence, GWAS and genome prediction are complimentary activities. However, effective implementation of genomic prediction or high-powered GWAS for complex and/or low heritability traits may require thousands of animals with phenotypes and genotypes. Individually genotyping seedstock populations is cost effective when the cost is spread among large suites of routinely recorded traits. However, individually genotyping can be prohibitively expensive when collecting novel traits on commercial animals, without recorded pedigree, for which only one trait is recorded.

DNA pooling can be an effective tool to reduce genotyping cost, and it captures greater than 80% of the power of individual genotyping in GWAS [2]. If costs of collecting phenotypes are much lower than genotyping costs, DNA pooling can reduce experimental costs by 90%. Our group has established the utility of GWAS using DNA pooling for novel complex traits such as fertility and disease resistance [3–5].

In addition to GWAS, early efforts have been proposed to apply pooling results to genomic prediction. Marker predictions from high and low phenotype groups could be used to accurately rank candidates for selection [6]. An alternative strategy could be the estimation of genomic relationships between pools and candidates for selection; for instance, genomic relationships could be used to derive estimated breeding values for sires of multiple-sire progeny in pools of extreme phenotypes [7]. Both techniques involve genomic prediction of animals that are known to have close relationships with animals in the pool. Commercial data capture is complicated by the fact that whole genome relationships between commercial cattle (source of data) and seedstock animals (selection candidates) are more distant and less variable compared to relationships between animals with data and animals being selected within the seedstock sector. The effectiveness of using SNP chip data from pools for genomic prediction depends on the signal to noise ratio, with the signal being allele frequency differences or haplotype frequency differences between pools of animals with extreme phenotypes, and noise being pooling errors. Greater genetic differences occur when the trait has higher heritability or when phenotypes are more extreme. Pooling errors include pool construction error and technical error. Pool construction error includes weighing error, errors in measuring or recording DNA concentration, incomplete sample mixing, pipetting error, variation in DNA content of the tissue, variation in DNA extraction efficiency, and DNA fragmentation. In this study, technical error is defined as variation in estimates of sub-pool, animal, or haplotype contributions among replicate arrays for identical pools (same animals in same proportions). However, replicate arrays were not run in this study because technical error can be estimated at a lower cost as the variance in estimated animal contribution among sub-samples of SNP, sampled without replacement. This is the approach that we took. Technical error depends on variation in in allelic ratio (Y/X) for intensity within genotype for each SNP and the number of SNP; specifically, we estimate pooling allele frequency using the same formula that Illumina uses, Illumina $\theta = 2 * \tan^{-1}(Y/X)/\pi$ [8], for calling genotypes and detecting copy number variation (or structural variants). The number of animals in the pool is inversely proportional to the average contribution from each animal; hence, we deliberately varied the contribution from each animal to approximate the influence of each pool's size on error without spending a lot of money genotyping individual animals. In this way, we can approximate error for pool sizes of 100 with only 16 animals. Our objectives were to evaluate pool construction, technical error, and the influence of factors that affect these errors, such as number of SNP and planned representation of animals within the pool.

2. Materials and Methods

Animal samples utilized in this study were recovered from abattoirs. Therefore, no Institutional and Animal Care and Use approval was obtained.

Sample pooling strategy: Pools comprising many animals are more economical because the cost of genotyping a pool is spread over more animals. Pools with many animals have small contributions from each animal. To approximate variable numbers of animals per pool at a reasonable cost, we selected an experimental design with variable contributions from individual animals. Liver tissue samples were obtained from a set of 50 Holstein steers with severely abscessed livers and 50 Holstein steers with no liver abscesses in close chain proximity from a commercial abattoir. From these pairs of matched samples, 16 samples from abscessed and 16 samples from non-abscessed animals were randomly selected. There were no common animals between the two phenotypes; however, there were likely steers with abscessed livers that were related to animals with non-abscessed livers. Disease status (abscessed and non-abscessed livers) was used as a basis for dividing samples in this experiment into biological replicates.

Four sub-pools were created for each liver phenotype (abscessed/non-abscessed). A sub-pool was comprised of four different liver tissue samples in parts of 1:2:3:4 (each part was 0.1 g) and placed in a 10 mL tube for the designated sub-pool with 2 mL phosphate buffered saline for homogenization. Each liver tissue sample only contributed to one sub-pool (4 sub-pools × 4 samples/sub-pool = 16 samples total per phenotype). Isolation of DNA from each sub-pool was performed by a standard phenol/chloroform extraction protocol. Two super pools were then derived from the extracted DNA of the sub-pools in two different arrangements, with the four sub-pools again representing parts of 1:2:3:4. As shown in Figure 1, four different sub-pools and two different super pools were formed, as previously described, for each liver phenotype. Thus, the planned representation of 16 individual animals are in Tables A1 and A2 based on the design presented in Figure 1. Individual DNA from the 32 animals used in the liver tissue pools was extracted using the QiAamp DNA Mini Kit following the manufacturer's instructions (Qiagen, Santa Clarita, CA, USA). All individual DNA samples and pools were then mixed, by placing individual samples on a rotator, and quantified by the DeNovix DS-11 FX+ Spectrophotometer/Fluorometer (DeNovix Inc., Wilmington, DE, USA) using 2 µL of sample and the dsDNA photometric setting. Quality of the DNA was evaluated for all individual samples and pools using gel electrophoresis to ensure that a high molecular weight DNA was present and intact.

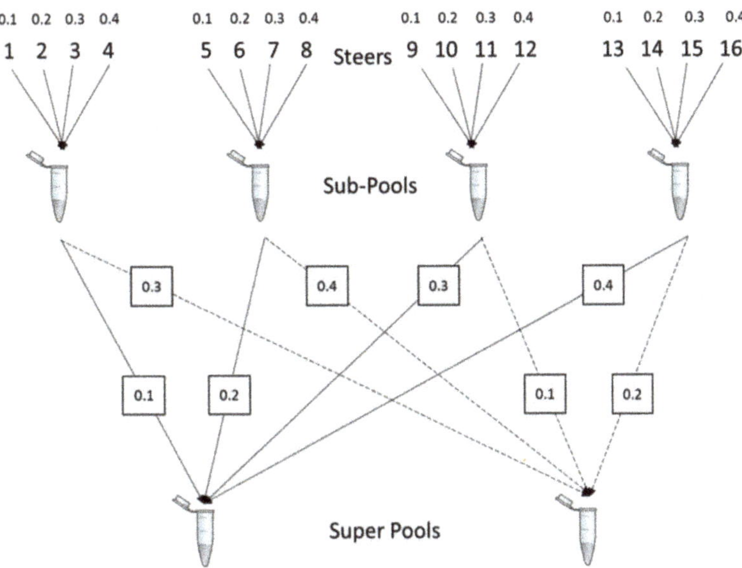

Figure 1. Graphical representation of the designed relative proportions of liver tissue from individual animals used to form sub-pools and DNA proportions used to form super pools. This design was replicated across two different liver phenotypes, abscessed and normal, to use a total of 32 animal samples. Planned contributions of animals to sub-pools ranged for 10 to 40 % and animal contributions to super pools ranged from 1 to 16 % (Table A2).

Our experimental design resulted in pools with a broad range of sample contributions ranging from 1 to 40% (Figure 1; Tables A1 and A2). Furthermore, there was planned proportionality in sample representation that was constant between sub-pools and super pools, allowing us to evaluate the stability of proportionality estimates at different dilutions. Discrepancy from proportionality would indicate instability of pool composition, changes in real animal proportions over time, or technical error. Realized or observed animal contributions were estimated from Illumina θ [8]. Illumina θ for super pools and sub-pools was computed by $2 * \tan^{-1}(Y/X)/\pi$ where X is the red intensity identifying the 'A' allele and Y is the green intensity representing the 'B' allele, using the Illumina

AB nomenclature [8]. We term pooling error sources as 'pool construction' (estimated DNA quantity does not match planned quantity), 'technical error' (caused by variation in Illumina θ within genotype or pools with the same animal representation or replicated arrays of the same pool construction), and error in estimating animal contributions. All 32 animals, eight sub-pools, and four super pools were genotyped using the Illumina BovineHD array (Illumina, Inc., San Diego, CA, USA) by Neogen Corporation (Lincoln, NE, USA).

Statistical analysis: For each sub-pool and super pool, we estimated the contribution of each of the 32 animals in the eight sub-pools (four animals that contribute to each sub-pool) and four super pools (16 animals in each) using quadratic programming to minimize the residual sums of squares, subject to the constraints that the estimated animal contributions were positive and summed to 1, using the solve.QP() within contributed package quadprog in R [9,10] with Illumina θ for super pool or sub-pool as the dependent variable and genotype (number of copies of B allele)/2 as the independent variable.

Theoretically, pool construction error should be greater for larger planned animal contributions compared to smaller planned animal contributions based on the Dirichlet distribution, which is commonly assumed for the probabilities underlying the multinomial distribution. We tested for equality of variance in pool construction error among groups of planned contributions using a Levene type test, which is robust to deviations from normality [11] using the levene test function in R [10]. There were four unique values among animal contributions to sub-pools and nine unique values among animal contributions to super pools. The Levene test requires greater replication within planned contribution level than our experiment allowed to achieve adequate power to detect differences in pool construction variance. In our analysis, we clustered similarly planned comparisons to overcome the small number of replicates within planned comparison level (Table A3) and looked at sensitivity to the level of granularity or aggregation to evaluate the robustness of our results. We used default parameters with the exception of correction.method = "zero.correction", kruskal.test = TRUE, bootstrap = TRUE, and num.bootstrap = 100,000, which implemented a bootstrap rank-based (Kruskal-Wallis) modified robust Brown-Forsythe Levene-type test based on the absolute deviations from the median with modified structural zero removal method and correction factor.

We evaluated technical error as influenced by the number of SNP by subsampling all 777,962 SNP without replacement, computing animal contributions for each subsample, estimating the standard deviation for each animal across subsample, and averaging the result across sub-pools and super pools within the number of subsamples. The number of subsamples and SNP per subsample are in Table 1.

Table 1. Standard deviation among technical errors for animals contributions estimated by bootstrapping sub-samples of SNP; sampled without replacement [1].

Number of SNP Samples	Number of SNP per Sample	Mean SD
3000	259	0.01960
2000	388	0.01610
1000	777	0.01140
100	7779	0.00360
50	15,559	0.00260
40	19,449	0.00230
30	25,932	0.00190
20	38,898	0.00160
10	77,962	0.00110
5	155,592	0.00069

[1] Standard deviation among SNP samples averaged over 4 super pools and 8 sub-pools.

To evaluate the consistency of proportionality between sub-pools and super pools, we regressed animal contribution to the super pool on animal contribution to the sub-pool. If the r^2 from this analysis is high, then there is a strong proportionality between animal contributions to sub-pools and super pools, and the estimated contribution of each animal is not affected much by being diluted in a pool of additional animals.

We estimated haplotypes without pedigree using Beagle version 5.2 [12] and hap-ibd version 1.0 [13] to identify shared identity by descent segments among the 32 liver samples plus hapmap animals [14].

Breed composition of the 32 liver samples was estimated using a multiple regression method [15], with the exception that we constrained the breed contributions to sum to 1 and be ≥ 0 using quadratic programming [10], and the breed SNP frequency reference data were derived from BovineHD 770 k data for multiple diverse breeds [14].

3. Results

The raw data produced in this study have been uploaded to Ag Data Commons (see Data Availability Statement).

3.1. Pool Construction Error

Pool construction error was 6.3-times greater when creating super pools from extracted sub-pools based on DNA concentration measurements compared to creating sub-pools from individual animals based on liver tissue mass ($p < 0.045$); variance for forming super pools from sub-pools was 0.0163 compared to 0.0026 when forming sub-pools from individual animals.

Pool construction error variation increased with the level of planned animal contribution, both within and between super pools and sub-pools (Figure 2; Table 2). Variation in pool construction error increased with larger planned contributions. Equality of variances was rejected for course and intermediate granularity ($p < 0.0154$; Table 2). At the finest level of granularity possible, equality of variance among distinct planned contributions was not rejected ($p = 0.121$), demonstrating the need to cluster planned contributions into bins to achieve sufficient replication within the bin of similar planned contributions to detect differences in variance. Significant results for rejecting equality of variance occurred for five levels of granularity, supporting the hypothesis that increasing variance with increasing planned contribution was not simply a function of lucky placement of bin boundaries.

Table 2. Testing equality of variance for pool construction error; granularity of tests.

			Number of Levels of Granularity				
Pool Type	**Planned Contribution**	**n [1]**	**2**	**3**	**4**	**5**	**6**
Super pool	0.01	4	1	1	1	1	1
Super pool	0.02	8	1	1	1	1	1
Super pool	0.03	8	1	1	1	1	2
Super pool	0.04	12	1	1	2	2	2
Super pool	0.06	8	1	2	2	2	3
Super pool	0.08	8	1	2	2	3	3
Super pool	0.09	4	2	2	3	3	4
Sub-pool	0.10	8	2	2	3	3	4
Super pool	0.12	8	2	3	3	4	4
Super pool	0.16	4	2	3	3	4	5
Sub-pool	0.20	8	2	3	4	5	5
Sub-pool	0.30	8	2	3	4	5	6
Sub-pool	0.40	8	2	3	4	5	6
p-value			0.0012	0.0056	0.0054	0.0098	0.0154

[1] n is the number of planned contributions per distinct value for super pools or sub-pools. The sum of the n column is 96 which is the total number of planned comparisons in all 4 super pools and 8 sub-pools.

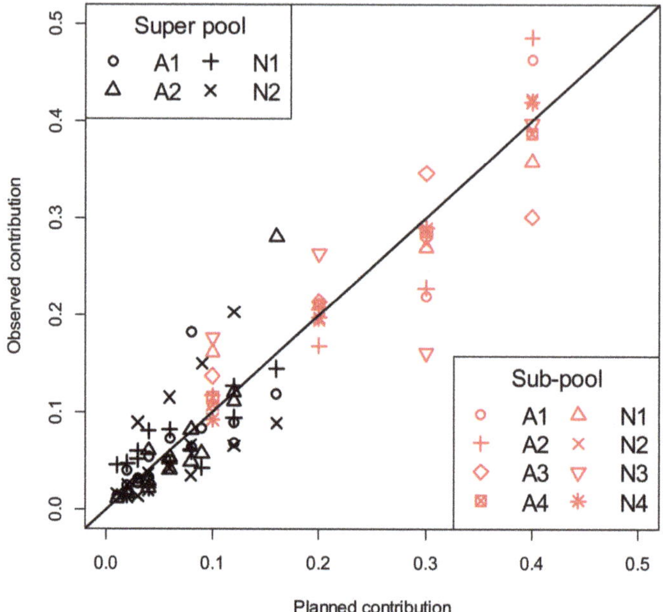

Figure 2. Pool construction errors are deviations of observed animal contributions from planned contributions for super and sub-pools. The solid black line depicts where observed and planned animal contributions are equal. Observed animal contributions were estimated by quadratic programming [9] minimizing error sums of squares subject to animal contributions summing to 1 and animal contributions ≥0. In legends, sub-pools or super pools starting with A or N are from cattle with liver abscess or normal livers, respectively.

3.2. Technical Error

In this section we characterize where technical error originates at the individual SNP level and evaluate the impact of individual SNP variation in Illumina θ on multiple SNP estimates of animal contribution to a pool. Mean Illumina θ was obtained from approximately 500 cattle of diverse ancestry by Illumina [16]. Technical error at the individual SNP level varied both within and between SNP, and the between SNP differences were consistent with differences in mean Illumina θ for heterozygotes [17] (Figure 3).

Technical error of estimated animal contributions for random samples of SNP decreased with increasing numbers of SNP per sample (Table 2).

3.3. Proportionality of Animal Contributions Conserved with Dilution

Estimated animal contributions to super pools were proportional to the contribution of the same animal to the sub-pool even though any given animal was diluted to different extents in the two super pools they were in (each animal was in one sub-pool and two super pools); linear regression of the super pool observed contribution on the sub-pool observed contribution yielded $r^2 = 0.99$ for super pool abscess 1, 0.98 for super pool abscess 2, 0.96 for super pool normal 1 and 0.99 for super pool normal 2.

3.4. Identity by Descent Sharing

Haplotypes within animals in sub-pools or super pools share identity by descent (IBD) with at least one animal not in a sub-pool or super pool for 85 to 95% of their genome. Each haplotype within each sub-pool was checked for shared IBD with 1492 haplotypes outside the sub-pool; two haplotypes for each of the 746 animals, comprising 718 animals from [14], 32 animals from the current study, minus four animals in each sub-pool. Each haplotype within each super pool was checked for shared IBD with 1468 haplotypes outside the super pool; two haplotypes for each of the 734 animals, comprising 718 animals from [14], 32 animals from the current study minus 16 animals in each super pool.

To further evaluate the structure of our population, we estimated breed composition of the liver samples using a regression analysis similar to [15], with the exception that we used BovineHD 770k data from [14]. One animal was a mix of Holstein and Jersey, four were crossbred beef females, and 27 were purebred Holstein. Hence, our assumption prior to analysis that all 32 animals were Holstein steers proved to be incorrect.

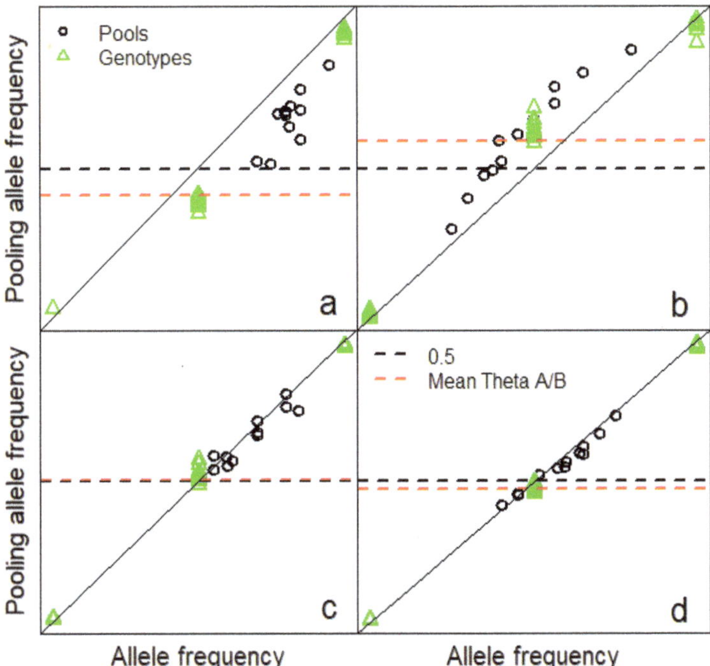

Figure 3. Technical error expressed as deviations of observed pooling allele frequency from allele frequency. All vertical and horizontal coordinates in this figure range from 0 to 1. Pooling allele frequency was estimated as Illumina $\theta = 2 * \tan^{-1}(Y/X)/\pi$ in this study where Y/X is the allelic ratio for green/red intensity for both pools and individuals. Considering pooling allele frequency for individuals is not ridiculous, because individuals are technically a pool of two haplotypes with equal representation. Technical error varies within and between SNP, and we present four examples, (**a**–**d**). (**a**) Pooling allele frequency was low compared to allele frequency for both pools and heterozygotes which was consistent with low mean Illumina θ for heterozygotes (depicted A/B in figure legend. (**b**) Pooling allele frequency and mean heterozygote Illumina θ were high compared to allele frequency. (**c**,**d**) Pooling allele frequency and mean heterozygote Illumina θ were similar to allele frequency.

4. Discussion

Pool construction error increased with planned animal contribution ($p \leq 0.0154$), which implies that pool construction error decreases with an increasing numbers of animals equally represented in a pool, because planned contribution is the reciprocal of the number of animals equally represented. Based on these results, we recommend more animals per pool, as also supported by previous literature [18].

Technical error in animal contributions decreases as more SNP are used to estimate animal contribution, which is consistent with the Central Limit Theorem coming into play and reducing technical error in estimating animal contributions as more SNP are included in the computation. An implication of this result is that we can accurately estimate haplotype contributions within a chromosomal region if there are adequate numbers of SNP within the region.

The four animals in each sub-pool are each represented in two super pools at two different dilutions. The proportion of the animals in the sub-pool were strongly correlated with the proportion in the two super pools regardless of the dilution; furthermore, the two dilutions in the super pools were strongly correlated. This finding suggests that pools of animals with extreme phenotypes from different breeds can be combined into larger pools to save money. Similarly, pools of animals with extreme phenotypes from different seasons, pens within a feedlot, feedlots, and pastures can be combined. Commercial feedlot cattle being collected from a packing plant are generally comprised of multiple breeds and crossbreeds, and phenotypically extreme animals in a particular pen of animals is likely to contain more than one breed; indeed, in most circumstances we do not know the breed makeup when we are processing the samples into pools. The phenotypic extremes of animals within a particular pen may not comprise of very many animals. For example, the top 5% of 200 animals in a pen is only 10 animals. Combining animals with extreme phenotypes from 10 pens to make one pool of 100 animals results in a savings of 90% relative to genotyping 10 pools with 10 animals each.

Although not typically thought of this way, the number of copies of a B allele for an individual divided by two is the allele frequency of two haplotypes in the individual, one of maternal and the other of paternal origin. When we regress Illumina θ for a pool of genotypes for individuals, we are estimating the representation of pools of two haplotypes in a larger pool context. If phenotypically extreme animals in the pool share chromosomal regions IBD with other animals not in the pool, then the distribution of haplotypes of animals not in the pool can be accurately estimated, and those haplotype contributions can be used to inform the estimated breeding value of other animals with shared IBD through the IBD sharing. Using 718 animals from 18 diverse breeds, we found that all 32 animals in our pools each shared between 85 and 95 % of their genomes in IBD with at least one reference haplotype from 746 or 734 animals not in the pool representing multiple diverse breeds. This demonstrates that reference haplotypes from approximately 750 diverse animals (1500 haplotypes) is sufficient to cover IBD for 85 to 95% of the genome for a purebred Holstein or crossbred beef animal; it all hinges on whether the haplotypes in the pool are covered by reference haplotypes, and they were in this case. It is unknown whether the high coverage in this case was due to a small sample of Holstein haplotypes or due to fairly large haplotype segments being ubiquitous across populations as a result of historical natural and artificial selection or random drift [19,20].

5. Conclusions

Pool construction error decreases as more animals are incorporated into the pool; hence, pools with more equally represented animals would be expected to have less pool construction error, that is, the actual contribution would be closer to the planned contribution compared to pools with fewer animals. Technical error decreases as more SNP are used to estimate haplotype contributions. Similar proportionality of animal contribution estimates in the sub-pool and after dilution to the super pool indicates that animals with extreme phenotypes of different breeds can be mixed into larger pools to save cost without much loss of information. Pools of phenotypically extreme animals can inform genetic evaluation if there is IBD sharing between animals in the pool and selection candidates outside the pool; hence, population distant IBD ensures the relevance of pools of phenotypically extreme animals for the purpose of genetic evaluation.

Author Contributions: Conceptualization, J.K., T.M., L.K., J.J. and T.L.; methodology, T.M., J.K. and T.L.; software, J.K.; validation, J.K. and L.K.; formal analysis, J.K.; investigation, J.J. and T.L.; resources, J.J. and T.L.; data curation, T.L., T.M. and J.K.; writing—original draft preparation, J.K. and L.K.; writing—review and editing, J.K., L.K.,T.M., T.L. and J.J.; visualization, J.K.; supervision, L.K., J.J. and

T.L.; project administration, L.K.; funding acquisition, L.K., J.J. and T.L. All authors have read and agreed to the published version of the manuscript.

Funding: This research received no external funding.

Institutional Review Board Statement: Animal samples utilized in this study were recovered *post mortem* from abattoirs. Therefore, no Institutional and Animal Care and Use approval was obtained.

Informed Consent Statement: Not applicable because study did not involve humans.

Data Availability Statement: Raw data are available at Ag Data Commons, https://doi.org/10.15482/USDA.ADC/1523112 (accessed on 15 September 2021); and Ag Data Commons, https://doi.org/10.15482/USDA.ADC/1523111 (accessed on 15 September 2021).

Acknowledgments: We acknowledge the technical contributions of Sandra Nejezchleb and Tammy Sorensen. Mention of trade names or commercial products in this publication is solely for the purpose of providing specific information and does not imply recommendation or endorsement by the U.S. Department of Agriculture.

Conflicts of Interest: The authors declare no conflict of interest. The funders had no role in the design of the study; in the collection, analyses, or interpretation of data; in the writing of the manuscript, or in the decision to publish the results.

Appendix A

Planned sub-pool contributions to super pools are in Table A1 as depicted in Figure 1. This design is replicated with sub-pools of animal with liver abscess or normal livers.

Table A1. Sub-pool contributions to super pools.

	Super Pools	
Sub-Pool	**1**	**2**
1	0.1	0.3
2	0.2	0.4
3	0.3	0.1
4	0.4	0.2

Planned animal contributions to super pools and sub-pools are in Table A2 for animals with abscessed on normal livers based on Figure 1. The same design was applied to each phenotype (abscessed or normal livers) with a different set of animals for each phenotype.

Table A2. Animal contributions to super pools and sub-pools.

	Sub-Pools				Super Pools	
Animal	**1**	**2**	**3**	**4**	**1**	**2**
1	0.1	0.0	0.0	0.0	0.01	0.03
2	0.2	0.0	0.0	0.0	0.02	0.06
3	0.3	0.0	0.0	0.0	0.03	0.09
4	0.4	0.0	0.0	0.0	0.04	0.12
5	0.0	0.1	0.0	0.0	0.02	0.04
6	0.0	0.2	0.0	0.0	0.04	0.08
7	0.0	0.3	0.0	0.0	0.06	0.12
8	0.0	0.4	0.0	0.0	0.08	0.16
9	0.0	0.0	0.1	0.0	0.03	0.01
10	0.0	0.0	0.2	0.0	0.06	0.02
11	0.0	0.0	0.3	0.0	0.09	0.03
12	0.0	0.0	0.4	0.0	0.12	0.04
13	0.0	0.0	0.0	0.1	0.04	0.02
14	0.0	0.0	0.0	0.2	0.08	0.04
15	0.0	0.0	0.0	0.3	0.12	0.06
16	0.0	0.0	0.0	0.4	0.16	0.08

Boundaries separating bins for planned contributions to enable testing for equality of variance among levels of planned contributions are int Table A3. Granularity increases with number of bins. The maximum number of bins would be the number of unique values for planned contributions.

Table A3. Boundaries between bins for planned contribution bins.

	Number of Bins				
Item [1]	**2**	**3**	**4**	**5**	**6**
Boundary	0.010	0.010	0.010	0.010	0.010
Label	1	1	1	1	1
Boundary	0.082	0.045	0.035	0.035	0.025
Label	2	2	2	2	2
Boundary	0.400	0.110	0.085	0.065	0.045
Label		3	3	3	3
Boundary		0.400	0.165	0.110	0.085
Label			4	4	4
Boundary			0.400	0.165	0.125
Label				5	5
Boundary				0.40	0.25
Label					6
Boundary					0.400

[1] Items are either boundaries between bins or labels for bins.

References

1. Visscher, P.M.; Hemani, G.; Vinkhuyzen, A.A.E.; Chen, G.B.; Lee, S.H.; Wray, N.R.; Goddard, M.E.; Yang, J. Statistical Power to Detect Genetic (Co)Variance of Complex Traits Using SNP Data in Unrelated Samples. *PLoS Genet.* **2014**, *10*, e1004269. [CrossRef] [PubMed]
2. MacGregor, S.; Zhao, Z.Z.; Henders, A.; Nicholas, M.G.; Montgomery, G.W.; Visscher, P.M. Highly cost-effective genome-wide association studies using DNA pools and dense SNP arrays. *Nucleic Acids Res.* **2008**, *36*, e35. [CrossRef] [PubMed]
3. McDaneld, T.G.; Kuehn, L.A.; Thomas, M.G.; Snelling, W.M.; Smith, T.P.; Pollak, E.J.; Cole, J.B.; Keele, J.W. Genomewide association study of reproductive efficiency in female cattle. *J. Anim. Sci.* **2014**, *92*, 1945–1957. [CrossRef] [PubMed]
4. Keele, J.W.; Kuehn, L.A.; McDaneld, T.G.; Tait, R.G., Jr.; Jones, S.A.; Smith, T.P.L.; Shackelford, S.D.; King, D.A.; Wheeler, T.L.; Lindholm-Perry, A.K.; et al. Genomewide association study of lung lesions in cattle using sample pooling. *J. Anim. Sci.* **2015**, *93*, 956–964. [CrossRef] [PubMed]
5. Keele, J.W.; Kuehn, L.A.; McDaneld, T.G.; Tait, R.G., Jr.; Jones, S.A.; Keel, B.N.; Snelling, W.M. Genomewide association of liver abscesses in beef cattle. *J. Anim. Sci.* **2016**, *94*, 490–499. [CrossRef] [PubMed]
6. Sonesson, A.K.; Meuwissen, T.H.; Goddard, M.E. The use of communal rearing of families and DNA pooling in aquaculture genomic selection schemes. *Genet. Sel. Evol.* **2010**, *42*, 41. [CrossRef] [PubMed]
7. Bell, A.M.; Henshall, J.M.; Porto-Neto, L.R.; Dominik, S.; McCulloch, R.; Kijas, J.; Lehnert, S.A. Estimating the genetic merit of sires by using pooled DNA from progeny of undetermined pedigree. *Genet. Sel. Evol.* **2017**, *49*, 28. [CrossRef] [PubMed]
8. Improved Genotype Clustering with Gentrain3.0. 370-2016-015-A. Available online: Illumina.com (accessed on 2 September 2021).
9. Goldfarb, D.; Idnani, A. Dual and Primal-Dual Methods for Solving Strictly Convex Quadratic Programs. In *Numerical Analysis*; Hennart, J.P., Ed.; Springer: Berlin, Germany, 1982; pp. 226–239.
10. R Core Team. *R: A Language and Environment for Statistical Computing*; R Foundation for Statistical Computing: Vienna, Austria, 2020. Available online: https://www.R-project.org/ (accessed on 15 September 2021).
11. Brown, M.B.; Forsythe, A.B. Robust tests for the equality of variances. *J. Am. Stat. Assoc.* **1974**, *69*, 364–367. [CrossRef]
12. Browning, B.L.; Zhou, Y.; Browning, S.R. A one-penny imputed genome from next generation reference panels. *Am. J. Hum. Genet.* **2018**, *103*, 338–348. [CrossRef]
13. Zhou, Y.; Browning, S.R.; Browning, B.L. A fast and simple method for detecting identity by descent segments in large-scale data. *Am. J. Hum. Genet.* **2020**, *106*, 426–437. [CrossRef] [PubMed]
14. Porto-Neto, L.R.; Sonstegard, T.S.; Liu, G.E.; Bickhart, D.M.; Da Silva, M.V.; Machado, M.A.; Utsunomiya, Y.T.; Garcia, J.F.; Gondro, C.; Van Tassell, C.P. Genomic divergence of zebu and taurine cattle identified through high-density SNP genotyping. *BMC Genom.* **2013**, *14*, 87. [CrossRef] [PubMed]
15. Kuehn, L.A.; Keele, J.W.; Bennett, G.L.; McDaneld, T.G.; Smith, T.P.; Snelling, W.M.; Sonstegard, T.S.; Thallman, R.M. Predicting breed composition using breed frequencies of 50,000 markers from the US Meat Animal Research Center 2000 Bull Project. *J. Anim. Sci.* **2011**, *89*, 1742–1750. [CrossRef] [PubMed]

16. BovineHD Genotyping BeadChip Datasheet. 370-2010-018, BovineHD Genotyping BeadChip. Available online: Illumina.com (accessed on 7 September 2021).
17. Peiris, B.L.; Ralph, J.; Lamont, S.J.; Dekkers, J.C. Predicting allele frequencies in DNA pools using high density SNP genotyping data. *Anim. Genet.* **2011**, *42*, 113–116. [CrossRef] [PubMed]
18. Vargas Jurado, N.; Kuehn, L.A.; Keele, J.W.; Lewis, R.M. Accuracy of GEBV of sires based on pooled allele frequency of their progeny. *G3* **2021**, *11*, jkab231. [CrossRef] [PubMed]
19. Nait Saada, J.; Kalantzis, G.; Shyr, D.; Cooper, F.; Robinson, M.; Gusev, A.; Palamara, P.F. Identity-by-descent detection across 487,409 British samples reveals fine scale population structure and ultra-rare variant associations. *Nat. Commun.* **2020**, *11*, 6130. [CrossRef] [PubMed]
20. Zhou, Y.; Browning, S.R.; Browning, B.L. IBDkin: Fast estimation of kinship coefficients from identity by descent segments. *Bioinformatics* **2020**, *36*, 4519–4520. [CrossRef] [PubMed]

 *agriculture* 

Article

Accounting for Missing Pedigree Information with Single-Step Random Regression Test-Day Models

Minna Koivula [1,*], Ismo Strandén [1], Gert P. Aamand [2] and Esa A. Mäntysaari [1]

[1] Natural Resources Institute Finland (Luke), 31600 Jokioinen, Finland; ismo.stranden@luke.fi (I.S.); esa.mantysaari@luke.fi (E.A.M.)
[2] NAV Nordic Cattle Genetic Evaluation, 8200 Aarhus, Denmark; gap@seges.dk
* Correspondence: Minna.koivula@luke.fi

Abstract: Genomic selection is widely used in dairy cattle breeding, but still, single-step models are rarely used in national dairy cattle evaluations. New computing methods have allowed the utilization of very large genomic data sets. However, an unsolved model problem is how to build genomic- (**G**) and pedigree- ($\mathbf{A}_{22}$) relationship matrices that satisfy the theoretical assumptions about the same scale and equal base populations. Incompatibility issues have also been observed in the manner in which the genetic groups are included in the model. In this study, we compared three approaches for accounting for missing pedigree information: (1) GT_H used the full Quaas and Pollak (QP) transformation for the genetic groups, including both the pedigree-based and the genomic-relationship matrices, (2) GT_A_{22} used the partial QP transformation that omitted QP transformation in $\mathbf{G}^{-1}$, and (3) GT_MF used the metafounder approach. In addition to the genomic models, (4) an official animal model with a unknown parent groups (UPG) from the QP transformation and (5) an animal model with the metafounder approach were used for comparison. These models were tested with Nordic Holstein test-day production data and models. The test-day data included 8.5 million cows with a total of 173.7 million records and 10.9 million animals in the pedigree, and there were 274,145 genotyped animals. All models used VanRaden method 1 in **G** and had a 30% residual polygenic proportion (RPG). The **G** matrices in GT_H and GT_A_{22} were scaled to have an average diagonal equal to that of $\mathbf{A}_{22}$. Comparisons between the models were based on Mendelian sampling terms and forward prediction validation using linear regression with solutions from the full- and reduced-data evaluations. Models GT_H and GT_A_{22} gave very similar results in terms of overprediction. The MF approach showed the lowest bias.

Keywords: ssGBLUP; ssGTBLUP; genomic evaluation; single-step; Holstein; genetic groups; metafounder

Citation: Koivula, M.; Strandén, I.; Aamand, G.P.; Mäntysaari, E.A. Accounting for Missing Pedigree Information with Single-Step Random Regression Test-Day Models. *Agriculture* **2022**, *12*, 388. https://doi.org/10.3390/agriculture12030388

Academic Editors: Heather Burrow and Michael Goddard

Received: 19 January 2022
Accepted: 8 March 2022
Published: 10 March 2022

1. Introduction

Meuwissen et al. [1] introduced the genome-wide marker-assisted prediction model called genomic selection. Many alternative prediction models have been developed to use genomic selection in dairy cattle breeding [2]. Theoretically, the best model is single-step genomic BLUP (ssGBLUP) when phenotypes are available from both genotyped and non-genotyped individuals. The single-step approach offers a unified method for the analysis of all animals simultaneously [3,4]. Even though a decade has passed since the introduction of ssGBLUP, it is still not widely implemented in national dairy cattle evaluations.

Practical implementation of ssGBLUP has encountered different computational and modeling challenges. Some of the computational challenges related to very large genomic data sets can be overcome by using an alternative expression of the inverse genomic-relationship matrix or a model equivalent to ssGBLUP [1,5,6]. However, an unsolved modeling problem is how to build genomic- (**G**) and pedigree- ($\mathbf{A}_{22}$) relationship matrices that satisfy the theoretical assumptions about the same scale and equal base populations [7]. Several methods were proposed that make **G** and $\mathbf{A}_{22}$ compatible. For example, base-population allele frequencies (AF) are used [8], and elements of **G** are scaled and centered to have, on average, the same diagonal and off-diagonal elements as in $\mathbf{A}_{22}$ [7,9]. Similar incompatibility issues were observed when genetic groups were included in the model [10].

Genetic groups can be included in a model as regression coefficients when the group proportions of an individual are traced to groups of unknown parents in the pedigree [11]. When the number of genetic groups is small, it is easy to explicitly include such groups in the model. However, this may lead to significant memory and computational costs especially in complicated models having many genetic groups [10]. A computationally more efficient approach with such models is to re-express genetic groups as unknown parent groups (UPG) resulting from the Quaas and Pollak (QP) transformation [12,13]. After the QP transformation, the genetic groups, by regression, are replaced by UPGs that extend the inverse-relationship matrix of all animals $\mathbf{A}^{-1}$ in the mixed-model equations (MME) of the pedigree-based animal model.

The ssGBLUP uses relationship matrix $\mathbf{H}$, which includes both pedigree-based and genomic-based relationship matrix information. UPGs can be accounted for in many ways in the MME of ssGBLUP. If UPGs are included in $\mathbf{A}^{-1}$ but not accounted for in $\mathbf{A}_{22}{}^{-1}$ and $\mathbf{G}^{-1}$, severe convergence problems may arise, suggesting an incorrect model [14]. In many cases, this problem can be solved by properly accounting for genetic groups. The full QP transformation as described in [14,15] includes UPGs in $\mathbf{A}^{-1}$, $\mathbf{A}_{22}{}^{-1}$, and $\mathbf{G}^{-1}$ in the MMEs of the single-step models. However, there is an alternative approach where the contributions of $\mathbf{G}^{-1}$ are ignored in the QP transformation, or the altered QP $\mathbf{H}$ inverse is used [16–18]. Ignoring $\mathbf{G}^{-1}$ in the QP transformation can be justified by considering that the $\mathbf{G}$ matrix contains all the information, whereas the pedigree information to build the $\mathbf{A}$ matrix is incomplete and requires a UPG.

An alternative to the genetic groups is the use of metafounders (MF). The MF approach proposed by Legarra et al. [19] attempts to make the $\mathbf{A}^{-1}$ and $\mathbf{A}_{22}{}^{-1}$ matrices compatible with the $\mathbf{G}$ matrix. The MF approach is based on the idea of using an allele frequency (AF) equal to 0.5 for all the markers when calculating the $\mathbf{G}$ matrix [7]. The unknown parents are replaced by MF or pseudo-individuals with relationships and self-relationships that augment the $\mathbf{A}$ matrix. The MFs are like UPGs but allow a related base- population or populations with nonzero inbreeding coefficients, e.g., as in [19,20]. The relationships within and between the MFs are modeled by a $\mathbf{\Gamma}$ matrix, which is used in forming the pedigree-based relationship matrix ($\mathbf{A}^{\mathbf{\Gamma}}$).

When MFs are defined as the same as the genetic groups, the large number of UPGs that are common in the breeding-value prediction models of dairy cattle can make the estimation of the $\mathbf{\Gamma}$ matrix challenging [21]. Because the estimation of the $\mathbf{\Gamma}$ matrix requires the base-population AF, the large number of UPGs increases the chances that a UPG is associated with some rare allele genotypes, which can make the base-population AF estimate very uncertain. Thus, instead of estimating the base-population AF for all UPGs, a well-estimated $\mathbf{\Gamma}$ submatrix can be extended to a full $\mathbf{\Gamma}$ matrix using a covariance function, [22] as suggested by Kudinov et al. [23].

The aim of this study was to test different options to handle genetic groups in single-step models with the Nordic Holstein test-day (TD) data. We applied three different single-step TD models: (1) UPG by full QP transformation (GT_H), (2) UPGs in $\mathbf{A}^{-1}$ and $\mathbf{A}_{22}{}^{-1}$ or partial QP (GT_A_{22}), and (3) the MF approach (GT_MF). In addition to the genomic models, (4) an official pedigree-based animal model with a UPG by QP transformation and (5) a pedigree-based animal model with the metafounder approach were run for comparison.

2. Materials and Methods

2.1. Data

We used the official Nordic Holstein (HOL) milk production evaluation data obtained from the Nordic Cattle Genetic Evaluation (NAV). The data included TD records of milk, fat, and protein production from Denmark (DNK), Finland (FIN), and Sweden (SWE). The TD data included 8.8 million cows with a total of 173.7 million test-day records and a total of 447.5 million observations. The pedigree file had 10.9 million animals.

There were 274,145 genotyped animals, of which 75,802 were genotyped bulls (also including Holstein bulls from the EuroGenomics genotype exchange (EuroGenomics, 2020) and young bull calves), and 198,343 genotyped cows and heifers. Until 2019, bulls were genotyped using the BovineLD Bead Chip (Illumina, San Diego, CA, USA) (25% of all genotyped) and cows with the Eurogenomics custom LD chip (51% of all genotyped). Since 2019, all animals have been genotyped using the Eurogenomics EG MD chip (24% of all genotyped) [24]. After applying editing criteria for a minor allele frequency of 0.01 and a locus average GenCall score of 0.60, a total of 46,342 SNP markers on the 29 bovine autosomes were chosen for the analyses, and all the genotypes were imputed to this density. Genotype imputation was carried out using a family-and-population-based approach implemented in the FImpute program [25].

2.2. Models

Instead of the original ssGBLUP model using the **H** matrix, we used the ssGTBLUP approach in the computations. The ssGTBLUP approach allows the key computations involving the $\mathbf{G}^{-1}$ matrix to be replaced by efficient multiplications with a Woodbury matrix identity [26]. Thus, a dense **T** matrix of size m by n is used instead of the dense $\mathbf{G}^{-1}$ matrix of size n by n, where n is the number of genotyped animals and m is the number of SNP markers. Three different single-step models named GT_H, GT_A$_{22}$, and MF were used in the comparisons. In all models, the genomic-relationship matrix was as in VanRaden method 1 and included 30% residual polygenic proportion (RPG) and an AF of 0.5 for all markers. Earlier studies indicated that 30% RPG reduces the inflation of genomic evaluations more than models with smaller or larger RPGs (unpublished data). In the GT_H model, full QP transformation of genetic groups for $\mathbf{A}^{-1}$ and $\mathbf{A}_{22}{}^{-1}$ and $\mathbf{G}^{-1}$ was used. The computations by ssGTBLUP were described in Koivula et al. [15]. In GT_A$_{22}$, the QP transformation was applied to $\mathbf{A}^{-1}$ and $\mathbf{A}_{22}{}^{-1}$. The QP part for $\mathbf{A}_{22}$ can be completed with an equivalent sparse formulation by reading the pedigree and including UPGs in the $\mathbf{A}_{22}$ as described in Koivula et al. [15]. In GT_MF, the metafounder approach was used. In addition to the single-step models, an animal model with UPGs by QP transformation (EBV) and an animal model with metafounders (EBV_MF) were used for comparison and to observe the changes in predictions due to genomic information only. The pedigree inbreeding coefficients were accounted for in $\mathbf{A}^{-1}$ and $\mathbf{A}_{22}{}^{-1}$, and in GT_MF, inbreeding coefficients were accounted for in corresponding inverse matrices $\mathbf{A}^{\Gamma}$ and $\mathbf{A}_{22}{}^{\Gamma}$. The **G** matrices in GT_H and GT_A$_{22}$ were scaled to have an average diagonal equal to the pedigree-based relationship matrix of the genotyped animals ($\mathbf{A}_{22}$).

Genetic groups and MFs were defined by the same logic. First, we defined fewer genetic groups than in the original NAV evaluation. The new groups were based on 4 breed groups (Holstein, red dairy cattle, Jersey, and 'other breeds') and 5 country-of-origin groups within the Holstein group (HOLDNK, HOLSWE, HOLFIN, HOLother, and HOLred). Second, each of these nine sources was further grouped by birth year decade and by selection path when appropriate. Thus, the original 446 groups were reduced to 176. These groups were considered random UPGs with variances equal to the genetic (co)variance in GT_H, GT _A$_{22,}$ and EBV. The 176 groups were used as metafounders in GT_MF and EBV_MF.

The MF approach needs a covariance matrix for the metafounders, i.e., the $\mathbf{\Gamma}$ matrix. The MF self-relationship matrix $\mathbf{\Gamma}$ was defined using a covariance function (CF) model [22]. The $\mathbf{\Gamma}_0$ matrix of nine base MFs used values from [21]. In the 176 groups, a breed-specific linear time trend of a decade was assumed in the self-relationships, which were estimated using the base $\mathbf{\Gamma}_0$ matrix in the CF model. The known regression coefficients and design matrices in the CF model allow describing $\mathbf{\Gamma}$ matrices of any size, such as for our 176 groups, $\mathbf{\Gamma}_{176} = \mathbf{\Phi}_{176}\mathbf{K}\mathbf{\Phi}_{176}{}'$. We chose heuristic values for **K** based on expectations from numerous descriptive analyses. For more formal analyses, see Kudinov et al. [23]. After solving the $\mathbf{\Gamma}_{176}$-matrix, we computed the $\mathbf{\Gamma}$-matrix-compliant inbreeding coefficients needed for the computations involving the inverses $\mathbf{A}^{\Gamma}$ and $\mathbf{A}_{22}{}^{\Gamma}$ when solving MMEs. As the MF approach changes the relationship structure, Legarra et al. [19] derived a correction factor to be used to change the unrelated base-population-variance components to related base-population components. For our derived $\boldsymbol{\Gamma}$ matrix, the correction factor was close to one. Therefore, the same genetic-variance components were used in all models.

The study used the Nordic multiple-trait, reduced-rank, random regression TD model from the routine genetic evaluations for milk, fat, and protein in Finland, Sweden, and Denmark (a detailed description of the model can be found in Lidauer et al. [27]). Production records from the first three lactations are considered as nine different traits, with each having its own lactation curve. Therefore, the model had 27 traits in total: 3 countries, 3 yield traits, and 3 lactations. In this-reduced rank random regression TD model, each animal received 15 solutions to the random regression effects in the same way as in the official NAV evaluations. The TD-model solutions of genetic random regression coefficients were used to compile the total yields for milk, protein, and fat for the 305 d lactation [27] where the first, second, and third lactations had weights of 0.3, 0.25, and 0.45, respectively.

Comparison of models was based on forward prediction validation with solutions from the full- and the reduced-data evaluations. The reduced data were extracted from the full data by removing the last four years of observations from the full data. The linear regression validation (LR) method [28] was used for validation. The LR method compares predictions based on reduced and full data, which results in estimates of accuracy and bias. Candidate animals in the validation were selected according to their effective daughter contributions (EDC). Danish, Finnish, and Swedish (DFS) Holstein bulls born between the years 2010 and 2016 and having their EBV based on an EDC $\geq$ 3.0 (corresponding to roughly 20 daughters) in the full-data set but an EDC of zero in the reduced-data set were defined as candidate bulls. This resulted in 524 candidate bulls for validation. The EDCs were calculated using the Interbull-EDC method tailored for the animal model by the ApaX99-program [29] for all bulls in the pedigree using both the full and the reduced data sets.

All MMEs of the TD models were solved by MiX99 software [30], which uses preconditioned conjugate gradient (PCG) iteration. The main computational cost in every iteration of the PCG method was due to the MME coefficient matrix times a vector product, where the most time-consuming computations were due to genomic information. To save memory and computing time, the inverse of the $\mathbf{A}_{22}$ matrix was not computed in advance, but instead, the computations used the method by Strandén et al. [31], which is based on sparse submatrices of $\mathbf{A}^{-1}$ by pedigree information. The PCG method was assumed to be converged when $C_r < 10^{-7}$, where C_r is defined as a Euclidean norm of the difference between the right-hand side (RHS) of the MME and the one predicted by the current solutions relative to the norm of RHS.

3. Results and Discussion

Average diagonal elements of the $\mathbf{A}_{22}$, $\mathbf{A}_{22}{}^{\boldsymbol{\Gamma}176}$, and $\mathbf{G}$ by birth year of genotyped animals are presented in Figure 1. The use of the $\boldsymbol{\Gamma}_{176}$ matrix lifted the diagonal elements of the $\mathbf{A}_{22}$ matrix close to those of the $\mathbf{G}$ matrix. Average inbreeding coefficients in $\mathbf{A}_{22}$ and $\mathbf{G}$ were 0.05 and 0.34, respectively. The difference is close to those reported earlier by VanRaden et al. [32] and Kudinov et al. [21]. The average inbreeding coefficient increased to 0.29 in $\mathbf{A}_{22}{}^{\boldsymbol{\Gamma}176}$.

Wall clock times for preprocessing and solving MMEs are given in Table 1. The preprocessing, i.e., the computing time and the peak memory used in the construction of the $\mathbf{T}$ matrix for the $\mathbf{G}^{-1}$ matrix computations, did not differ considerably between the single-step models. For the MF model, there was an additional step of building the self-relationship matrix $\boldsymbol{\Gamma}$. However, this step only marginally increased the total time.

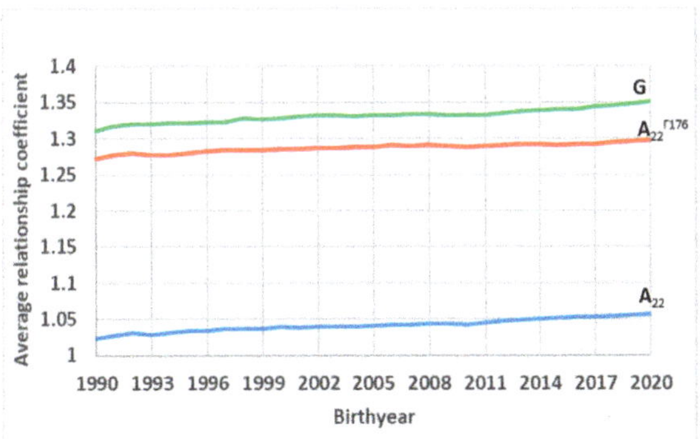

Figure 1. Average diagonal elements of the pedigree-relationship matrix of the genotyped animals ($\mathbf{A}_{22}$), the genomic-relationship matrix constructed assuming that all allele frequencies were 0.5 (**G**), and the pedigree-relationship matrix of the genotyped animals augmented by $\mathbf{\Gamma}_{176}$ ($\mathbf{A}_{22}{}^{\mathbf{\Gamma}176}$) presented by the birth year of the animal.

The number of PCG iterations to solve (G)EBV was 1227 for the animal model (EBV) and 1264 for the animal model with metafounders (EBV_MF). The total solving time to calculate EBV_MF was longer than for the animal model with UPGs by the QP transformation. Of the single-step models, GT_H with full QP needed 1019 iteration rounds for convergence, GT_A_{22} with QP only in $\mathbf{A}^{-1}$ and $\mathbf{A}_{22}$ needed 1051 iteration rounds, and GT_MF with metafounders needed 1307 iteration rounds. Thus, the MF model required more iterations. However, because the time per iteration round in the GT_MF model was less than that in the GT_H and GT_A22 models, the total computing time by GT_MF was lowest among the single-step models. Compared with the pedigree-based animal models, the single-step models required about 40–50% more time to calculate the solutions using the PCG iteration.

Table 1. Computing time (wall clock time in hours) and peak memory use (in GB) for an animal model with UPGs (EBV), an animal model with metafounders (EBV_MF), and different single-step models. The single-step models had unknown parent groups (UPG) with full QP (GT_H), UPGs with partial QP (GT_A_{22}), and metafounders (GT_MF). Convergence was assumed when $C_r < 10^{-7}$ (the norm of relative error in predicted RHS of MME).

	EBV	EBV_MF	GT_H	GT_A_{22}	GT_MF
Building **T** matrix time (h)			23	20	20
Peak memory GB			208	207	207
Solving					
Iterations	1227	1264	1019	1051	1307
Seconds/PCG round	101	125	173	178	125
Time (h)	34	44	49	52	45
Peak memory GB	14.9	15.0	114.5	114.3	114.5
Total computing time (h)	34	44	72	72	65

Table 2 illustrates the LR-validation results from the different models for 524 DFS Holstein validation bulls. The level differences in the (G)EBV predictions were corrected by standardizing the (G)EBV so that the mean (G)EBV for cows born in 2007 was the same in all models. The b_0 column in Table 1 shows the mean difference (in kg) between the full- and reduced-data (G)EBV evaluations. This illustrates the realized bias. GT_MF showed a slightly smaller difference than the UPG models. The regression coefficients (b_1) showed the same trend as b_0. The largest b_1, i.e., the smallest overdispersion, was observed for GT_MF for all traits. Although there were no large differences in the coefficients of correlation (R^2) between the models, GT_MF had the highest R^2 values. The R^2 from the LR validation can be interpreted as a reciprocal of the increase in reliability from the reduced-data evaluations to the full-data evaluations. Thus, the results indicate that, both in terms of bias and reliability, the MF model was slightly better than the other models. This conclusion is similar to that of other studies where the use of the MF model improved the single-step evaluations, e.g., [17,18]. Similarly, the MF approach seems to give better b_1 and R^2 values than the UPG model for the animal model without genomic information. Our results indicate that when the QP transformation is used, GT_H is as practical an alternative as GT_A_{22}. This result is different from the results from other studies [16,18], where GT_A_{22}, which was called altered **H**-inverse, had better predictive abilities than the full QP model GT_H.

Table 2. Bull linear regression validation (number of bulls = 524) results. Regression coefficients (b_1) and coefficients of correlation (R^2) from the models. The b_0 = mean (Full_(G)EBV—reduced_(G)EBV). The different models are an animal model with UPGs (EBV), an animal model with metafounders (EBV_MF), and different single-step models. The single-step models used UPGs with full QP (GT_H), UPGs with partial QP (GT_A_{22}), and metafounders (GT_MF).

	Model	b_0	b_1	R^2
Milk	EBV	−101.7	0.84	0.32
	EBV_MF	−141.36	0.89	0.35
	GT_H	−319.8	0.87	0.67
	GT_A_{22}	−315.2	0.87	0.67
	GT_MF	−272.3	0.89	0.68
Protein	EBV	0.80	0.74	0.24
	EBV_MF	0.58	0.82	0.27
	GT_H	−11.10	0.81	0.63
	GT_A_{22}	−10.99	0.81	0.62
	GT_MF	−9.71	0.83	0.64
Fat	EBV	−2.18	0.73	0.23
	EBV_MF	−2.24	0.80	0.26
	GT_H	−16.16	0.82	0.64
	GT_A_{22}	−15.61	0.81	0.63
	GT_MF	−14.67	0.85	0.65

In a genomic-selection program, the genomic pre-selection of bull calves [9,32] conflicts with the usual assumptions in the pedigree-based evaluations, which do not use genomic information. The pre-selected bulls are no longer a random sample of the progeny of their parents. This leads to an inflated mean for Mendelian sampling (MS) terms for these bulls and to a violation of normal assumptions in the pedigree-based animal model. Additionally, the MS term for the genotyped animals is likely to be different from zero with selective genotyping. In contrast, MS is expected to be zero when genotyping involves all young animals or is random.

Genomic selection allows selecting animals with superior MS. Consequently, the mean GEBV of all candidate animals is lower than for selected animals and their progeny [33]. The selected animals with many genotyped progenies are also more likely to have MS greater than zero [34]. The larger MS has an impact on genetic trends when animals are selected based on genomic information, especially if the selection happens before phenotypes are recorded [35].

Figure 2 shows the mean MS of genotyped bulls by birth year for protein using the reduced-data GEBV. Means are for DFS bulls and include all genotyped young bulls, those without daughters, and those that never entered AI service. In the reduced-data model, the bulls born after 2011 only have genomic information. There were no significant differences in the mean MS terms between the single-step models. The figure shows that for the youngest age classes, the difference is about 4 kg for different single-step models, whereas in both animal models, the MS term is zero, as expected. Before the start of genomic selection, the mean MS terms were quite stable. The mean MS was presumably below zero because of the overprediction of bull dam EBVs. After genomic selection began to take effect, the mean MS also started to increase. In animal models, both approaches seem to give zero MS for the youngest age classes, but in the older bulls, the MS term in EBV_MF was not as negative as in the other models.

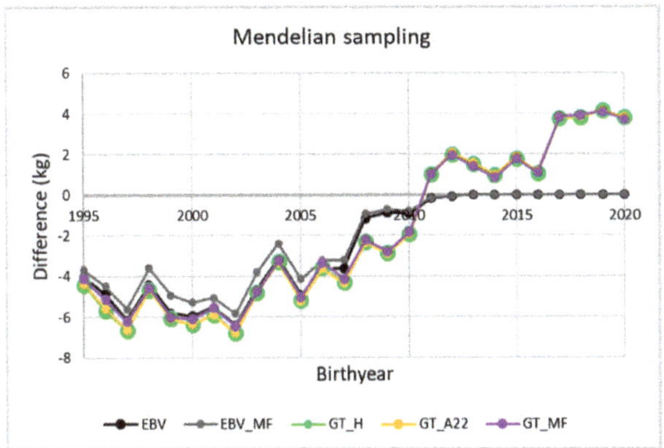

Figure 2. Mendelian sampling term means for protein for all genotyped DFS bulls by birth year calculated from (G)EBV from the reduced data. The different models are animal model (EBV), animal model with metafounders (EBV_MF), and three different single-step models by ssGTBLUP. The single-step models had unknown parent groups (UPG) with full QP (GT_H), UPGs with partial QP (GT_A_{22}), and metafounders (GT_MF).

Figure 3 shows the genetic trend and yearly SD of protein (G)EBV for DFS Holstein bulls. Solid lines are from the full-data runs and dashed lines from the reduced-data runs. Except for the lower trend for animal-model EBVs, the trends from the single-step models were quite similar. However, for the MF model in the reduced-data set, the trend was slightly lower than in the other single-step models, indicating lower overprediction in the MF model compared with the UPG models. The same can be observed in estimates of b_0 in Table 2. The genetic trends in the official animal model (EBV) and the animal model using the MF approach (MF_EBV) were similar, as were their SD trends.

Based on all our comparisons, it seems that the traditional genetic group model and the MF model are both feasible options for handling genetic groups in single-step evaluations. The single-step MF model can be a more sophisticated way to combine pedigree and genomic information than the traditional single-step model with UPGs because genomic information affects both the genomic- and the pedigree-based relationship matrices in the MF model. Moreover, it seems that the MF model does not increase the trend of young, genotyped animals as much as the UPG-based single-step methods. The MF model also gives marginally better validation results compared with the other models. However, the current MF approach might still require some further development in building the $\mathbf{\Gamma}$ matrix.

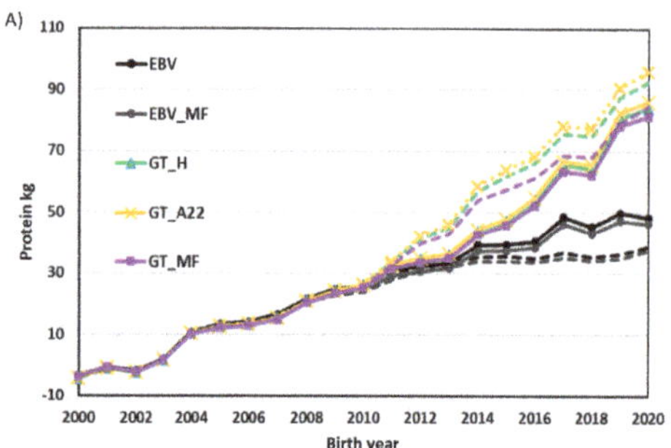

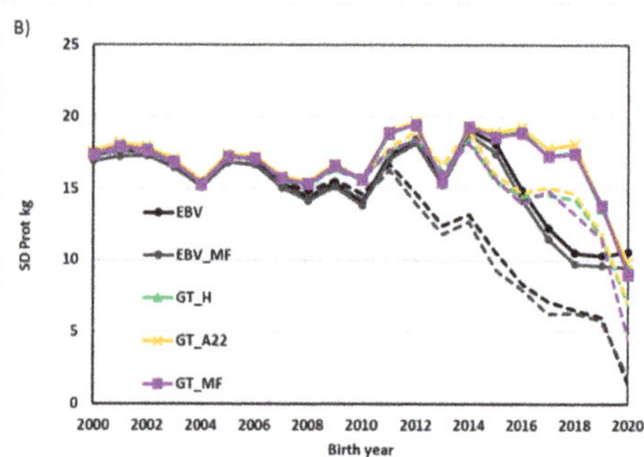

Figure 3. (**A**) Genetic trends for protein (G)EBV (kg) for the bulls presented by birth year averages. (**B**) SD for protein (G)EBVs (kg) by birth year. The different models are animal model (EBV), animal model with metafounders (EBV_MF), and three different single-step models by ssGTBLUP. The single-step model had UPGs with full QP (GT_H), UPGs with partial QP (GT_A_{22}), and metafounders (GT_MF). Solid lines are means and SDs for full-data trends, and dashed lines are for reduced-data trends.

4. Conclusions

Both the traditional UPG models and the MF approach can be implemented efficiently in single-step models with large genomic data sets. In our study, the MF approach had a lower bias than the UPG models. Moreover, when the QP transformation was used to arrive at a UPG model, the results from the full QP transformation were similar to the partial QP transformation in which the $\mathbf{G}^{-1}$ contributions were not included in the UPG computations. The mean MS by birth year was positive for the genotyped bulls during the last decade according to the single-step models but close to zero in the pedigree-based animal model. Selective genotyping can explain some of the positive mean MS values, but the most recent years need further investigation.

Author Contributions: Conceptualization and original draft preparation, M.K.; writing, review, and editing, E.A.M., I.S. and G.P.A. All authors have read and agreed to the published version of the manuscript.

Funding: This study was funded by the following Nordic cattle-breeding organizations: Viking Genetics and Nordic Cattle Genetic Evaluation.

Institutional Review Board Statement: Not applicable.

Data Availability Statement: Not applicable.

Acknowledgments: This work was a part of the Genomics in BLUP project. The Nordic cattle-breeding organizations: Viking Genetics (Randers, Denmark), Nordic Cattle Genetic Evaluation (NAV, Aarhus, Denmark), and Faba (Hollola, Finland) are acknowledged for providing the genotype data and the test-day data.

Conflicts of Interest: The authors declare no conflict of interest.

References

1. Meuwissen, T.H.E.; Hayes, B.J.; Goddard, M.E. Prediction of total genetic value using genome-wide dense marker maps. *Genetics* **2001**, *157*, 1819–1829. [CrossRef] [PubMed]
2. VanRaden, P.M. Symposium review: How to implement genomic selection. *J. Dairy Sci.* **2020**, *103*, 5291–5301. [CrossRef] [PubMed]
3. Aguilar, I.; Misztal, I.; Johnson, D.-L.; Legarra, A.; Tsuruta, S.; Lawlor, T.J. Hot topic: A unified approach to utilize phenotypic, full pedigree, and genomic information for genetic evaluation of Holstein final score. *J. Dairy Sci.* **2010**, *93*, 743–752. [CrossRef]
4. Christensen, O.F.; Lund, M.S. Genomic prediction when some animals are not genotyped. *Genet. Sel. Evol.* **2010**, *42*, 2. [CrossRef] [PubMed]

5. Mäntysaari, E.A.; Koivula, M.; Strandén, I. Symposium review: Single-step genomic evaluations in dairy cattle. *J. Dairy Sci.* **2020**, *103*, 5314–5326. [CrossRef]
6. Misztal, I.; Lourenco, D.; Legarra, A. Current status of genomic evaluation. *J. Anim. Sci.* **2020**, *98*, 1–14. [CrossRef]
7. Christensen, O.F. Compatibility of pedigree-based and marker-based relationship matrices for single-step genetic evaluation. *Genet. Sel. Evol.* **2012**, *44*, 37. [CrossRef]
8. VanRaden, P.M. Efficient methods to compute genomic predictions. *J. Dairy Sci.* **2008**, *91*, 4414–4423. [CrossRef]
9. Vitezica, Z.G.; Aguilar, I.; Misztal, I.; Legarra, A. Bias in genomic predictions for populations under selection. *Genet. Res.* **2011**, *93*, 357–366. [CrossRef]
10. Misztal, I.; Vitezica, Z.G.; Legarra, A.; Aguilar, I.; Swan, A.A. Unknown-parent groups in single-step genomic evaluation. *J. Anim. Breed. Genet.* **2013**, *130*, 252–258. [CrossRef]
11. Thompson, R. The estimation of heritability with unbalanced data: II. Data available on more than two generations. *Biometrics* **1979**, *33*, 497–504. [CrossRef]
12. Westell, R.A.; Quaas, R.L.; Van Vleck, L.D. Genetic groups in an animal model. *J. Dairy Sci.* **1988**, *71*, 1310–1318. [CrossRef]
13. Quaas, R.L.; Pollak, E.J. Modified equations for sire models with groups. *J. Dairy Sci.* **1981**, *64*, 1868–1872. [CrossRef]
14. Matilainen, K.; Strandén, I.; Aamand, G.P.; Mäntysaari, E.A. Single step genomic evaluation for female fertility in Nordic Red dairy cattle. *J. Anim. Breed. Genet.* **2018**, *135*, 337–348. [CrossRef] [PubMed]
15. Koivula, M.; Strandén, I.; Aamand, G.P.; Mäntysaari, E.A. Practical implementation of genetic groups in single-step genomic evaluations with Woodbury matrix identity based genomic relationship inverse. *J. Dairy Sci.* **2021**, *104*, 10049–10058. [CrossRef]
16. Tsuruta, S.; Lourenco, D.A.L.; Masuda, Y.; Misztal, I.; Lawlor, T.J. Controlling bias in genomic breeding values for young genotyped bulls. *J. Dairy Sci.* **2019**, *102*, 9956–9970. [CrossRef] [PubMed]
17. Cesarani, A.; Masuda, Y.; Tsuruta, S.; Nicolazzi, E.K.; VanRaden, P.M.; Lourenco, D.; Misztal, I. Genomic predictions for yield traits in US Holsteins with unknown parent groups. *J. Dairy Sci.* **2021**, *104*, 5843–5853. [CrossRef]
18. Masuda, Y.; Tsuruta, S.; Bermann, M.; Bradford, H.L.; Misztal, I. Comparison of models for missing pedigree in single-step genomic prediction. *J. Anim. Sci.* **2021**, *99*, 1–10. [CrossRef]
19. Legarra, A.; Christensen, O.F.; Vitezica, Z.G.; Aguilar, I.; Misztal, I. Ancestral relationships using metafounders: Finite ancestral populations and across population relationships. *Genetics* **2015**, *200*, 455–468. [CrossRef]
20. Garcia-Baccino, C.; Legarra, A.A.; Christensen, O.F.; Misztal, I.; Pocrnic, I.; Vitezica, Z.G.; Cantet, R.J. Metafounders are related to Fst fixation indices and reduce bias in single-step genomic evaluations. *Genet. Sel. Evol.* **2017**, *49*, 34. [CrossRef]
21. Kudinov, A.A.; Mäntysaari, E.A.; Aamand, G.P.; Uimari, P.; Strandén, I. Metafounder approach for single-step genomic evaluations of Red Dairy cattle. *J. Dairy Sci.* **2020**, *103*, 6299–6310. [CrossRef] [PubMed]
22. Kirkpatrick, M.; Hill, W.G.; Thompson, R. Estimating the covariance structure of traits during growth and ageing, illustrated with lactation in dairy cattle. *Genet. Res.* **1994**, *64*, 57–69. [CrossRef] [PubMed]
23. Kudinov, A.A.; Koivula, M.; Strandén, I.; Aamand, G.P.; Mäntysaari, E.A. Single-step genomic predictions of a minor breed concurrently with a main breed large national genomic evaluation. *Interbull Bull.* **2021**, *56*, 174–179.
24. EuroGenomics. A European Network for a Reliable Cattle Breeding. 2022. Available online: https://www.eurogenomics.com/?rub=88&unce_contenus_webclient=0&view=afficher_elasticsearch_results&submitted=1&q=A+European+Network+for+a+Reliable+Cattle+Breeding.+2022 (accessed on 10 January 2022).
25. Sargolzaei, M.; Chesnais, J.P.; Schenkel, F.S. A new approach for efficient genotype imputation using information from relatives. *BMC Genom.* **2014**, *15*, 478. [CrossRef]
26. Mäntysaari, E.A.; Evans, R.D.; Strandén, I. Efficient single-step genomic evaluation for a multibreed beef cattle population having many genotyped animals. *J. Anim. Sci.* **2017**, *95*, 4728–4737. [CrossRef]
27. Lidauer, M.; Pösö, J.; Pederson, J.; Lassen, J.; Madsen, P.; Mäntysaari, E.A.; Nielsen, U.; Eriksson, K.-Å.; Johansson, K.; Pitkänen, T.; et al. Across-country test-day model evaluations for Nordic Holstein, Red Cattle and Jersey. *J. Dairy Sci.* **2015**, *98*, 1296–1309. [CrossRef]
28. Legarra, A.; Reverter, A. Semi-parametric estimates of population accuracy and bias of predictions of breeding values and future phenotypes using the LR method. *Genet. Sel. Evol.* **2018**, *50*, 53. [CrossRef]
29. Stranden, I.; Lidauer, M.; Mäntysaari, E.A.; Pösö, J. Calculation of Interbull weighting factors for the Finnish test day model. *Interbull Bull.* **2001**, *26*, 78–79.
30. Strandén, I.; Lidauer, M. Solving large mixed models using preconditioned conjugate gradient iteration. *J. Dairy Sci.* **1999**, *82*, 2779–2787. [CrossRef]
31. Strandén, I.; Matilainen, K.; Aamand, G.P.; Mäntysaari, E.A. Solving efficiently large single-step genomic best linear unbiased prediction models. *J. Anim. Breed. Genet.* **2017**, *134*, 264–274. [CrossRef]
32. VanRaden, P.M.; Olson, K.M.; Wiggans, G.R.; Cole, J.B.; Tooker, M.E. Genomic inbreeding and relationships among Holsteins, Jerseys, and Brown Swiss. *J. Dairy Sci.* **2011**, *94*, 5673–5682. [CrossRef] [PubMed]
33. Tyrisevä, A.-M.; Mäntysaari, E.A.; Jakobsen, J.; Aamand, G.P.; Dürr, J.; Fikse, W.F.; Lidauer, M.H. Detection of evaluation bias caused by genomic preselection. *J. Dairy Sci.* **2018**, *101*, 3155–3163. [CrossRef]

34. Masuda, Y.; VanRaden, P.M.; Misztal, I.; Lawlor, T.J. Differing genetic trend estimates from traditional and genomic evaluations of genotyped animals as evidence of preselection bias in US Holsteins. *J. Dairy Sci.* **2018**, *101*, 5194–5206. [CrossRef] [PubMed]
35. Abdollahi-Arpanahi, R.; Lourenco, D.; Misztal, I. Detecting effective starting point of genomic selection by divergent trends from best linear unbiased prediction and single-step genomic best linear unbiased prediction in pigs, beef cattle, and broilers. *J. Anim. Sci.* **2021**, *99*, 1–11. [CrossRef] [PubMed]

 agriculture

Article

Time-Course Transcriptome Landscape of Bursa of Fabricius Development and Degeneration in Chickens

Lan Huang [1], Yaodong Hu [2], Qixin Guo [1], Guobin Chang [1] and Hao Bai [1,*]

[1] Joint International Research Laboratory of Agriculture and Agri-Product Safety, the Ministry of Education of China, Yangzhou University, Yangzhou 225009, China
[2] College of Animal Science, Xichang University, Xichang 615000, China
* Correspondence: bhowen1027@yzu.edu.cn; Tel.: +86-18796608824

Abstract: The bursa of Fabricius (BF) is a target organ for various pathogenic microorganisms; however, the genes that regulate BF development and decline have not been fully characterized. Therefore, in this study, histological sections of the BF were obtained from black-boned chickens at 7 (N7), 42 (N42), 90 (N90) and 120 days (N120) of age, and the differential expression and expression trends of the BF at different stages were analyzed by transcriptome analysis. The results showed that the growth of the BF progressively matured with age, followed by gradual shrinkage and disappearance. Transcriptome differential analysis revealed 5914, 5513, 4575, 577, 530 and 66 differentially expressed genes (DEG) in six different comparison groups: N7 vs. N42, N7 vs. N90, N7 vs. N120, N42 vs. N90, N42 vs. N120 and N90 vs. N120, respectively. Moreover, we performed transcriptomic analysis of the time series of BF development and identified the corresponding stages of biological process enrichment. Finally, quantitative real-time polymerase chain reaction (qRT-PCR) was used to validate the expression of the 16 DEGs during bursal growth and development. These results were consistent with the transcriptome results, indicating that they reflect the expression of the BF during growth and development and that these genes reflect the characteristics of the BF at different times of development and decline. These findings reflect the characteristics of the BF at different time intervals.

Keywords: bursa of Fabricius; development; degradation; chicken; transcriptomic analysis

Citation: Huang, L.; Hu, Y.; Guo, Q.; Chang, G.; Bai, H. Time-Course Transcriptome Landscape of Bursa of Fabricius Development and Degeneration in Chickens. *Agriculture* **2022**, *12*, 1194. https://doi.org/10.3390/agriculture12081194

Academic Editors: Heather Burrow and Michael Goddard

Received: 14 July 2022
Accepted: 8 August 2022
Published: 10 August 2022

1. Introduction

The bursa is the central organ of humoral immunity in birds and is a target organ for a variety of pathogenic microorganisms. The bursa is a unique immune tissue in birds that secretes immune cells (B lymphocytes) to produce specific antibodies and complete a specific immune response. In most poultry production, the bursa is mainly affected by immune diseases (e.g., Infectious Bursal Disease (IBD), Malignant Disease (MD), Avian Leukemia, etc.), resulting in a decrease in the immune status of the bird, which may even lead to the death of the bird as a result of the immune disease [1,2]. The BF was first discovered in 1621 by Italian anatomist Hieronymus Fabricius [3,4]. For a long time, it was thought that the BF was an organ associated with reproduction, until 1956, when Glick [5] discovered that it is a gut-associated lymphoid tissue with immune functions. As a primary immune organ, the BF provides the microenvironment necessary for the development and maturation of B cells in birds [6]. In both humans and mice, B-cell development occurs in the bone marrow. Bird B-cells develop in the BF, a unique organ located dorsal to the cloaca in birds [7] that is critical to early B-lymphocyte proliferation and differentiation [8–10]. Additionally, the BF contains a variety of polypeptides that improve both innate and acquired immune responses. The body's immunological response is a requirement for normal BF growth [11].

The BF originates from hematopoietic stem cells [12], appears in the embryo, develops at a young age, reaches a peak of development at sexual maturity and then gradually degenerates until it disappear [13]. For example, in chickens, at approximately 8 day after embryonic development, the BF forms and B cells colonize it. Approximately 1 week after emergence, the medulla of the BF begins to proliferate. Until approximately 30 day of age, lymphocytes begin to transfer in large numbers to the peripheral lymphoid organs to perform their immune functions. The growth of the BF reaches its peak at approximately 60 day of age, and the adult BF is atrophied and the lymphocytes in the follicular medulla are largely empty at 130 day of age. In contrast to the biological process of growth and maturation followed by organ failure in many organisms, the BF naturally atrophies after maturation in the absence of pathology until it disappears. It is thought that BF degeneration is related to the apoptosis of lymphocytes and secretion of hormones after sexual maturation in birds [14]. To date, the molecular mechanisms underlying BF degeneration are poorly understood.

Most current research on the BF has focused on the relationship between avian diseases, attack toxins, and organs at the phenological and molecular levels, along with a small number of studies on the BF morphology. However, the immune function of the BF is closely related to its structural development, and apoptosis of the BF leads to the destruction of immune function and causes immune deficiency in the animal, which is detrimental to the economic efficiency and healthy development of poultry farming. Therefore, in order to fill the gaps in previous studies on the molecular mechanisms of BF degeneration, we performed transcriptomic analyses of the early developmental period (N7), the middle developmental period (N42), the peak developmental period, the early degeneration period (N90) and the late degeneration period (N120) of BFs to determine the changes in gene expression during the differentiation to atrophy of BF. The degenerative process of BF was dissected from a structural developmental perspective, and the unique developmental degenerative mechanisms of BFs were elucidated.

2. Materials and Methods

2.1. Ethics Statement

All samples were collected in accordance with the guidelines proposed by the China Council on Animal Care and Ministry of Agriculture of the People's Republic of China. The study was approved by the Institutional Animal Care and Use Committee and the School of Animal Experiments Ethics Committee (license number: SYXK [Su] IACUC 2012–0029) of Yangzhou University.

2.2. Sampling

Jiuyuan Black chickens were obtained from the Laboratory of Poultry Genetic Resources Evaluation and Germplasm Utilization at Yangzhou University. All test individuals were hatched and raised under the same conditions. The chickens were sacrificed at 7, 42, 90 or 120 d by severing the jugular vein after anesthesia and were bled out for 5 min before dissection. The BFs were quickly isolated, washed twice with fresh ice-cold phosphate-buffered saline (PBS) and cut in half along the sagittal plane. One-half was fixed in 4% paraformaldehyde and the other was stored in liquid nitrogen.

2.3. Histological Observation

After fixation in a 4% paraformaldehyde solution for 24 h at room temperature, the BFs were trimmed, dehydrated with alcohol and embedded in paraffin. Then, 5 mm serial sections were prepared and stained with hematoxylin and eosin. Sections were mounted with neutral balsam and histopathological changes were observed and photographed under a Nikon Eclipse 90i microscope (Nikon, Tokyo, Japan).

2.4. RNA Extraction, cDNA Library Preparation and RNA Sequencing (RNA-Seq)

The total RNA was extracted using the RNAprep Pure Tissue Kit (TianGen, Beijing, China), according to the manufacturer's protocol. RNA degradation and contamination were visualized on 1% agarose gels, RNA purity was checked using a NanoPhotometer spectrophotometer (Implen, Munich, Germany) and concentrations were determined using the Qubit RNA Assay Kit in a Qubit 2.0 Fluorometer (Life Technologies, Shanghai, China). The total RNA was depleted of ribosomal RNA (rRNA) using the Epicentre Ribo-Zero rRNA Removal Kit (Epicentre, Madison, WI, USA), and fragmented, purified and sequenced using the Illumina HiSeq system (Illumina, Inc., San Diego, CA, USA). Briefly, the mRNA was fragmented and strand cDNA synthesis was primed with random hexamers. After second-strand cDNA synthesis, the transcripts were poly A-tailed for ligation of the sequencing adaptors. The library fragments were size-selected for cDNA fragments 150–200 bp in length, and the pool of cDNA libraries was sequenced by paired-end sequencing on the Illumina HiSeq sequencer (Illumina, San Diego, CA, USA).

2.5. Bioinformatics Analysis

Bioinformatics analysis was performed on the OmicShare platform, a free online platform for data analysis (https://www.omicshare.com/tools, accessed on 11 May 2022). Raw reads were processed using Trimmomatic software. Reads containing poly-N and low-quality reads were removed to obtain clean reads. The clean reads were mapped to the *Gallus gallus* genome GRCg6a (https://asia.ensembl.org/Gallus_gallus/Info/Index, accessed on 11 April 2022) using HISAT2 (version: 2.0.2-beta, Baltimore, MD, USA) software [15]. Transcript abundances were determined using the fragments per kilobase of transcript per million mapped reads (FPKM) values, and genes with FPKM values ≥ 0.1 were retained for further analysis. Differentially expressed genes (DEG) were identified using DESeq2 (version: 3.15, http://www.bioconductor.org/packages/release/bioc/html/DESeq2.html, accessed on 19 April 2022) [16]. Values of $p < 0.05$ and log2 (fold change) > 1.5 were set as the thresholds for significantly differential expression levels. Hierarchical cluster analysis of the DEGs was performed to explore gene expression patterns.

2.6. Gene Expression Pattern Analysis

The Mfuzz R package was used to apply the fuzzy c-means algorithm to profile rhythmic genes according to their expression patterns [17]. The average FPKM (fragments per kilobase per million) value for each gene at different time points was clustered using the Mfuzz package. After standardization, each gene was assigned to a unique cluster according to its membership value. Additionally, ImpulseDE2, which is a Bioconductor R package specifically designed for time series data, was employed in the case-only mode to discern steadily increasing or decreasing expression trajectories from transiently up- or downregulated genes (false discovery rate-adjusted $p < 0.01$) [18].

2.7. Functional Annotation

Gene Ontology (GO) enrichment analysis of DEGs was implemented using clusterProfiler 4.0, in which gene length bias was corrected [19]. GO terms with corrected *p*-values less than 0.05 were considered significantly enriched among the DEGs. Additionally, we used KOBAS software to test the statistical enrichment of DEGs in KEGG pathways [20]. The OmicShare platform was also used for REACTOME enrichment.

2.8. Quantitative Real-Time Polymerase Chain Reaction (qRT-PCR)

The total RNA was isolated using the RNAprep Pure Tissue Kit (TianGen, Beijing, China), according to the manufacturer's protocol. For quantification of gene expression, qRT-PCR was conducted using One-step RT-qPCR kit (TianGen, Beijing, China). The relative mRNA expression was quantified using the $2^{-\Delta\Delta CT}$ method. Beta-actin was used as an internal control. All experiments were repeated thrice. The primers used for qRT-PCR are listed in Table S1.

2.9. Statistical Analysis

Data are expressed as the mean ± standard error (SE). Significance was determined using one-way analysis of variance (ANOVA), as implemented in SPSS (version: 25, New York, NY, USA) software. Differences were considered statistically significant at $p < 0.05$.

3. Results

3.1. Histomorphological Changes in the BF at Different Developmental Times

To determine the changes during BF development, histological observations were performed at 7, 42, 90 and 120 d. The results show that the BFs at 7 d displayed thin and empty reticular submucosa (Figure 1a). The number of smooth muscle cells in the muscle layer was lower, the serosa was thinner, the volume of the lymph nodes was significantly increased and the cells were closely arranged. At 42 d, the follicular cortex was obvious, the boundary between the cortex and medulla was clear, and the medulla was well developed and filled with lymphocytes (Figure 1b). At 90 d, the lymphoid follicles of the BFs were trapped in the interstitium, the medullary lymphocytes decreased in number and the cortical cell walls were dense (Figure 1c). At 120 d, the lymphoid follicles of the BFs were trapped in the interstitium, and the medullary lymphocytes had almost disappeared completely. Significant fibrosis was observed in the foam component (Figure 1d).

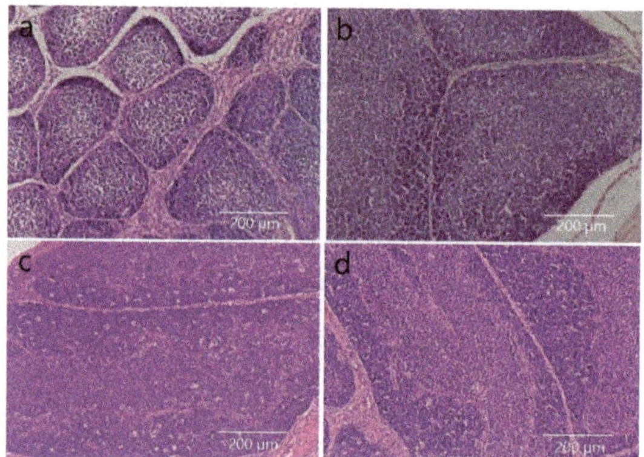

Figure 1. Histomorphological image of a BF at 4 different stages: 7 d (**a**), 42 d (**b**), 90 d (**c**) and 120 d (**d**). Magnification = 200×.

3.2. Sequencing Quality Analysis

To ensure the quality and reliability of the data analysis, the raw data were first filtered, and then the reads with joints, reads containing N and low-quality reads were removed. After filtering the original data, the sequencing error rate and GC content distribution were determined. A total of 487,181,964 clean reads were obtained, accounting for 99.5% of the raw reads. The data are summarized in Table S2. In the 12 samples, the proportion of high-quality reads to original reads was greater than 97%, with 72.7 Gb of high-quality reads, and the GC content per sample was greater than 49.81%. Thus, the reads obtained were of high quality. High-quality reads with a quality score of Q20 exceeded 97.48%, and those with Q30 exceeded 93.86%. These data revealed the sequencing quality of the transcriptome. High-quality reads (high Q20 percentage) were selected from the 12 samples. After quality control, clean reads were compared with the Gallus gallus genome GRCg6a. HISAT2 (version: 2.0.2-beta, MD, USA) software was used to compare clean reads quickly and accurately with the reference genome to obtain the locational information of the reads on the reference genome. To calculate the respective mapping rates of read1 and read2, the total read number was calculated as the sum of read1 and read2, which are shown as clean reads in Table S2. A comparison between the samples and reference genomes is also shown in Table S2.

3.3. Identification and Functional Annotation of Differentially Expressed Genes (DEGs) during BF Development

To determine the changes in gene expression during BF development, we comparatively analyzed the expression levels of genes at different developmental stages. A total of 7605 DEGs were identified in six different comparison groups (N7 vs. N42, N7 vs. N90, N7 vs. N120, N42 vs. N90, N42 vs. N120, N90 vs. N120). According to the results, the number of DEGs tended to decrease as the BF developed. The numbers of DEGs for N7 vs. N42, N90 and N120 were 5914 (3293 upregulated genes and 2621 downregulated genes), 5513 (2852 and 2661) and 4575 (3004 and 1753), respectively (Figure 2a–c, Table S3). The results of Upset (Figure 2g) showed that N7 and the other three time points shared 3375 DEGs, while the number of DEGs for N42 vs. N90 and N120 was 530 (178 upregulated and 352 downregulated genes) and 577 (343 and 234), respectively (Figure 2d,e, Table S3). Moreover, 31 of these co-differentially expressed genes were in the N42, N90 and N120 comparison groups. Finally, we identified 66 DEGs (46 upregulated genes and 20 downregulated genes) in the N90 and N120 comparison groups (Figure 2f, Table S3). In all comparison groups, only one DEG showed differential expression between all comparison groups (Figure 2g).

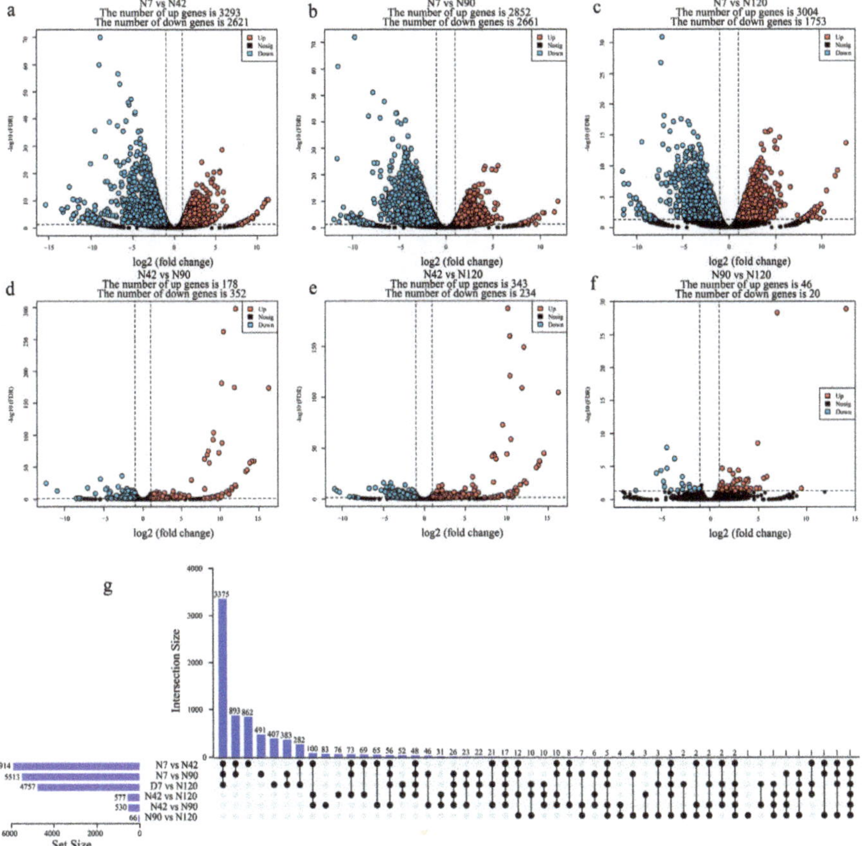

Figure 2. Volcano plot of differentially expressed genes (DEG) identified by comparison groups. (**a**) D7 vs. D42; (**b**) D7 vs. D90, (**c**) D7 vs. D120, (**d**) D42 vs. D90, (**e**) D42 vs. D120, and (**f**) D90 vs. D120. "Up" and "Down" indicate that the expression levels of the DEGs were significantly (FDR $p < 0.05$ and

| log2FoldChange | > 1) higher and lower in the different comparison groups, respectively. "Nosig" indicates that the expression levels of the genes are not significantly different in the different comparison groups; (**g**) UpSet plot of the differentially expressed genes (DEGs) of RNA-Seq.

3.4. Functional Analysis of DEGs Using the GO Database

To investigate the functional associations of common DEGs, we performed GO database analysis using the R clusterProfiler package. In the comparison group of N7 and the other three time points separately, we found that the cell cycle, chromosome, supramolecular complex and chromosomal region were significantly enriched (Figure 3a–c, Table S4). Thus, we suggest that at N7, BFs develop and the rapid development of lymphocytes during this period results in the formation of a complete BF structure. In the comparison groups of N42, N90 and N120, the terms extracellular region, response to endogenous stimulus and extracellular space were significantly enriched, respectively (Figure 3d,e, Table S4). Additionally, in the N90 and N120 comparison groups, we found that DEGs were significantly enriched in cell surface, external side of plasma membrane, extracellular region, extracellular region part, extracellular space, lipase inhibitor activity, phospholipase inhibitor activity, specific granule and other terms (Figure 3f, Table S4).

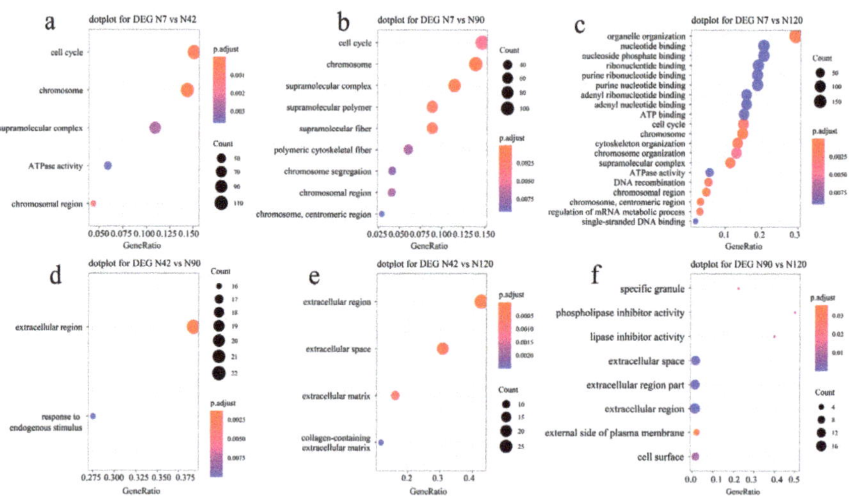

Figure 3. Dot plot of DEGs enriched by the GO database. (**a**) Dot plot of DEGs in the N7 vs. N42 comparison group; (**b**) dot plot of DEGs in the N7 vs. N90 comparison group; (**c**) dot plot of DEGs in the N7 vs. N120 comparison group; (**d**) dot plot of DEGs in the N42 vs. N90 comparison group; (**e**) dot plot of DEGs in the N42 and N120 comparison group; and (**f**) dot plot of DEGs in the N90 vs. N120 comparison group.

3.5. Functional Analysis of DEGs Using the Kyoto Encyclopedia of Genes and Genomes (KEGG) Database

To elucidate the pathways and metabolic pathways involved in DEGs in each comparator group, clusterProfiler was used for KEGG enrichment. Enrichment results based on DEGs in the N7 and three other time point comparison groups showed that the cell cycle, cellular senescence, focal adhesion, MAPK signaling pathway, AGE-RAGE signaling pathway in diabetic complications, C-type lectin receptor signaling pathway and VEGF signaling pathway were significantly enriched in tissue development-related pathways (Figure 4a–c, Table S5). The differential genes N42, N90 and N120 were mainly enriched in the relaxin signaling, cell adhesion molecules, ECM–receptor interaction, hematopoietic cell lineage and other pathways (Figure 4d,e, Table S5). In addition, differential genes for N90 and N120 were significantly enriched in the rheumatoid arthritis, epithelial cell signaling in Helicobacter pylori infection, NF-kappa B signaling, phagosome and phospholipase D signaling pathways (Figure 4f, Table S5). The IL-8, IGH and LBP, which are included in the NF-kappa B signaling pathway, show upregulation in the later stages of bursal development.

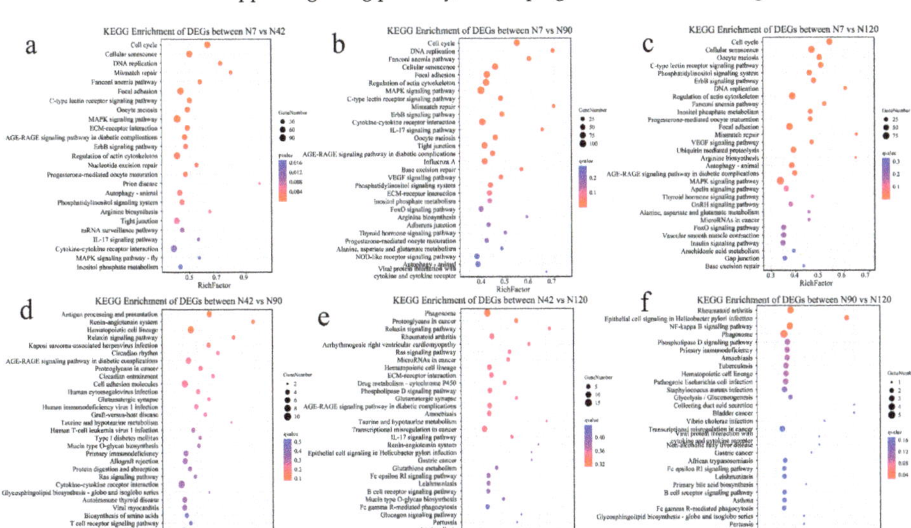

Figure 4. Dot plots of DEGs enriched by KEGG. (**a**) Dot plot of DEGs in the N7 vs. N42 comparison group; (**b**) dot plot of DEGs in the N7 vs. N90 comparison group; (**c**) dot plot of DEGs in the N7 vs. N120 comparison group; (**d**) dot plot of DEGs in the N42 vs. N90 comparison group; (**e**) dot plot of DEGs in the N42 vs. N120 comparison group; (**f**) dot plot of DEGs in the N90 vs. N120 comparison group.

3.6. Functional Analysis of DEGs Using the REACTOME Pathway

REACTOME enrichment analysis of the up- and downregulated DEGs from the N7 and the other three time point comparison groups revealed that the enriched pathways were largely consistent with GO, with most of the upregulated genes involved in mitotic prometaphase, mitotic metaphase and anaphase, resolution of sister chromatid cohesion and other cell and tissue development-related pathways (Figure 5a–c, Table S6). The DEGs identified in N42, N90 and N120 were mainly enriched in the extracellular matrix organization, metabolism of angiotensinogen to angiotensins, activation of matrix metalloproteinases, degradation of the extracellular matrix and release of endostatin-like peptides pathways (Figure 5d,e, Table S6). The DEGs identified by N90 and N120 were enriched in the lactoferrin scavenges iron ions. BPI binds lipopolysaccharides (LPS) on the bacterial surface, complement factor H binds to C3b, factor H displaces Bb in the Cb:Bb complex and complement factor H binds to surface-bound C3b pathways. Factor H binds to the host cell surface and other immune system-related pathways (Figure 5f, Table S6).

3.7. Gene Expression during Different BF Development Stages

To investigate the gene expression patterns during BF development, we performed c-means clustering analysis for 24,357 expressed genes and generated 20 co-expression clusters. Genes in the same cluster showed similar expression patterns. The genes in clusters 1, 10, 3 and 6 were highly expressed at only one of the four developmental stages, indicating that they might have specific functions at the corresponding stages (Figure 6a). Moreover, to understand the kinetics of gene expression during BF development, we used the ImpulseDE2 model to identify the differential expression of all expressed genes at the four time points. This model produced results similar to the differential gene identification;

the BFs tended to be similar over time, indicating that they were relatively more stable during the later stages of development (Figure 6b).

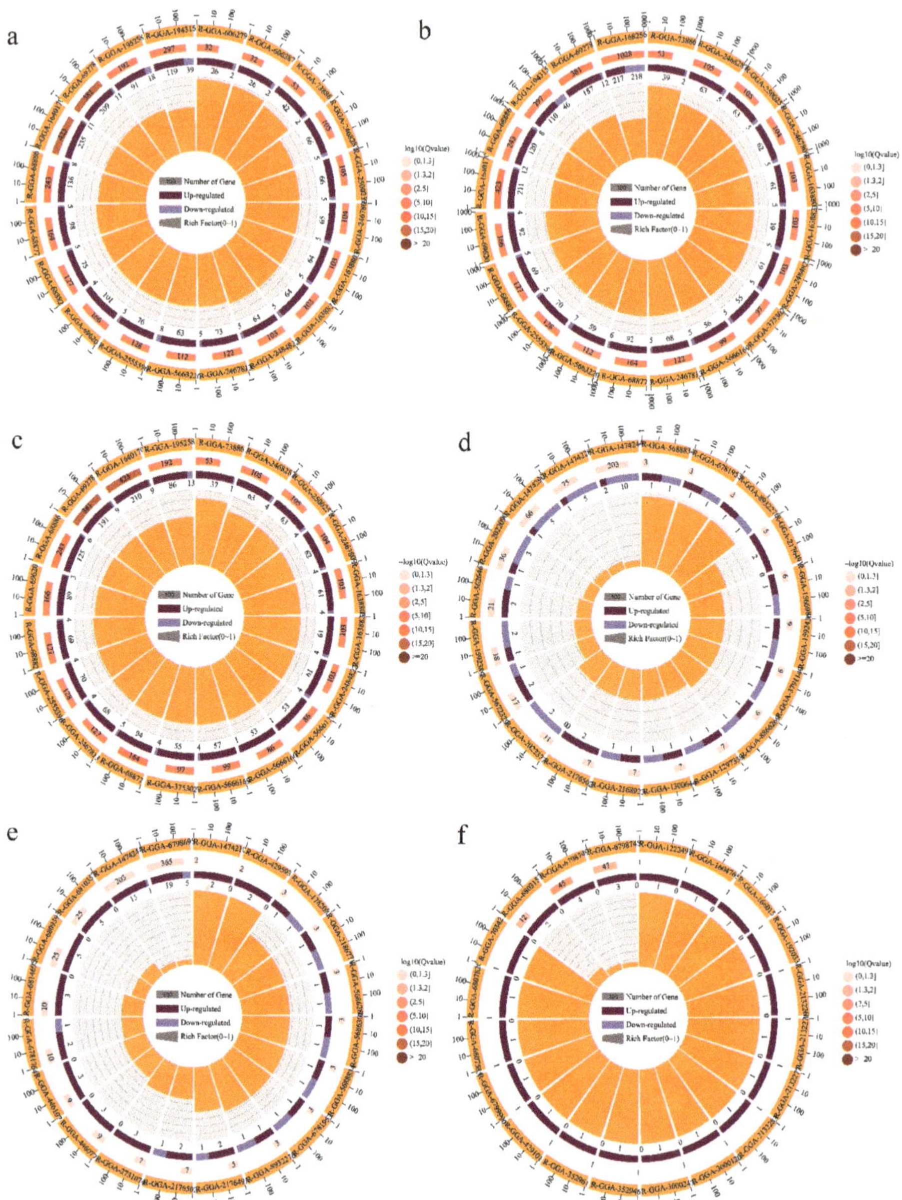

Figure 5. Circular plot of DEGs enriched by REACTOME database analysis. (**a**) Circular plot of DEGs in the N7 vs. N42 comparison group; (**b**) circular plot of DEGs in the N7 vs. N90 comparison group; (**c**) circular plot of DEGs in the N7 vs. N120 comparison group; (**d**) circular plot of DEGs in the N42 vs. N90 comparison group; (**e**) circular plot of DEGs in the N42 vs. N120 comparison group; (**f**) circular plot for DEGs in the N90 vs. N120 comparison group.

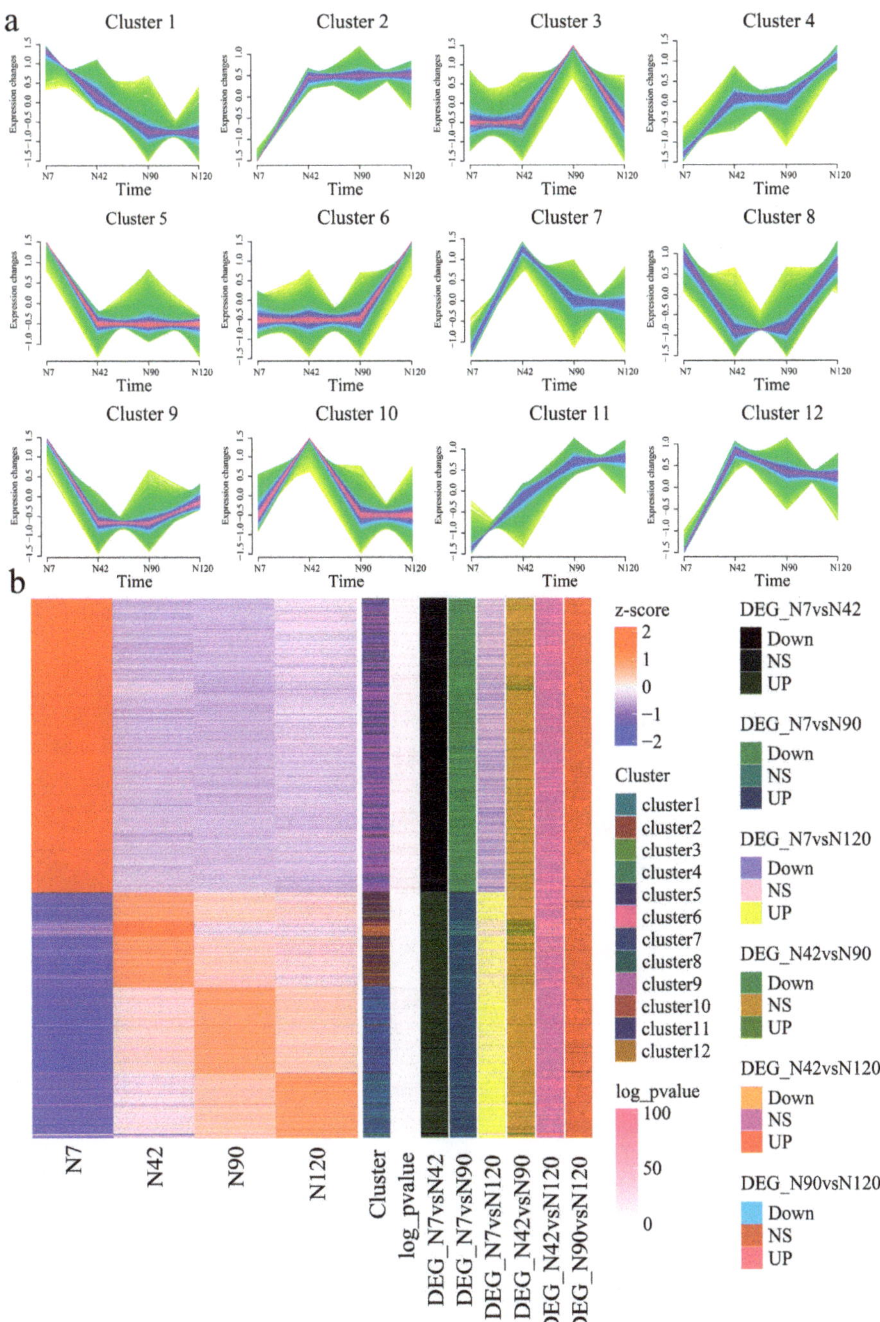

Figure 6. Transcriptome-wide time series cluster of DEGs. (**a**) Cluster analysis of DEGs based on Mfuzz. (**b**) The dynamic changes in RNA sequencing (RNA-seq) are sequentially correlated.

3.8. Validation of DEGs by QRT-PCR

To validate the accuracy of RNA-seq, we selected RT-PCR for 16 genes, of which *SPP1* [21–23], *CTNNB1* [24], *BMF* [25,26], *IL10* [27,28], *TUBB3* [29], *TUBA8* [30], *TUBA3E* [31], *LMNB2* [32,33], *MYL9* [34,35], *LITAF* [36], *CDH11* [37,38], *MYBL1* [39] and *TRAIL* [40,41] were shown to be significantly associated with cell proliferation and apoptosis. In addition, *BF2*, *B2M* and *BF1* are important members of the MHC, which has been shown to be significantly associated with the immune competence of the organism [42–44]. The qRT-PCR and RNA-seq results were consistent (Figure 7). Although the measured gene expression patterns differed slightly from those obtained from transcriptome analysis, the trends were essentially the same.

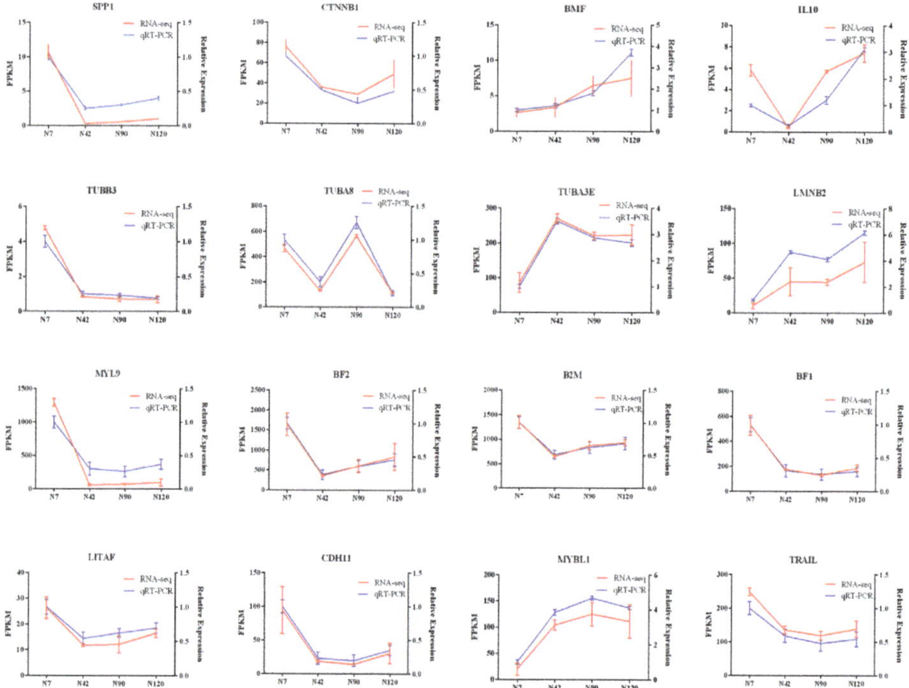

Figure 7. Expression levels of immune genes verified by both quantitative real-time polymerase chain reaction (qRT-PCR) and RNA sequencing (RNA-seq).

4. Discussion

Understanding physiological changes in the development of the BF is an important step in exploring its developmental process. In a previous study, the medullary lymphocytes of the BFs were largely empty at 4.5 months [13]. The results of this experiment were in accordance with this; the medullary lymphocytes of the BFs were largely empty at 120 d of age. In this study, observations of BF tissue at 7, 42, 90 and 120 d revealed the developmental state of the BF at the four time points.

RNA-seq allows the rapid exploration of key genes associated with specific phenotypes or important biological processes [45–48]. Therefore, RNA-seq was used in this study to identify a large number of DEGs that play an important role in the regulation of cell development during BF formation. The extracellular matrix (ECM) pathway plays an important role in tissue and organ morphogenesis and in the maintenance of cellular and tissue structure and function. The interaction of substances in the ECM signaling pathway leads to the direct or indirect control of cellular activity [49,50]. Moreover, the

MAPK signaling pathway plays an important role in B-cell development [51–54], along with the Wnt signaling pathway [55]. Genes in the Wnt signaling pathway, including Wnt protein, Frizzled (Fzd) receptor and lymphatic enhancer factor (LEF), have been found to be highly expressed in the progenitor cells of B cells [56–58]. Additionally, Fzd-9 knockout mice exhibited a significant reduction in B cells [59]. In this study, the DEGs of the transcriptome by time series revealed that the key signaling pathways involved in cell growth, development and adhesion, such as the Wnt signaling pathway, MAPK signaling pathway and ECM receptor interaction, were significantly enriched. In addition, the category of cell cycle was significant enriched in the early stage of BF development. The previous study showed that the relationship between cell cycle processes and development is complex and characterized by interdependence. At the level of the individual cell, this interrelationship has an impact on pattern formation and cell morphogenesis. At the supracellular level, this interrelationship affects hyphal tissue function and organ growth. In general, developmental signals not only guide cell cycle progression, but also set the framework for cell cycle regulation by identifying cell type-specific cell cycle patterns [60]. At the same time, the enrichment of differentially expressed genes showed that the category of NF-kappa B pathway was significantly enriched in the later stages of bursal development, a result that suggests that the NF-kappa B pathway plays an important role in the degenerative stage of bursal development. Analysis of the genes in the NF-kappa B pathway revealed that IL-8 [61,62], IGH and LBP [63–65] showed upregulated expression in the later stages. In addition, the key cell cycle pathway related to development [66] was significantly enriched in the comparison group at N7 compared to other time points, demonstrating that the development of the BF predominantly occurs early in the bursa.

Interestingly, the expression levels of some genes, including *SPP1*, *BMP*, *IL10*, *TUBB3* and *MYL9*, also appear to be different between the four BF development stages. Among the DEGs, *SPP1*, which is a secreted protein, may mediate the expression of interferon and interleukin-12 [67,68]. Moreover, *SPP1*, also known as early type 1 T lymphocyte activating protein (ETA-1), may be involved in early BF development through the Toll-like receptor signaling pathway [69,70]. Furthermore, we found that the MHC superfamily members *B2M*, *BF2* and *BF1* are also differentially expressed during BF development; however, the exact role they play in BF development requires further study.

The degeneration of the BF is gradually initiated after sexual maturation in birds and is caused by the mature differentiation of B lymphocytes in the BF. *SPP1*, a gene involved in early BF development, is mainly involved in BF development through the Toll-like receptor signaling pathway [71,72]. It is involved in the regulation of BF development by mediating the expression of interferon and interleukin-12 [73]. In this study, *SPP1* was also found to be involved in BF development, mainly through the Toll-like receptor signaling pathway in the early stage of BF development. In addition, the *BMF* gene, a member of the Bcl-2 family, is an important regulatory factor. The protein includes a BH3-only structural domain, which binds to and releases the anti-apoptotic proteins Bax and Bak, which in turn activate apoptosis [74]. In this study, we found that the expression of *BMF* continued to increase as the developmental time progressed, and *BMF* may induce apoptosis in BF cells by participating in the biological process of anoikis.

In conclusion, in this study, we observed the tissue structure of the BF at different developmental stages at the tissue level and found that the growth of the BF showed a process of gradual shrinkage after continued maturation with increasing age. Second, transcriptomic analysis of the time sequences identified a series of genes associated with the development and decline of the BF. However, the specific regulatory mechanisms require further research, and this study lays the foundation for the development and decline of the BF. Overall, this study elucidates the regulatory role of differential genes throughout the process of BF development and atrophy and provides a theoretical basis for selecting more immunocompetent birds through molecular breeding.

Supplementary Materials: The following supporting information can be downloaded at: https://www.mdpi.com/article/10.3390/agriculture12081194/s1, Table S1: Information of RNA-seq data; Table S2: The list of primers used in the present study; Table S3: All DEG expression changes in each comparison group; Table S4: The list of DEGs enriched by KEGG; Table S5: The list of DEGs enriched by GO; Table S6: The list of DEGs enriched by REACTOME.

Author Contributions: L.H., formal analysis, writing—original draft; Y.H., H.B. and G.C., writing—review and editing; Q.G., visualization; H.B. and G.C., funding acquisition. All authors submitted comments on the draft. All authors have read and agreed to the published version of the manuscript.

Funding: This research was supported by the earmarked fund for CARS (grant no. CARS-41-G26) and Open Project of Key Laboratory for Poultry Genetics and Breeding of Jiangsu Province (grant no. JQLAB-KF-202102).

Institutional Review Board Statement: All samples were collected in accordance with the guidelines proposed by the China Council on Animal Care and Ministry of Agriculture of the People's Republic of China. The study was approved by the Institutional Animal Care and Use Committee and the School of Animal Experiments Ethics Committee (license number: SYXK [Su] IACUC 2012-0029) of Yangzhou University.

Acknowledgments: The authors thank to the supported by the earmarked fund for CARS (grant no. CARS-41-G26) and Open Project of Key Laboratory for Poultry Genetics and Breeding of Jiangsu Province (grant no. JQLAB-KF-202102).

Conflicts of Interest: The authors declare there was no conflict of interest.

References

1. Paramithiotis, E.; Ratcliffe, M.J. B cell emigration directly from the cortex of lymphoid follicles in the bursa of Fabricius. *Eur. J. Immunol.* **1994**, *24*, 458–463. [CrossRef] [PubMed]
2. Islam, M.N.; Khan, M.Z.I.; Jahan, M.; Fujinaga, R.; Yanai, A.; Kokubu, K.; Shinoda, K. Histomorphological study on prenatal development of the lymphoid organs of native chickens of Bangladesh. *Pak. Vet. J.* **2012**, *32*, 175–178.
3. Schat, K.A.; Kaspers, B.; Kaiser, P. *Avian Immunology*, 2nd ed.; Elsevier: Amsterdam, The Netherlands, 2012.
4. Ribatti, D.; Porzionato, A.; Emmi, A.; de Caro, R. The bursa of Hieronymus Fabricius ab Aquapendente: From original iconography to most recent research. *Rom. J. Morphol. Embryol.* **2020**, *61*, 583–585. [CrossRef] [PubMed]
5. Glick, B.; Chang, T.S.; Jaap, R.G. The Bursa of Fabricius and Antibody Production. *Poult. Sci.* **1956**, *35*, 224–225. [CrossRef]
6. Monson, M.S.; van Goor, A.G.; Ashwell, C.M.; Persia, E.M.; Rothschild, M.F.; Schmidt, C.J.; Lamont, S.J. Immunomodulatory effects of heat stress and lipopolysaccharide on the bursal transcriptome in two distinct chicken lines. *BMC Genom.* **2018**, *19*, 643. [CrossRef]
7. Kaspers, B.; Schat, K.; Göbel, T.; Vervelde, L. *Avian Immunology*, 3rd ed.; Elsevier: Amsterdam, The Netherlands, 2021.
8. Cooper, M.D.; Raymond, D.A.; Peterson, R.D.; South, M.A.; Good, R.A. The functions of the thymus system and the bursa system in the chicken. *J. Exp. Med.* **1966**, *123*, 75–102. [CrossRef]
9. Salvi, E.; Buffa, R.; Renda, T.G. Ontogeny, distribution and amine/peptide colocalization of chromogranin A- and B-immunoreactive cells in the chicken gizzard and antrum. *Anat. Embryol.* **1995**, *192*, 547–555. [CrossRef]
10. Lassila, O.; Eskola, J.; Toivanen, P. Prebursal stem cells in the intraembryonic mesenchyme of the chick embryo at 7 days of incubation. *J. Immunol.* **1979**, *123*, 2091–2094.
11. Otsubo, Y.; Chen, N.; Kajiwara, E.; Horiuchi, H.; Matsuda, H.; Furusawa, S. Role of bursin in the development of B lymphocytes in chicken embryonic Bursa of Fabricius. *Dev. Comp. Immunol.* **2001**, *25*, 485–593. [CrossRef]
12. Le Douarin, N.M.; Houssaint, E.; Jotereau, F.V.; Belo, M. Origin of Hemopoietic Stem Cells in Embryonic Bursa of Fabricius and Bone Marrow Studied through Interspecific Chimeras. *Proc. Natl. Acad. Sci. USA* **1975**, *72*, 2701–2705. [CrossRef]
13. Pike, K.A.; Baig, E.; Ratcliffe, M.J. The avian B-cell receptor complex: Distinct roles of Igalpha and Igbeta in B-cell development. *Immunol. Rev.* **2004**, *197*, 10–25. [CrossRef] [PubMed]
14. Makino, K.; Omachi, R.; Suzuki, H.; Tomobe, K.; Kawashima, T.; Nakajima, T.; Kawashima-Ohya, Y. Apoptosis Occurs during Early Development of the Bursa of Fabricius in Chicken Embryos. *Biol. Pharm. Bull.* **2014**, *37*, 1982–1985. [CrossRef] [PubMed]
15. Kim, D.; Langmead, B.; Salzberg, S.L. HISAT: A fast spliced aligner with low memory requirements. *Nat. Methods* **2015**, *12*, 357–360. [CrossRef] [PubMed]
16. Love, M.I.; Huber, W.; Anders, S. Moderated estimation of fold change and dispersion for RNA-seq data with DESeq2. *Genome Biol.* **2014**, *15*, 550. [CrossRef] [PubMed]
17. Kumar, L.; Futschik, M.E. Mfuzz: A software package for soft clustering of microarray data. *Bioinformation* **2007**, *2*, 5–7. [CrossRef]
18. Fischer, D.S.; Theis, F.J.; Yosef, N. Impulse model-based differential expression analysis of time course sequencing data. *Nucleic Acids Res.* **2018**, *46*, e119. [CrossRef]

19. Wu, T.; Hu, E.; Xu, S.; Chen, M.; Guo, P.; Dai, Z.; Feng, T.; Zhou, L.; Tang, W.; Zhan, L.; et al. clusterProfiler 4. *0: A universal enrichment tool for interpreting omics data. Innovation* **2021**, *2*, 100141.
20. Bu, D.; Luo, H.; Huo, P.; Wang, Z.; Zhang, S.; He, Z.; Wu, Y.; Zhao, L.; Liu, J.; Guo, J.; et al. KOBAS-i: Intelligent prioritization and exploratory visualization of biological functions for gene enrichment analysis. *Nucleic Acids Res.* **2021**, *49*, W317–W325. [CrossRef]
21. Tu, Y.; Chen, C.; Fan, G. Association between the expression of secreted phosphoprotein-related genes and prognosis of human cancer. *BMC Cancer* **2019**, *19*, 1230. [CrossRef]
22. Saleh, S.; Thompson, D.E.; McConkey, J.; Murray, P.; Moorehead, R.A. Osteopontin regulates proliferation, apoptosis, and migration of murine claudin-low mammary tumor cells. *BMC Cancer* **2016**, *16*, 359. [CrossRef]
23. Dalal, S.; Zha, Q.; Singh, M.; Singh, K. Osteopontin-stimulated apoptosis in cardiac myocytes involves oxidative stress and mitochondrial death pathway: Role of a pro-apoptotic protein BIK. *Mol. Cell Biochem.* **2016**, *418*, 22683–22690. [CrossRef]
24. Ming, M.; Wang, S.; Wu, W.; Senyuk, V.; le Beau, M.M.; Nucifora, G.; Qian, Z. Activation of Wnt/β-catenin protein signaling induces mitochondria-mediated apoptosis in hematopoietic progenitor cells. *J. Biol. Chem.* **2012**, *287*, 22683–22690. [CrossRef] [PubMed]
25. Grespi, F.; Soratroi, C.; Krumschnabel, G.; Sohm, B.; Ploner, C.; Geley, S.; Hengst, L.; Häcker, G.; Villunger, A. BH3-only protein Bmf mediates apoptosis upon inhibition of CAP-dependent protein synthesis. *Cell Death Differ.* **2010**, *17*, 1672–1683. [CrossRef] [PubMed]
26. Piñon, J.D.; Labi, V.; Egle, A.; Villunger, A. Bim and Bmf in tissue homeostasis and malignant disease. *Oncogene* **2008**, *27*, S41–S52. [CrossRef] [PubMed]
27. Schmidt, M.; Lügering, N.; Pauels, H.G.; Schulze-Osthoff, K.; Domschke, W.; Kucharzik, T. IL-10 induces apoptosis in human monocytes involving the CD95 receptor/ligand pathway. *Eur. J. Immunol.* **2000**, *30*, 1769–1777. [CrossRef]
28. Wang, Z.Q.; Bapat, A.S.; Rayanade, R.J.; Dagtas, A.S.; Hoffmann, M.K. Interleukin-10 induces macrophage apoptosis and expression of CD16 (FcgammaRIII) whose engagement blocks the cell death programme and facilitates differentiation. *Immunology* **2001**, *102*, 331–337. [CrossRef]
29. Sadeghi, I.; Behmanesh, M.; Chashmi, N.A.; Sharifi, M.; Soltani, B.M. 6-Methoxy Podophyllotoxin Induces Apoptosis via Inhibition of TUBB3 and TOPIIA Gene Expressions in 5637 and K562 Cancer Cell Lines. *Cell J.* **2015**, *17*, 502–509.
30. Di Donato, N.; Timms, A.E.; Aldinger, K.A.; Mirzaa, G.M.; Bennett, J.T.; Collins, S.; Olds, C.; Mei, D.; Chiari, S.; Carvill, G.; et al. Analysis of 17 genes detects mutations in 81% of 811 patients with lissencephaly. *Genet. Med.* **2018**, *20*, 1354–1364. [CrossRef]
31. Wu, I.; Shin, S.C.; Cao, Y.; Bender, I.K.; Jafari, N.; Feng, G.; Lin, S.; Cidlowski, J.A.; Schleimer, R.P.; Lu, N.Z. Selective glucocorticoid receptor translational isoforms reveal glucocorticoid-induced apoptotic transcriptomes. *Cell Death Dis.* **2013**, *4*, e453. [CrossRef]
32. Zhao, C.-C.; Chen, J.; Zhang, L.-Y.; Liu, H.; Zhang, C.-G.; Liu, Y. Lamin B2 promotes the progression of triple negative breast cancer via mediating cell proliferation and apoptosis. *Biosci. Rep.* **2021**, *41*, BSR20203874. [CrossRef]
33. Dong, C.-H.; Jiang, T.; Yin, H.; Song, H.; Zhang, Y.; Geng, H.; Shi, P.-C.; Xu, Y.-X.; Gao, H.; Liu, L.-Y.; et al. LMNB2 promotes the progression of colorectal cancer by silencing p21 expression. *Cell Death Dis.* **2021**, *12*, 331. [CrossRef] [PubMed]
34. Feng, M.; Dong, N.; Zhou, X.; Ma, L.; Xiang, R. Myosin light chain 9 promotes the proliferation, invasion, migration and angiogenesis of colorectal cancer cells by binding to Yes-associated protein 1 and regulating Hippo signaling. *Bioengineered* **2022**, *13*, 96–106. [CrossRef] [PubMed]
35. Martin, K.; Pritchett, J.; Llewellyn, J.; Mullan, A.F.; Athwal, V.S.; Dobie, R.; Harvey, E.; Zeef, L.; Farrow, S.; Streuli, C.; et al. PAK proteins and YAP-1 signalling downstream of integrin beta-1 in myofibroblasts promote liver fibrosis. *Nat. Commun.* **2016**, *7*, 12502. [CrossRef] [PubMed]
36. Shi, Y.; Kuai, Y.; Lei, L.; Weng, Y.; Berberich-Siebelt, F.; Zhang, X.; Wang, J.; Zhou, Y.; Jiang, X.; Ren, G.; et al. The feedback loop of LITAF and BCL6 is involved in regulating apoptosis in B cell non-Hodgkin's-lymphoma. *Oncotarget* **2016**, *7*, 77444–77456. [CrossRef]
37. Alimperti, S.; Andreadis, S.T. CDH2 and CDH11 act as regulators of stem cell fate decisions. *Stem. Cell Res.* **2015**, *14*, 270–282. [CrossRef] [PubMed]
38. Li, L.; Ying, J.; Li, H.; Zhang, Y.; Shu, X.; Fan, Y.; Tan, J.; Cao, Y.; Tsao, S.W.; Srivastava, G.; et al. The human cadherin 11 is a pro-apoptotic tumor suppressor modulating cell stemness through Wnt/β-catenin signaling and silenced in common carcinomas. *Oncogene* **2012**, *31*, 3901–3912. [CrossRef]
39. Bolcun-Filas, E.; Bannister, L.A.; Barash, A.; Schimenti, K.J.; Hartford, S.A.; Eppig, J.J.; Handel, M.A.; Shen, L.; Schimenti, J.C. A-MYB (MYBL1) transcription factor is a master regulator of male meiosis. *Development* **2011**, *138*, 3319–3330. [CrossRef]
40. Allen, J.E.; El-Deiry, W.S. Regulation of the human TRAIL gene. *Cancer Biol. Ther.* **2012**, *13*, 1143–1151. [CrossRef]
41. Dai, X.; Zhang, J.; Arfuso, F.; Chinnathambi, A.; Zayed, M.E.; Alharbi, S.A.; Kumar, A.P.; Ahn, K.S.; Sethi, G. Targeting TNF-related apoptosis-inducing ligand (TRAIL) receptor by natural products as a potential therapeutic approach for cancer therapy. *Exp. Biol. Med.* **2015**, *240*, 760–773. [CrossRef]
42. Jongsma, M.L.M.; Guarda, G.; Spaapen, R.M. The regulatory network behind MHC class I expression. *Mol. Immunol.* **2019**, *113*, 16–21. [CrossRef]
43. Dunon, D.; Salomonsen, J.; Skjødt, K.; Kaufman, J.; Imhof, B.A. Ontogenic appearance of MHC class I (B-F) antigens during chicken embryogenesis. *Dev. Immunol.* **1990**, *1*, 127–135. [CrossRef]

44. Cheng, S.; Liu, X.; Mu, J.; Yan, W.; Wang, M.; Chai, H.; Sha, Y.; Jiang, S.; Wang, S.; Ren, Y.; et al. Intense Innate Immune Responses and Severe Metabolic Disorders in Chicken Embryonic Visceral Tissues Caused by Infection with Highly Virulent Newcastle Disease Virus Compared to the Avirulent Virus: A Bioinformatics Analysis. *Viruses* **2022**, *14*, 911. [CrossRef] [PubMed]
45. Jax, E.; Wink, M.; Kraus, R.H.S. Avian transcriptomics: Opportunities and challenges. *J. Ornithol.* **2018**, *159*, 599–629. [CrossRef]
46. Shah, A.U.; Li, Y.; Ouyang, W.; Wang, Z.; Zuo, J.; Shi, S.; Yu, Q.; Lin, J.; Yang, Q. From nasal to basal: Single-cell sequencing of the bursa of Fabricius highlights the IBDV infection mechanism in chickens. *Cell Biosci.* **2021**, *11*, 212. [CrossRef] [PubMed]
47. Zhang, Y.; Zhou, Y.; Sun, G.; Li, K.; Li, Z.; Su, A.; Liu, X.; Li, G.; Jiang, R.; Han, R.; et al. Transcriptome profile in bursa of Fabricius reveals potential mode for stress-influenced immune function in chicken stress model. *BMC Genom.* **2018**, *19*, 918. [CrossRef]
48. Rai, M.F.; Cai, L.; Tycksen, E.D.; Chamberlain, A.; Keener, J. RNA-Seq analysis reveals sex-dependent transcriptomic profiles of human subacromial bursa stratified by tear etiology. *J. Orthop. Res.* **2022**. [CrossRef]
49. Lu, P.; Takai, K.; Weaver, V.M.; Werb, Z. Extracellular matrix degradation and remodeling in development and disease. *Cold Spring Harb. Perspect. Biol.* **2011**, *3*, a005058. [CrossRef]
50. Poltavets, V.; Kochetkova, M.; Pitson, S.M.; Samuel, M.S. The Role of the Extracellular Matrix and Its Molecular and Cellular Regulators in Cancer Cell Plasticity. *Front. Oncol.* **2018**, *8*, 431. [CrossRef]
51. Khiem, D.; Cyster, J.G.; Schwarz, J.J.; Black, B.L. A p38 MAPK-MEF2C pathway regulates B-cell proliferation. *Proc. Natl. Acad. Sci. USA* **2008**, *105*, 17067–17072. [CrossRef]
52. Kotelnikova, E.; Kiani, N.A.; Messinis, D.; Pertsovskaya, I.; Pliaka, V.; Bernardo-Faura, M.; Rinas, M.; Vila, G.; Zubizarreta, I.; Pulido-Valdeolivas, I.; et al. MAPK pathway and B cells overactivation in multiple sclerosis revealed by phosphoproteomics and genomic analysis. *Proc. Natl. Acad. Sci. USA* **2019**, *116*, 9671–9676. [CrossRef]
53. McLaurin, J.D.; Weiner, O.D. Multiple sources of signal amplification within the B-cell Ras/MAPK pathway. *Mol. Biol. Cell* **2019**, *30*, 1610–1620. [CrossRef] [PubMed]
54. Zhang, W.; Liu, H.T. MAPK signal pathways in the regulation of cell proliferation in mammalian cells. *Cell Res.* **2002**, *12*, 9–18. [CrossRef] [PubMed]
55. Yu, Q.; Quinn, W.J., 3rd; Salay, T.; Crowley, J.E.; Cancro, M.P.; Sen, J.M. Role of beta-catenin in B cell development and function. *J. Immunol.* **2008**, *181*, 3777–3783. [CrossRef] [PubMed]
56. Wu, Q.L.; Zierold, C.; Ranheim, E.A. Dysregulation of Frizzled 6 is a critical component of B-cell leukemogenesis in a mouse model of chronic lymphocytic leukemia. *Blood* **2009**, *113*, 3031–3039. [CrossRef] [PubMed]
57. Ljungberg, J.K.; Kling, J.C.; Tran, T.T.; Blumenthal, A. Functions of the WNT Signaling Network in Shaping Host Responses to Infection. *Front. Immunol.* **2019**, *10*, 2521. [CrossRef]
58. Pai, S.G.; Carneiro, B.A.; Mota, J.M.; Costa, R.; Leite, C.A.; Barroso-Sousa, R.; Kaplan, J.B.; Chae, Y.K.; Giles, F.J. Wnt/beta-catenin pathway: Modulating anticancer immune response. *J. Hematol. Oncol.* **2017**, *10*, 101. [CrossRef]
59. Ranheim, E.A.; Kwan, H.C.K.; Reya, T.; Wang, Y.-K.; Weissman, I.L.; Francke, U. Frizzled 9 knock-out mice have abnormal B-cell development. *Blood* **2005**, *105*, 2487–2494. [CrossRef]
60. Coffman, J.A. Cell Cycle Development. *Dev. Cell* **2004**, *6*, 321–327. [CrossRef]
61. Hagimoto, N.; Kuwano, K.; Kawasaki, M.; Yoshimi, M.; Kaneko, Y.; Kunitake, R.; Maeyama, T.; Tanaka, T.; Hara, N. Induction of interleukin-8 secretion and apoptosis in bronchiolar epithelial cells by Fas ligation. *Am. J. Respir. Cell Mol. Biol.* **1999**, *21*, 436–445. [CrossRef]
62. Kettritz, R.; Gaido, M.L.; Haller, H.; Luft, F.C.; Jennette, C.J.; Falk, R.J. Interleukin-8 delays spontaneous and tumor necrosis factor-alpha-mediated apoptosis of human neutrophils. *Kidney Int.* **1998**, *53*, 84–91. [CrossRef]
63. Yu, Y.; Wu, X.; Pu, J.; Luo, P.; Ma, W.; Wang, J.; Wei, J.; Wang, Y.; Fei, Z. Lycium barbarum polysaccharide protects against oxygen glucose deprivation/reoxygenation-induced apoptosis and autophagic cell death via the PI3K/Akt/mTOR signaling pathway in primary cultured hippocampal neurons. *Biochem. Biophys. Res. Commun.* **2018**, *495*, 1187–1194. [CrossRef] [PubMed]
64. Li, D.; Wang, W.J.; Wang, Y.Z.; Wang, Y.B.; Li, Y.L. Lobaplatin promotes (125)I-induced apoptosis and inhibition of proliferation in hepatocellular carcinoma by upregulating PERK-eIF2α-ATF4-CHOP pathway. *Cell Death Dis.* **2019**, *10*, 744. [CrossRef] [PubMed]
65. Lou, L.; Chen, G.; Zhong, B.; Liu, F. Lycium barbarum polysaccharide induced apoptosis and inhibited proliferation in infantile hemangioma endothelial cells via down-regulation of PI3K/AKT signaling pathway. *Biosci. Rep.* **2019**, *39*, BSR20191182. [CrossRef] [PubMed]
66. Duronio, R.J.; Xiong, Y. Signaling pathways that control cell proliferation. *Cold Spring Harb. Perspect. Biol.* **2013**, *5*, a008904. [CrossRef]
67. Shan, M.; Yuan, X.; Song, L.Z.; Roberts, L.; Zarinkamar, N.; Seryshev, A.; Zhang, Y.; Hilsenbeck, S.; Chang, S.H.; Dong, C.; et al. Cigarette smoke induction of osteopontin (SPP1) mediates T(H)17 inflammation in human and experimental emphysema. *Sci. Transl. Med.* **2012**, *4*, 117ra9. [CrossRef] [PubMed]
68. Poole, D.H.; Ndiaye, K.; Pate, J.L. Expression and regulation of secreted phosphoprotein 1 in the bovine corpus luteum and effects on T lymphocyte chemotaxis. *Reproduction* **2013**, *146*, 527–537. [CrossRef]
69. Ashkar, S.; Weber, G.F.; Panoutsakopoulou, V.; Sanchirico, M.E.; Jansson, M.; Zawaideh, S.; Rittling, S.R.; Denhardt, D.T.; Glimcher, M.J.; Cantor, H. Eta-1 (osteopontin): An early component of type-1 (cell-mediated) immunity. *Science* **2000**, *287*, 860–864. [CrossRef]
70. Castello, L.M.; Raineri, D.; Salmi, L.; Clemente, N.; Vaschetto, R.; Quaglia, M.; Garzaro, M.; Gentilli, S.; Navalesi, P.; Cantaluppi, V.; et al. Osteopontin at the Crossroads of Inflammation and Tumor Progression. *Mediat. Inflamm.* **2017**, *2017*, 4049098. [CrossRef]

71. Prince, C.W.; Oosawa, T.; Butler, W.T.; Tomana, M.; Bhown, A.S.; Bhown, M.; Schrohenloher, R.E. Isolation, characterization, and biosynthesis of a phosphorylated glycoprotein from rat bone. *J. Biol. Chem.* **1987**, *262*, 2900–2907. [CrossRef]
72. van Rooij, E.; Sutherland, L.B.; Thatcher, J.E.; DiMaio, J.M.; Naseem, R.H.; Marshall, W.S.; Hill, J.A.; Olson, E.N. Dysregulation of microRNAs after myocardial infarction reveals a role of miR-29 in cardiac fibrosis. *Proc. Natl. Acad. Sci. USA* **2008**, *105*, 13027–13032. [CrossRef]
73. Heinemeier, K.M.; Skovgaard, D.; Bayer, M.L.; Qvortrup, K.; Kjaer, A.; Kjaer, M.; Magnusson, S.P.; Kongsgaard, M. Uphill running improves rat Achilles tendon tissue mechanical properties and alters gene expression without inducing pathological changes. *J. Appl. Physiol.* **2012**, *113*, 827–836. [CrossRef] [PubMed]
74. Sjöblom, B.; Salmazo, A.; Djinović-Carugo, K. Alpha-actinin structure and regulation. *Cell Mol. Life Sci.* **2008**, *65*, 2688–2701. [CrossRef] [PubMed]

 agriculture

Article

Comprehensive Profiling of Circular RNAs in Goat Dermal Papilla Cells and Prediction of Their Modulatory Roles in Hair Growth

Sen Ma [1,2,3], Xiaochun Xu [4], Xiaolong Wang [5], Yuxin Yang [5], Yinghua Shi [1,2,3,*] and Yulin Chen [5,*]

1. College of Animal Science and Technology, Henan Agricultural University, Zhengzhou 450002, China
2. Henan Key Laboratory of Innovation and Utilization of Grassland Resources, Zhengzhou 450002, China
3. Henan Engineering Research Center for Forage, Zhengzhou 450002, China
4. Collaborative Innovation Center for Food Production and Safety, North Minzu University, Yinchuan 750000, China
5. Key Laboratory of Animal Genetics, Breeding and Reproduction of Shaanxi Province, College of Animal Science and Technology, Northwest A&F University, Xianyang 712100, China

* Correspondence: annysyh@henau.edu.cn (Y.S.); chenyulin@nwafu.edu.cn (Y.C.)

Abstract: Circular RNAs (circRNAs) are capable of finely modulating gene expression at transcriptional and post-transcriptional levels; however, their characters in dermal papilla cells (DPCs)—the signaling center of hair follicle—are still obscure. Herein, we established a comprehensive atlas of circRNAs in DPCs and their skin counterparts—dermal fibroblasts (DFs)—from cashmere goats. In terms of the results, a sum of 3706 circRNAs were bioinformatically identified. Subsequent analysis suggested that the detected transcripts exhibited several prominent genomic features, including exons as their main sources. Compared with DFs, 76 circRNAs significantly displayed higher abundances in goat DPCs, with 45 transcripts markedly exhibiting adverse trends ($p < 0.05$). Furthermore, potential roles and underlying molecular mechanisms of circRNAs in goat DPCs were speculated through constructing their possible regulatory networks with mRNAs and microRNAs (miRNAs). We found that the circRNAs may serve as miRNA sponges to alleviate three hair growth-related functional genes (*HOXC8*, *RSPO1*, and *CCBE1*) of DPCs from miRNAs-imposed post-transcriptional modulation, further facilitating two critical processes (*HOXC8* and *RSPO1*: hair follicle stem cell activation; *CCBE1*: follicular angiogenesis) closely involved in hair growth. In addition, we also speculated that two intron-derived circRNAs (chi_circ_0005569 and chi_circ_0005570) possibly affect the expression of their host gene *CCBE1* at a transcriptional level in the nucleus. The above results demonstrated that circRNAs are abundantly expressed in goat DPCs, and certain circRNAs are potential participators in hair growth via the effects on the levels of related functional genes. Our study offers a preliminary clue for researchers hoping to untangle the roles of non-coding RNAs in hair growth.

Keywords: circRNAs; DPCs; cashmere goats; *HOXC8*; *RSPO1*; *CCBE1*

Citation: Ma, S.; Xu, X.; Wang, X.; Yang, Y.; Shi, Y.; Chen, Y. Comprehensive Profiling of Circular RNAs in Goat Dermal Papilla Cells and Prediction of Their Modulatory Roles in Hair Growth. *Agriculture* **2022**, *12*, 1306. https://doi.org/10.3390/agriculture12091306

Academic Editors: Heather Burrow and Michael Goddard

Received: 31 May 2022
Accepted: 5 August 2022
Published: 25 August 2022

1. Introduction

Dermal papilla cells (DPCs) are a group of specialized fibroblasts located at the base of the hair follicle (HF), the mini-organ responsible for the continuous production of mammalian hair in the skin [1]. Previous studies have validated that the DPCs are the signaling center of the HF and determine the growth of hair via modulating several key biological processes in hair growth [2,3]. Meanwhile, a few reports have demonstrated that such a unique capacity of DPCs is decided by the intrinsic expressions of signature genes in the cells. For example, the initial step of hair growth—activation of hair follicle stem cells (HFSCs)—is under the genetic control of *Hoxc8* and *RSPO1* [4,5], whose overexpression results in precious HF development and hair overgrowth. In addition, follicular angiogenesis, a critical event closely related to active hair growth, is stimulated by several potent angiogenic factors (e.g., VEGF) highly expressed in DPCs [6–8]. Although these studies highlighted the functional importance of the related genes in hair biology, how their expressions are precisely regulated remains as yet unknown.

In recent years, a class of enclosed non-coding RNAs called circular RNAs (circRNAs) gradually emerged as important participators in a wide array of biological processes, including organogenesis [9], pathogenesis [10], and carcinogenesis [11]. At the same time, extensive studies have found that circRNAs exert a regulatory character on the expressions of the protein-coding genes at transcriptional and post-transcriptional levels, through a series of unique molecular mechanisms [12,13]. Functioning as competing endogenous RNAs (ceRNAs) to bind the microRNAs (miRNAs) and modulate the activity of the cognate miRNAs and their target mRNAs is one of the widely accepted approaches. Therefore, the utilization of the ceRNAs theory to deduce the functionality of circRNAs, and the subsequent experimental verification of the proposed hypothesis are the common methods in the livestock field of study. In fiber-producing animals, such as goats and sheep, several studies have shown that the differentially expressed circRNAs in skin tissues are closely related to HF formation [14,15], and key fiber traits (e.g., fineness and quality) [16,17]. Acting as ceRNAs to finely regulate the levels of mRNAs through the circRNAs–miRNA–mRNAs axis seems to be the molecular basis of circRNAs in the above processes. Apart from validating the relationships, some of the researchers have also explored the roles of circRNAs hair biology, using in vitro cellular models. In goats, the promotive effect of circRNA-1926 on directing the committed differentiation of HFSCs towards the follicular cells via titrating miR-148a/b-3p to alleviate their inhibitory roles on the target gene, *CDK19*, was observed [18]. In addition to acting as ceRNAs, some of the circRNAs have been demonstrated as adjusting the gene expression via modulating the gene transcription in the nucleus [19,20], or competing with mRNAs for transcript alternative splicing in the cytosol [21]. Although mounting evidence suggests that the circRNAs should be unneglected players in modulating the gene expression and functionality of DPCs, their information currently remains scare.

In the present study, we established a genome-wide profile of circRNAs expressed in goat DPCs and screened the functional circRNAs via comparing the transcriptomes between goat DPCs and DFs. We also reported that circRNAs might affect the expression of the signature genes of DPCs at transcriptional and post-transcriptional levels, highlighting the possible characteristics of circRNAs in key events (i.e., HFSCs' activation and angiogenesis) in hair biology. Our study shines new light on a deeper exploration of the roles of non-coding RNAs in hair biology.

2. Materials and Methods

2.1. Animals and Cell Culture

Three healthy female Shanbei white cashmere goats (~2 years old, ~35 kg weight) with independent genetic lineage background were selected from a private farm located in Yangling District, Shannxi, China (34°28′ N and 108°07′ E). The rearing and management of the animals were performed under the recommended guidelines provided by the regional standard (DB61/T 584-2013). The skin samples harvested from the lateral backsides on cashmere goats were used for the cell lines' acquisition. The primary culture of the dermal papilla cells (DPCs) was obtained using a canonical microdissection-based method [22]. At the same time, an explants-based protocol was adopted to acquire goat dermal fibroblasts (DFs) from the skin samples [23]. All of the cells were maintained in a sterile incubator at 37 °C temperature, 100% humidity, and 5% CO_2/95% air atmosphere. A conventional DMEM/F12, with the addition of 10% FBS (*v*/*v*), 100 UI/mL penicillin, and 100 μg/mL streptomycin, was chosen as the culture medium. At the fourth passage, the cell samples from three lines of DPCs and DFs were collected and subjected to downstream analysis. All of the reagents used in the present study were purchased from Sigma-Aldrich (Shanghai, China). The entire experimental procedure was approved and supervised by the Animal Care Commission of Northwest A and F University under the forced guideline (2013-31101684).

2.2. Sequencing Library Construction and Reference Genome Mapping

The total RNA extraction was implemented according to the classic Trizol-based RNA extraction method. The RNA degradation and contamination were monitored on 1% agarose gels, and their purity was checked using the NanoPhotometer® spectrophotometer (IMP LEN, CA, USA). Furthermore, the RNA concentration was measured using a Qubit ® RNA Assay Kit in a Qubit ®2.0 Fluorometer (Life Technologies, CA, USA), and the RNA integrity was assessed using the RNA Nano 6000 Assay Kit of the Bioanalyzer 2100 system (Agilent Technologies, CA, USA). After the extraction and quality examination of the total RNA from all of the samples, 5 μg RNA from each sample was used for sequencing the library construction. In brief, the ribosomal RNA (rRNA) was eliminated by the Epicentre Ribo-zero™ rRNA Removal Kit (Epicentre, WI, USA) and the sequencing libraries were generated using the rRNA-depleted RNA by NEBNext® Ultra™ Directional RNA Library Prep Kit for Illumina® (NEB, USA), following the manufacturer's recommendations. In addition, the products were purified (AMPure XP system) and the library quality was assessed on the Bioanalyzer 2100 system. Finally, all of the libraries were sequenced on the Illumina HiSeq 4000 platform and 150 bp paired-end reads were generated.

Thereafter, a series of quality control procedures, including the removal of the invalid reads and evaluation of the Q20 index of clean reads, were carried out, according to pipeline established by Novogene, as reported in a previous study [24]. Further, the clean reads were mapped and aligned to the representative goat reference genome (Code: ARS1), using Bowtie 2 [25]. The mapped reads were selected to identify the existence and relative expression of the mRNAs as we already reported in published literature [26]; meanwhile, the reads unmapped to the goat reference genome were chosen for the subsequent circRNAs' identification. In addition, the construction of the sequencing libraries for the miRNAs and the subsequent bioinformatic analysis were executed, as stated before [26].

2.3. CircRNAs Identification

The two mainstream programs find_circ and CIRI2 were picked to predict the existence of the circular RNAs (circRNAs) in both of the cell types, according to their respective bioinformatic algorithms [27,28]. To overcome the problem that a high false-positive probability of circRNAs' detection exists, the transcripts detected by both pieces of software were deemed as reliable circular candidates.

2.4. Normalization of circRNAs Abundance and Differential Expression Analysis

The raw circRNAs counts were first normalized using standard TPM (transcripts per million clean reads) as in the following equation: normalized expression level = (read counts per 1,000,000 reads)/libsize (libsize is the sum of circRNAs counts). Principal component analysis (PCA) and cluster dendrogram construction were performed using FactoMineR 2.4 [29] and Cluster 2.1.2 [30] R packages, respectively. The differential expression analysis of the circRNAs between the two sample sets was performed, using DESeq R package (1.10.1), which provides statistical routines for determining the differential expression in digital gene expression data, using a model based on the negative binomial distribution [31]. The transcripts with a p value <0.05 determined by the DESeq tool were thought of as differentially expressed.

2.5. Gene Ontology and KEGG Enrichment Analysis

The gene ontology (GO) enrichment for the host genes of the differentially expressed circRNAs was carried out using the GOseq R package, in which the gene length bias was corrected [32]. The GO terms with a corrected p value < 0.05 were deemed as significantly enriched by the genes. The KEGG pathway enrichment analysis of the host genes was performed, using a web server called KOBAS 3.0 [33]. The results were visualized by ggplot2 package.

2.6. Prediction of miRNA Binding Sites on circRNAs and mRNAs

MiRanda-3.3a was used to predict the target sites for the miRNA binding on the linear 3′ untranslated regions and the overall segments of mRNAs and circRNAs [34], respectively.

2.7. CircRNAs-miRNAs-mRNAs Interplay Network Construction

The interplay network among the circRNAs, miRNAs, and mRNAs was constructed according to the miRNAs target sites prediction on mRNAs and circRNAs. The theory of competing endogenous RNAs (ceRNAs), in which the circRNAs compete with the mRNAs for miRNAs binding, to ensure the stability of the mRNAs, was used to infer the mutual regulatory relationships [35]. R package ggalluvial 0.12.3 was used to generate the Sankey graphics.

3. Results

3.1. Bioinformatic Identification and Genomic Feature Characterization of Circular RNAs

To deeply explore the potential roles of the circRNAs in hair growth, we utilized transcriptomic data generated in our experiment and performed a series of bioinformatic analysis to predict the functioning pathways of circRNAs specifically expressed by the goat DPCs. As shown in Figure 1, Step 1 and part of the work of Step 2 (i.e., the analysis of the mRNAs and miRNAs) were finished for a previous study, and the remaining work belongs to the present study.

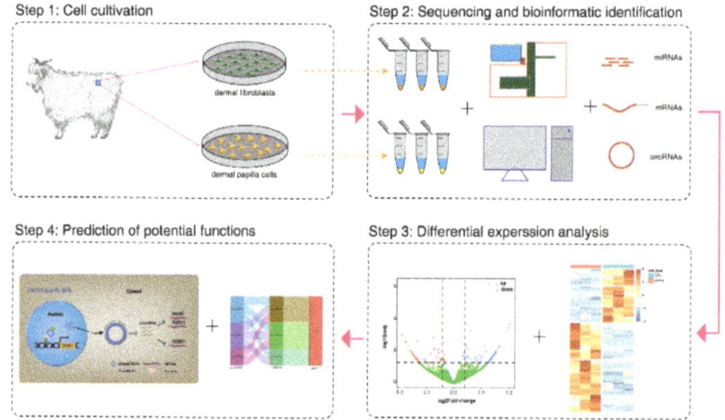

Figure 1. Workflow of present study. Cell cultivation and profiling of miRNAs and mRNAs were finished in a previous study [26].

Through combining the prediction results of find_circ and CIRI2, a total of 3706 circular RNAs (circRNAs) were bioinformatically identified (Figure 2a). Among these transcripts, 94.6% of them originated from the exons of coding genes; only a small proportion derived from the introns (3.78%) and intergenic regions (1.62%) (Figure 2b). Next, we found that all of the types of the circRNAs share a similar length distribution, in which the majority of them are less than 600 nt. Moreover, the average lengths of the exonic, intronic, and intergenic circRNAs are 280, 266, and 253 nt, respectively (Figure 2c). We also discovered that most of the circRNAs were made up of one–four genomic segments, and the percentage of the circRNAs containing two segments ranked first for exonic and intronic transcripts (Figure 2d). In addition, we demonstrated that the average length of the circRNAs visually increased along with more of the segments (Figure 2e). Meanwhile, the average length per segment was inversely related to the component counts (Figure 2f). Finally, we demonstrated that 62% of the coding genes only produced one circularized transcript, but the minority of the genes were more prolific than that. For instance, the genes generating two and three circRNAs accounted for 22% and 8% of the total genes, respectively (Figure 2g). The above findings suggested that the presently detected circRNAs from the DPCs and DFs possess outstanding genomic features, which could be utilized to judge the fidelity of the circRNAs. The detailed sequence information of all of the transcripts is provided in Table S1 (Supplementary Materials).

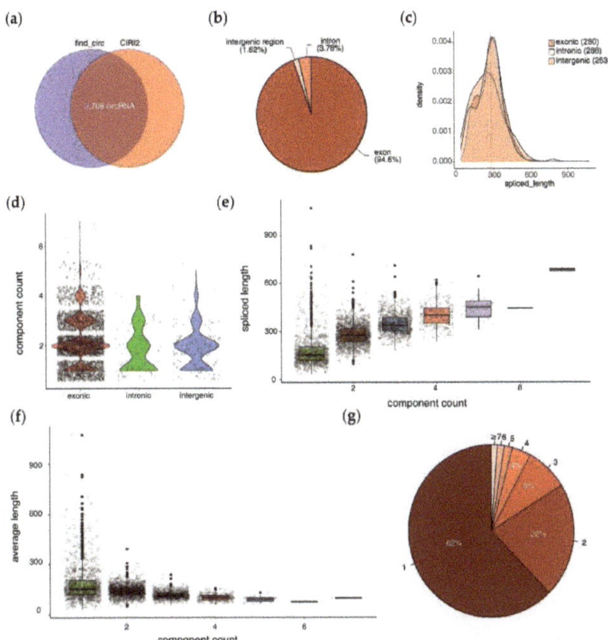

Figure 2. Genomic feature characterization of circular RNAs from goat DPCs and DFs. (**a**) The counts of circRNAs detected by both tools; (**b**) The proportion of each type of circRNAs; (**c**) The length distribution and average length of circRNAs; (**d**) Component counts distribution of circRNAs; (**e**,**f**) The spliced length and average length of circRNAs with different component counts; (**g**) The percentage of genes producing one or more circRNAs.

3.2. Global Expression Pattern and Differential Expression Analysis of circRNAs

As shown by the boxplots in Figure 3a, the TPM values of all of the transcripts in the six samples displayed a similar distribution mode, suggesting that the circRNAs' transcripts of all of the cells shared a nearly identical expression pattern at a global level. Next, we found a significant segregation of the samples at the first dimension on PCA figure (Figure 3b). This result is in high accordance with the sample clustering analysis, in which the cellular samples clustered into two independent clades (Figure 3c). Altogether, these findings hinted that the expression of some of the transcripts were highly cell-type-dependent. Subsequently, we identified a sum of 121 differentially expressed circRNAs between two of the sample sets ($p < 0.05$), including 76 upregulated and 45 downregulated in the DPCs when the DFs were set as controls (Figure 3d,e). We listed the top 20 of the differentially expressed circRNAs ranked with a p value in Table 1, and found that a group of transcripts (e.g., chi_circ_0001124 and chi_circ_0005862) were exclusively expressed in one cell type. Furthermore, we also found that the three ciRNAs (e.g., chi_circ_0005569) and the three intergenic region-generated circRNAs (e.g., chi_circ_0000835) displayed distinct abundances between the DPCs and DFs (Table 2). Collectively, the above results indicated that the DPCs and DFs possessed a featured transcriptional profile and the signature transcripts might underpin their functional heterogeneity. Overall, the expression levels and the differential expression analysis result are provided in Table S2 (Supplementary Materials).

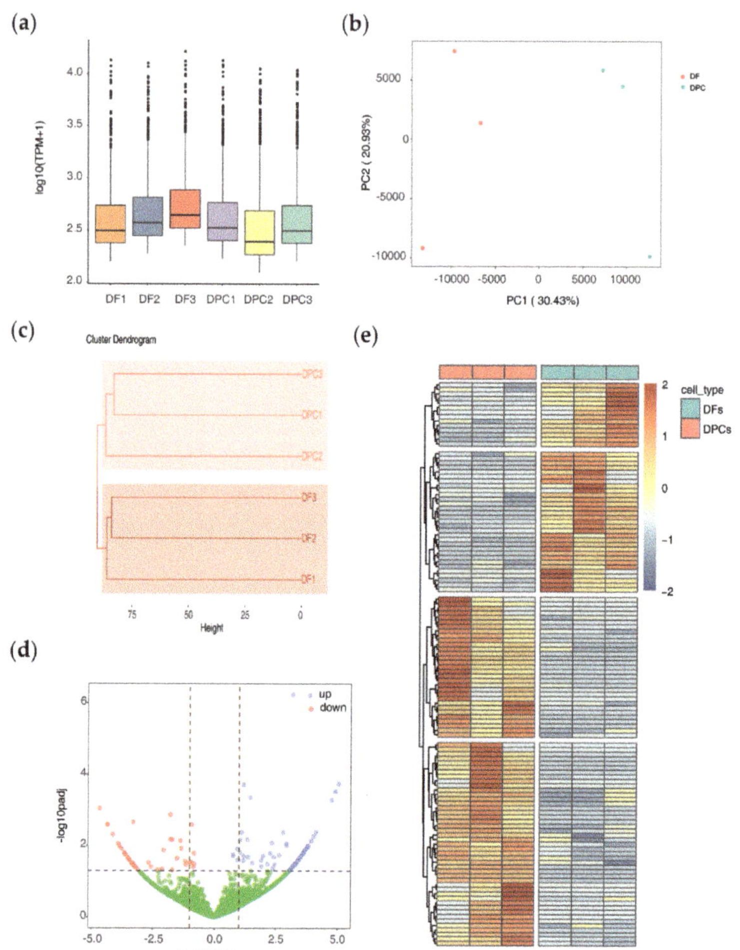

Figure 3. Global expression pattern and differential expression analysis of circRNAs. (**a**) Box-plot showing the distribution pattern of circRNAs based on normalized abundances; (**b**,**c**) Principal component analysis (PCA) and cluster dendrogram analysis of samples based on TPM values of each transcript; (**d**,**e**) Volcano plot and heatmap showing relative expression levels and statistical significances of circRNAs between goat DPCs and DFs. $p < 0.05$ was thought of as statistically significant.

Table 1. Top 20 differentially expressed circRNAs between goat DPCs and DFs ranked by statistical significance.

CircRNAs ID	DPCs TPM	DFs TPM	$\log_2$ (FC)	*p* Value	Host Genes
chi_circ_0001237	27.66	2.60	3.19	5.83×10^{-7}	*EPDR1*
chi_circ_0001124	10.42	0	5.06	0.00018033	*RBM33*
chi_circ_0005517	77.94	33.43	1.20	0.00019107	*MTCL1*
chi_circ_0005255	9.48	0	4.93	0.00029672	*ADAMTS9*
chi_circ_0003390	36.68	12.84	1.46	0.00044071	*FNDC3A*
chi_circ_0002616	8.49	0	4.78	0.00052082	*SMOC2*

Table 1. *Cont.*

CircRNAs ID	DPCs TPM	DFs TPM	$\log_2$ (FC)	*p* Value	Host Genes
chi_circ_0005862	0	7.36	−4.65	0.00091454	*HTRA1*
chi_circ_0002154	5.49	19.05	−1.76	0.0013659	*PLIN2*
chi_circ_0004069	0.67	8.68	−3.27	0.0022086	*ENAH*
chi_circ_0003353	0	5.74	−4.33	0.0025604	*CDK8*
chi_circ_0005056	56.58	106.67	−0.91	0.0025953	*CEMIP*
chi_circ_0004639	0.00	5.69	−4.32	0.0025977	*APPBP2*
chi_circ_0005733	5.61	0	4.15	0.0041094	*FGFR2*
chi_circ_0003643	35.55	16.39	1.12	0.0041723	*TASP1*
chi_circ_0001957	9.21	0.94	2.81	0.0041946	*DPYSL3*
chi_circ_0003371	0	5.05	−4.11	0.0048414	*DCLK1*
chi_circ_0004078	5.29	0	4.05	0.0053418	*KIF26B*
chi_circ_0002817	3.91	13.87	−1.75	0.006652	*PTGDR*
chi_circ_0003391	21.04	8.06	1.35	0.0067238	*FNDC3A*
chi_circ_0000379	4.27	14.53	−1.70	0.0068278	*ZMAT3*

Note: DPCs, dermal papilla cells; DFs, dermal fibroblasts; circRNAs, circular RNAs; TPM, transcripts per million clean reads; FC, fold change.

Table 2. Differentially expressed ciRNAs and intergenic region-generated circRNAs between goat DPCs and DFs.

CircRNAs ID	DPCs_TPM	DFs_TPM	$\log_2$ (FC)	*p* Value	Host Gene
CiRNAs					
chi_circ_0005569	4.95	0	3.95	0.0070649	*CCBE1*
chi_circ_0003704	2.93	0	3.04	0.049869	*ZFAND1*
chi_circ_0005570	48.90	24.95	0.95	0.012148	*CCBE1*
Intergenic region-derived circRNAs					
chi_circ_0000835	8.21	0	3.30	0.034524	NA
chi_circ_0004524	129.36	11.39	2.95	0.0098427	NA
chi_circ_0006241	19.51	34.13	−0.80	0.032297	NA

Note: CiRNAs, cNote: CiRNAs, circular intronic RNAs; NA, not available.

3.3. Relationship of circRNAs Expression with Their Host Genes

To determine the relationship between the expression levels of the circRNAs with their host genes, we correlated their relative abundances between the cell types and the performed statistical analysis. As shown by the heatmap in Figure 4a, the trend of the circRNAs expression pattern is not strictly consistent with that of the mRNAs. Notably, the fold changes of the two host genes, *ZMYM6* and *RPS6KC1*, both display inverse modes with the relative abundances of the circRNAs derived from them. In addition, the statistical results using Pearson's correlation analysis indicated that the relative levels of the circRNAs and mRNAs are correlated at a medium level (Figure 4b: $R = 0.41$; $p = 3.9 \times 10^{-5}$), suggesting that no more than a weak association exists. Furthermore, we summarized the expression status of the circRNAs and their host genes between DPCs and DFs (Table S2, Supplementary Materials), and found that more than half of the differentially expressed circRNAs derive from genes with equal abundances in DPCs and DFs. At the same time, less than 50% percent of the differentially expressed transcripts showed similar expression patterns with that of their host genes. In addition, the abundances of a small percentage of the circRNAs showed reverse trends compared to their host genes. These results demonstrated that the physiological and cellular functions of circRNAs may not strictly depend on the expression status of their host genes.

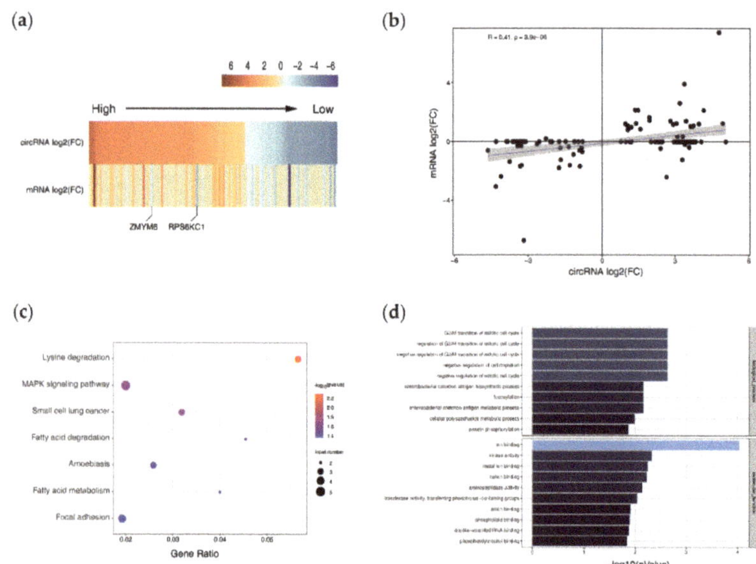

Figure 4. Relationship of circRNAs expression with their host genes and functional enrichment. (**a**) Heatmaps showing the relative abundances of circRNAs and corresponding mRNAs; (**b**) Pearson's correlation analysis of the relative level of circRNAs and their host genes; (**c**,**d**) KEGG pathways and gene ontology (GO) items enriched by host genes.

3.4. Functional Enrichment Analysis of the Host Genes

A total of 104 host genes were used as input for the functional enrichment analysis. As shown in Figure 4c, several signaling pathways, including Lysine degradation, MAPK signaling pathway, small-cell lung cancer, and four other signaling pathways were significantly enriched ($p < 0.05$). The GO analysis results showed that 76 of the items were significantly enriched, most of which belonged to the molecular function (MF) and biological process (BP) categories (Figure 4d). The most significantly enriched GO items in MF comprised ion binding (GO: 0043167), kinase activity (GO: 0016301), and others. At the same time, several cell cycle-related items, including the G2/M transition of mitotic cell cycle (GO: 0000086) and the regulation of the G2/M transition of the mitotic cell cycle (GO: 0010389) were significantly highlighted in BP. Detailed information is provided in Table S3 (Supplementary Materials).

3.5. Screening of circRNAs Acting as ceRNAs and Their Regulatory Relationships with Functional Genes in Goat DPCs

To infer the possible characteristics of the circRNAs in cells, we screened the circRNAs acting as ceRNAs and constructed the regulatory relationships of the circRNAs with miRNAs and mRNAs. As a result, a total of 48,676 circRNAs–miRNA–mRNAs interactive lines were bioinformatically identified (Table S4, Supplementary Materials). In our previous study, we defined the core signatures of the goat DPCs through a comparative transcriptomic analysis of the goat DPCs and DPCs [26]. Among the signature genes, *HOXC8* and *RSPO1* were shown to govern the activation of the hair follicle stem cells, the most critical path of the DPCs in controlling hair growth [4,5]. Thus, we filtered the circRNAs functioning as ceRNAs to modulate the transcripts' abundances of *HOXC8* and *RSPO1*, and exhibited the corresponding relationship between the three types of transcripts. As suggested by the heatmap in Figure 5a, the circRNAs' candidates and the genes showed reverse expression patterns compared to the trends of the miRNAs between the samples, which fitted the theoretical basis of the ceRNAs [35]. Furthermore, we demonstrated that the circRNAs might indirectly adjust the transcript abundances of the mRNAs via sponging miRNAs (Figure 5b). For example, the adverse effect of the miRNA-145 on the *HOXC8* mRNA abundance or translation could be specifically ameliorated by a set of circRNAs, including chi-circ_0001956, and others (Figure 4b). Moreover, the inhibitive actions of novel_624 and other six miRNAs on *RSPO1* mRNAs may be eased by cognate circRNAs, such as chi_circ_0004422 via chi_circ_0004422-novel_624-RSPO1 interacting line. These results implied that the circRNAs possibly participate in hair-follicle-stem cells vitalization via finely adjusting the expressions of the pivotal genes in DPCs.

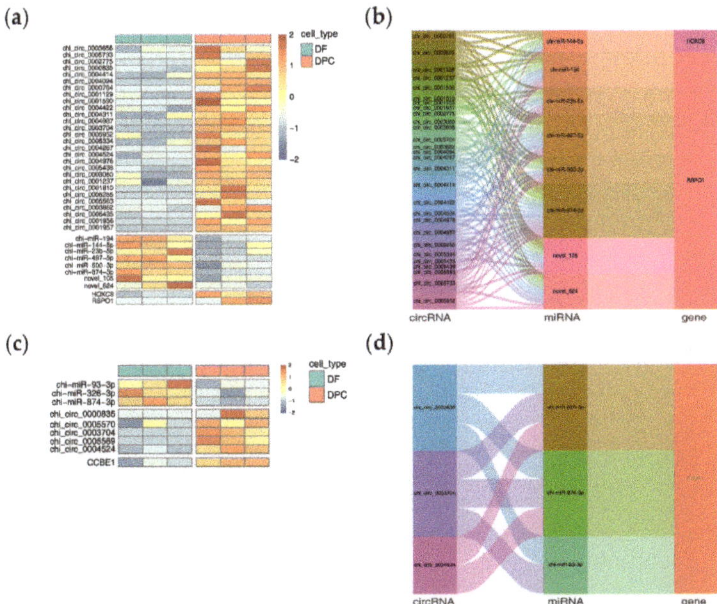

Figure 5. CircRNAs acting as ceRNAs in regulating the abundances of coding genes. (**a**) Heatmap showing the relative expression of transcripts between two cell samples; (**b**) Sankey graph showing the relationship between circRNAs, miRNAs, and coding genes (*HOXC8* and *RSPO1*); (**c**) Heatmap showing the relative levels of intronic circRNAs (i.e., chi_circ_0005569 and chi_circ_0005570) and other transcripts between goat DPCs and DFs; (**d**) Sankey graph showing the interactive lines of circRNAs, miRNAs, and *CCBE1*.

In addition, we observed the elevated abundances of the two intron-derived circRNAs chi_circ_0005569 and chi_circ_0005570 in goat DPCs compared to DFs (Figure 5c). Previous studies reported that their host genes *CCBE1* were involved in tissue angiogenesis [36], a critical physiological process related to hair follicle development. At the same time, several findings pointed out that intronic circRNAs mostly reside in the nucleus and regulate the transcriptions of their host genes in *cis* [19,20]. Based on the facts of the above studies, we reasonably reckoned that the two ciRNAs possibly execute similar roles on their parental gene. Moreover, we also found that two intergenic segment-formed circRNAs chi_circ_0000835 and chi_circ_0004524 possess binding sites for the miRNAs, targeting the matured *CCBE1* transcripts (Figure 5d), suggesting their strong potential in enhancing *CCBE1* transcript stability or protein output. The above results suggested that the circRNAs might take part in modulating the follicular angiogenesis process via acting as ceRNAs or transcriptional regulators.

4. Discussion

Serving as the signaling center of HF, DPCs are essential for a wide range of developmental events during the entire phase of hair growth [1,2]. Previous studies have identified a group of DPCs' signature genes involved in key events (e.g., HFSCs' activation) in HF growth [4,5]; however, how the expressions of the genes are regulated at transcriptional or translational level still remains elusive. In recent years, circRNAs gradually emerged as the key participators in various physiological and pathological processes via the regulating gene transcription, decoying miRNAs and other functioning mechanisms [12]. Meanwhile, the potential functions of the circRNAs in key aspects of hair biology (i.e., HF development and fiber traits) have been proposed, even though most of these studies were carried out at the tissue level [14–17]. Our present study points out that the circRNAs might participate in the pivotal events of hair growth via affecting the abundances of related functional genes in DPCs. We established a comprehensive genome-wide profile of circular transcripts in goat DPCs and DFs, and constructed the modulatory relationships between the circRNAs and coding genes.

We demonstrated that the presently identified circRNAs possess several prominent features regarding their sources, length, and other sequence characteristics. These genomic properties are highly consistent with the circular transcripts found in the tissues of goats [37], sheep [38], humans [39], and yaks [40]. The phenomenon not only validates the fidelity of our circRNAs identification, but also fits the evolutionary conservation of the biogenesis and functioning mechanisms of the circRNAs. In addition, we also found that DPCs and DFs possess a distinct transcriptional profile and the expressions of the circRNAs are highly cell-context dependent. Numerous studies have confirmed that the specific expressions of the transcripts is the hallmark and indicator of their special functionality in cells or tissues [12,13]. For example, circTshz2-1, a uniquely expressed transcript in differentiated adipocytes, exerts a promotive effect on mouse adipogenesis via upregulating the genes critical for lipid accumulation [10]. Similarly, the significant upregulation of circRNA-0100 expression positively drives the committed differentiation of HFSCs towards their progenies in cashmere goats [41]. Thus, it is reasonably to speculate that these differentially expressed circRNAs should perform functions of importance related to the characters of these cells in tissue development.

Next, we demonstrated that the expressions of the circRNAs are not tightly coupled to the levels of linear transcripts of their host genes. This discovery is in high accordance with the results found in the tissues of other animals [19,42]. Previous studies validated that both the circRNAs and mRNAs are alternative splicing products of primary mRNAs [21], confirming that a mutually competitive relationship exists during their biogenesis. This also partially explains why the expressions of a subset of circRNAs are independent of their linear isoforms in the present and other studies. Researchers have often performed GO and KEGG enrichment analysis of the host genes to reckon the functionalities of the circRNAs [37,40,43]. In human adipocytes, the targeted knockdown of linear or circular transcripts of the gene-*Arhgap5* obviously caused adverse trends in the abundance of the genes that determine adipogenesis [10]. A similar case occurs with circSMARCA5, which decreases the expression level of its parental gene via exclusively binding to the gene locus and pausing transcription in breast cancer tissue [44]. The above cases implied that the linear and circular isoforms of a gene can exert opposite effects on the same biological processes. However, our results imply that the enriched terms (e.g., MAPK signaling pathway and cell cycle-related terms) frequently appear in transcriptomic studies involving hair growth and reflect the differences in cellular identity of the cells [45,46]. We highly recommend that additional precautions should be taken when the bioinformatic deduction of the potential roles of circRNAs using functional enrichment of their host genes occurs.

Finally, we observed that the circRNAs might serve as ceRNAs, and constructed the interaction network of circRNAs–miRNAs–mRNAs, in which we highlighted the modulatory roles of circRNAs on genes (e.g., *HOXC8*, *RSPO1*, and *CCBE1*) concerning two key events in hair growth. Functioning as sponges that bind miRNAs and thus prevent them from binding and suppressing their target mRNAs is one of the most important and extensively explored approaches through which the circRNAs exert their functionalities. A large quantity of related circRNAs have been identified in the skin tissues of goats and sheep, and some of the circRNAs have been experimentally authenticated [15–17]. The circRNAs–miRNAs–mRNAs axis was gradually recognized as a pivotal avenue through which the circRNAs exert their important roles in several aspects of hair biology.

For example, the circRNA-1926–miR-148a/b-3p–CDK19 axis was exhibited to participate in the goat HFSCs' activation [18]. In the present study, we proposed that the circRNAs might participate in the regulation of the signature genes of goat DPCs in the same manner. *HOXC8* and *RSPO1* have been verified as the key drivers in the DPCs-stimulated activation of HFSCs and the subsequent hair regrowth via vitalization of the Wnt signaling pathway [4,5]. Therefore, it is possible that an interactive axis, such as chi-circ_0001956–miRNA-145–*HOXC8* could perform regulatory roles of crucial importance in HFSCs' activation via the post-transcriptional modification of genes involved.

In addition, we also discovered that the abundances of two intron-derived circRNAs chi_circ_0005569 and chi_circ_0005570 are consistent with that of their host gene *CCBE1*. A few studies have demonstrated that circRNAs are capable of regulating gene transcription via interacting with RNA Pol II, or transcription factors [19,20]. Notably, an intronic circRNA named ci-ankrd52 could drive its host gene expression via locating at the genomic sites of transcription and interacting with the transcriptional machinery [20]. Thus, it is possible that the expression of *CCBE1* is under the transcriptional control of intronic circRNAs, derived from itself. *CCBE1* encodes an extracellular matrix protein that implicates angiogenesis [36], which is a physiological process closely associated with active hair growth [7]. A decreased mRNA level of *CCBE1* was associated with the impaired hair growth-stimulatory capacity of DPCs under the treatment of dihydrotestosterone; the hormone causes undesired androgenic alopecia [47]. Therefore, the two intronic circRNAs perhaps exert regulatory characteristics in follicular angiogenesis via an adjustment of the abundances of *CCBE1* in goat DPCs. Moreover, we also found that some circRNAs might serve as ceRNAs to modulate the expression of *CCBE1*, further confirming the complexity of gene expression regulation at transcriptional and translational levels.

5. Conclusions

In present study, we established a comprehensive global profile of circRNAs in goat DPCs and DFs. Through comparative analysis, we identified a group of circRNAs specifically expressed in each cell type. Further, we predicted the potential roles of circRNAs in DPCs via constructing their regulatory relationships with the genes involved in key events in hair growth. The validation of such relationships will provide new insight into how the functionality of DPCs is maintained by circRNAs through regulating the gene expression at transcriptional and translational levels.

Supplementary Materials: The following are available online at https://www.mdpi.com/article/10.3390/agriculture12091306/s1, Table S1: Sequence information of circRNAs; Table S2: Normalized and differential expression of circRNAs, Table S3: GO and KEGG pathway enrichment of host genes; Table S4: Interactive lines of circRNAs-miRNAs-mRNAs.

Author Contributions: Conceptualization, S.M.; methodology, S.M.; software, S.M.; validation, Y.S.; formal analysis, S.M. and X.X.; investigation, S.M. and X.X.; resources, S.M.; data curation, S.M.; writing—original draft preparation, S.M.; writing—review and editing, Y.Y., Y.S. and X.W.; visualization, Y.S.; supervision, Y.C.; project administration, Y.C.; funding acquisition, Y.S. and Y.C. All authors have read and agreed to the published version of the manuscript.

Funding: This work was supported by National Natural Science Foundation of China (No. 31872332) and China Agriculture Research System (CARS-34). Financial support for this research was provided China Agriculture Research System (CARS-39).

Institutional Review Board Statement: Not applicable.

Informed Consent Statement: Not applicable.

Data Availability Statement: All the relevant data are presented. Other data are available from the corresponding author on reasonable request.

Acknowledgments: Thanks to all members of our labs for their help in the whole experiment process and life.

Conflicts of Interest: The authors declare no conflict of interest.

Abbreviation

DPCs	dermal papilla cells
DFs	dermal fibroblasts
HOXC8	homeobox C8
RSPO1	R-spondin 1
CCBE1	collagen and calcium-binding EGF domains 1
PTEN	phosphatase and tensin homolog
CDK19	cyclin-dependent kinase 19
Arhgap5	Rho GTPase activating protein 5

References

1. Yang, C.-C.; Cotsarelis, G. Review of hair follicle dermal cells. *J. Dermatol. Sci.* **2010**, *57*, 2–11. [CrossRef] [PubMed]
2. Morgan, B.A. The Dermal Papilla: An Instructive Niche for Epithelial Stem and Progenitor Cells in Development and Regeneration of the Hair Follicle. *Cold Spring Harb. Perspect. Med.* **2014**, *4*, a015180. [CrossRef] [PubMed]
3. Ramos, R.; Guerrero-Juarez, C.F.; Plikus, M.V. Hair Follicle Signaling Networks: A Dermal Papilla–Centric Approach. *J. Investig. Dermatol.* **2013**, *133*, 2306–2308. [CrossRef] [PubMed]
4. Yu, Z.; Jiang, K.; Xu, Z.; Huang, H.; Qian, N.; Lu, Z.; Chen, D.; Di, R.; Yuan, T.; Du, Z.; et al. Hoxc-Dependent Mesenchymal Niche Heterogeneity Drives Regional Hair Follicle Regeneration. *Cell Stem Cell* **2018**, *23*, 487–500.e6. [CrossRef] [PubMed]
5. Li, N.; Liu, S.; Zhang, H.-S.; Deng, Z.-L.; Zhao, H.-S.; Zhao, Q.; Lei, X.-H.; Ning, L.-N.; Cao, Y.-J.; Wang, H.-B.; et al. Exogenous R-Spondin1 Induces Precocious Telogen-to-Anagen Transition in Mouse Hair Follicles. *Int. J. Mol. Sci.* **2016**, *17*, 582. [CrossRef] [PubMed]
6. Yano, K.; Brown, L.F.; Detmar, M. Control of hair growth and follicle size by VEGF-mediated angiogenesis. *J. Clin. Investig.* **2001**, *107*, 409–417. [CrossRef] [PubMed]
7. Mecklenburg, L.; Tobin, D.J.; Müller-Röver, S.; Handjiski, B.; Wendt, G.; Peters, E.M.J.; Pohl, S.; Moll, I.; Paus, R. Active Hair Growth (Anagen) is Associated with Angiogenesis. *J. Investig. Dermatol.* **2000**, *114*, 909–916. [CrossRef]
8. Odorisio, T.; Cianfarani, F.; Failla, C.M.; Zambruno, G. The placenta growth factor in skin angiogenesis. *J. Dermatol. Sci.* **2006**, *41*, 11–19. [CrossRef]
9. Lee, E.C.S.; Elhassan, S.A.M.; Lim, G.P.L.; Kok, W.H.; Tan, S.W.; Leong, E.N.; Tan, S.H.; Chan, E.W.L.; Bhattamisra, S.K.; Rajendran, R.; et al. The roles of circular RNAs in human development and diseases. *Biomed. Pharmacother.* **2019**, *111*, 198–208. [CrossRef]
10. Arcinas, C.; Tan, W.; Fang, W.; Desai, T.P.; Teh, D.C.S.; Degirmenci, U.; Xu, D.; Foo, R.; Sun, L. Adipose circular RNAs exhibit dynamic regulation in obesity and functional role in adipogenesis. *Nat. Metab.* **2019**, *1*, 688–703. [CrossRef]
11. Lu, Q.; Liu, T.; Feng, H.; Yang, R.; Zhao, X.; Chen, W.; Jiang, B.; Qin, H.; Guo, X.; Liu, M.; et al. Circular RNA circSLC8A1 acts as a sponge of miR-130b/miR-494 in suppressing bladder cancer progression via regulating PTEN. *Mol. Cancer* **2019**, *18*, 111. [CrossRef] [PubMed]
12. Li, X.; Yang, L.; Chen, L.-L. The Biogenesis, Functions, and Challenges of Circular RNAs. *Mol. Cell* **2018**, *71*, 428–442. [CrossRef] [PubMed]
13. Chen, L.-L. The expanding regulatory mechanisms and cellular functions of circular RNAs. *Nat. Rev. Mol. Cell Biol.* **2020**, *21*, 475–490. [CrossRef] [PubMed]
14. Zhao, B.; Luo, H.; He, J.; Huang, X.; Chen, S.; Fu, X.; Zeng, W.; Tian, Y.; Liu, S.; Li, C.-J.; et al. Comprehensive transcriptome and methylome analysis delineates the biological basis of hair follicle development and wool-related traits in Merino sheep. *BMC Biol.* **2021**, *19*, 197. [CrossRef] [PubMed]
15. Shang, F.; Wang, Y.; Ma, R.; Di, Z.; Wu, Z.; Hai, E.; Rong, Y.; Pan, J.; Liang, L.; Wang, Z.; et al. Expression Profiling and Functional Analysis of Circular RNAs in Inner Mongolian Cashmere Goat Hair Follicles. *Front. Genet.* **2021**, *12*, 678825. [CrossRef]
16. Hui, T.; Zheng, Y.; Yue, C.; Wang, Y.; Bai, Z.; Sun, J.; Cai, W.; Zhang, X.; Bai, W.; Wang, Z. Screening of cashmere fineness-related genes and their ceRNA network construction in cashmere goats. *Sci. Rep.* **2021**, *11*, 21977. [CrossRef]
17. Wang, J.; Wu, X.; Kang, Y.; Zhang, L.; Niu, H.; Qu, J.; Wang, Y.; Ji, D.; Li, Y. Integrative analysis of circRNAs from Yangtze River Delta white goat neck skin tissue by high-throughput sequencing (circRNA-seq). *Anim. Genet.* **2022**, *53*, 405–415. [CrossRef]
18. Yin, R.H.; Zhao, S.J.; Jiao, Q.; Wang, Z.Y.; Bai, M.; Fan, Y.X.; Zhu, Y.B.; Bai, W.L. CircRNA-1926 Promotes the Differentiation of Goat SHF Stem Cells into Hair Follicle Lineage by miR-148a/b-3p/*CDK19* Axis. *Animals* **2020**, *10*, 1552. [CrossRef]

19. Stoll, L.; Rodríguez-Trejo, A.; Guay, C.; Brozzi, F.; Bayazit, M.B.; Gattesco, S.; Menoud, V.; Sobel, J.; Marques, A.C.; Venø, M.T.; et al. A circular RNA generated from an intron of the insulin gene controls insulin secretion. *Nat. Commun.* **2020**, *11*, 5611. [CrossRef]
20. Zhang, Y.; Zhang, X.-O.; Chen, T.; Xiang, J.-F.; Yin, Q.-F.; Xing, Y.-H.; Zhu, S.; Yang, L.; Chen, L.-L. Circular Intronic Long Noncoding RNAs. *Mol. Cell* **2013**, *51*, 792–806. [CrossRef]
21. Ashwal-Fluss, R.; Meyer, M.; Pamudurti, N.R.; Ivanov, A.; Bartok, O.; Hanan, M.; Evantal, N.; Memczak, S.; Rajewsky, N.; Kadener, S. circRNA Biogenesis Competes with Pre-mRNA Splicing. *Mol. Cell* **2014**, *56*, 55–66. [CrossRef] [PubMed]
22. Jahoda, C.; Oliver, R.F. The growth of vibrissa dermal papilla cells in vitro. *Br. J. Dermatol.* **1981**, *105*, 623–627. [CrossRef] [PubMed]
23. Bratka-Robia, C.B.; Mitteregger, G.; Aichinger, A.; Egerbacher, M.; Helmreich, M.; Bamberg, E. Primary cell culture and morphological characterization of canine dermal papilla cells and dermal fibroblasts. *Vet. Dermatol.* **2002**, *13*, 1–6. [CrossRef] [PubMed]
24. Zhou, G.; Kang, D.; Ma, S.; Wang, X.; Gao, Y.; Yang, Y.; Wang, X.; Chen, Y. Integrative analysis reveals ncRNA-mediated molecular regulatory network driving secondary hair follicle regression in cashmere goats. *BMC Genom.* **2018**, *19*, 222. [CrossRef]
25. Langmead, B.; Salzberg, S.L. Fast gapped-read alignment with Bowtie 2. *Nat. Methods* **2012**, *9*, 357–359. [CrossRef]
26. Ma, S.; Wang, Y.; Zhou, G.; Ding, Y.; Yang, Y.; Wang, X.; Zhang, E.; Chen, Y. Synchronous profiling and analysis of mRNAs and ncRNAs in the dermal papilla cells from cashmere goats. *BMC Genom.* **2019**, *20*, 512. [CrossRef]
27. Memczak, S.; Jens, M.; Elefsinioti, A.; Torti, F.; Krueger, J.; Rybak, A.; Maier, L.; Mackowiak, S.D.; Gregersen, L.H.; Munschauer, M.; et al. Circular RNAs are a large class of animal RNAs with regulatory potency. *Nature* **2013**, *495*, 333–338. [CrossRef]
28. Gao, Y.; Wang, J.; Zhao, F. CIRI: An efficient and unbiased algorithm for de novo circular RNA identification. *Genome Biol.* **2015**, *16*, 4. [CrossRef]
29. Lê, S.; Josse, J.; Husson, F. FactoMineR: An R Package for Multivariate Analysis. *J. Stat. Softw.* **2008**, *25*, 1–18. [CrossRef]
30. Mouselimis, L. ClusterR: Gaussian Mixture Models, K-Means, Mini-Batch-Kmeans, K-Medoids and Affinity Propagation Clustering, version 1.2.6; R Package: 2022. Available online: https://CRAN.R-project.org/package=ClusterR (accessed on 10 April 2022).
31. Love, M.I.; Huber, W.; Anders, S. Moderated estimation of fold change and dispersion for RNA-seq data with DESeq2. *Genome Biol.* **2014**, *15*, 550. [CrossRef]
32. Young, M.D.; Wakefield, M.J.; Smyth, G.K.; Oshlack, A. Gene ontology analysis for RNA-seq: Accounting for selection bias. *Genome Biol.* **2010**, *11*, R14. [CrossRef] [PubMed]
33. Bu, D.; Luo, H.; Huo, P.; Wang, Z.; Zhang, S.; He, Z.; Wu, Y.; Zhao, L.; Liu, J.; Guo, J.; et al. KOBAS-i: Intelligent prioritization and exploratory visualization of biological functions for gene enrichment analysis. *Nucleic Acids Res.* **2021**, *49*, W317–W325. [CrossRef] [PubMed]
34. Enright, A.J.; John, B.; Gaul, U.; Tuschl, T.; Sander, C.; Marks, D.S. MicroRNA targets in Drosophila. *Genome Biol.* **2003**, *5*, R1. [CrossRef] [PubMed]
35. Tay, Y.; Rinn, J.; Pandolfi, P.P. The multilayered complexity of ceRNA crosstalk and competition. *Nature* **2014**, *505*, 344–352. [CrossRef]
36. Tian, G.-A.; Zhu, C.-C.; Zhang, X.-X.; Zhu, L.; Yang, X.-M.; Jiang, S.-H.; Li, R.-K.; Tu, L.; Wang, Y.; Zhuang, C.; et al. CCBE1 promotes GIST development through enhancing angiogenesis and mediating resistance to imatinib. *Sci. Rep.* **2016**, *6*, 31071. [CrossRef]
37. Zheng, Y.; Hui, T.; Yue, C.; Sun, J.; Guo, D.; Guo, S.; Guo, S.; Li, B.; Wang, Z.; Bai, W. Comprehensive analysis of circRNAs from cashmere goat skin by next generation RNA sequencing (RNA-seq). *Sci. Rep.* **2020**, *10*, 516. [CrossRef]
38. Lv, X.; Chen, W.; Sun, W.; Hussain, Z.; Chen, L.; Wang, S.; Wang, J. Expression profile analysis to identify circular RNA expression signatures in hair follicle of Hu sheep lambskin. *Genomics* **2020**, *112*, 4454–4462. [CrossRef]
39. Li, Y.; Zheng, Q.; Bao, C.; Li, S.; Guo, W.; Zhao, J.; Chen, D.; Gu, J.; He, X.; Huang, S. Circular RNA is enriched and stable in exosomes: A promising biomarker for cancer diagnosis. *Cell Res.* **2015**, *25*, 981–984. [CrossRef]
40. Zhang, Y.; Guo, X.; Pei, J.; Chu, M.; Ding, X.; Wu, X.; Liang, C.; Yan, P. CircRNA Expression Profile during Yak Adipocyte Differentiation and Screen Potential circRNAs for Adipocyte Differentiation. *Genes* **2020**, *11*, 414. [CrossRef]
41. Zhao, J.; Shen, J.; Wang, Z.; Bai, M.; Fan, Y.; Zhu, Y.; Bai, W. CircRNA-0100 positively regulates the differentiation of cashmere goat SHF-SCs into hair follicle lineage via sequestering miR-153-3p to heighten the KLF5 expression. *Arch. Anim. Breed.* **2022**, *65*, 55–67. [CrossRef]
42. Rybak-Wolf, A.; Stottmeister, C.; Glažar, P.; Jens, M.; Pino, N.; Giusti, S.; Hanan, M.; Behm, M.; Bartok, O.; Ashwal-Fluss, R.; et al. Circular RNAs in the Mammalian Brain Are Highly Abundant, Conserved, and Dynamically Expressed. *Mol. Cell* **2015**, *58*, 870–885. [CrossRef] [PubMed]
43. Zhao, B.; Chen, Y.; Hu, S.; Yang, N.; Wang, M.; Liu, M.; Li, J.; Xiao, Y.; Wu, X. Systematic Analysis of Non-coding RNAs Involved in the Angora Rabbit (Oryctolagus cuniculus) Hair Follicle Cycle by RNA Sequencing. *Front. Genet.* **2019**, *10*, 407. [CrossRef] [PubMed]
44. Xu, X.; Zhang, J.; Tian, Y.; Gao, Y.; Dong, X.; Chen, W.; Yuan, X.; Yin, W.; Xu, J.; Chen, K.; et al. CircRNA inhibits DNA damage repair by interacting with host gene. *Mol. Cancer* **2020**, *19*, 128. [CrossRef] [PubMed]

45. Zhang, Y.; Wu, K.; Wang, L.; Wang, Z.; Han, W.; Chen, D.; Wei, Y.; Su, R.; Wang, R.; Liu, Z.; et al. Comparative study on seasonal hair follicle cycling by analysis of the transcriptomes from cashmere and milk goats. *Genomics* **2020**, *112*, 332–345. [CrossRef] [PubMed]
46. Katsuoka, K.; Schell, H.; Hornstein, O.P.; Deinlein, E.; Wessel, B. Comparative morphological and growth kinetics studies of human hair bulb papilla cells and root sheath fibroblasts in vitro. *Arch. Dermatol. Res.* **1986**, *279*, 20–25. [CrossRef]
47. Chew, E.G.Y.; Tan, J.H.J.; Bahta, A.W.; Ho, B.S.-Y.; Liu, X.; Lim, T.C.; Sia, Y.Y.; Bigliardi, P.L.; Heilmann, S.; Wan, A.C.A.; et al. Differential Expression between Human Dermal Papilla Cells from Balding and Non-Balding Scalps Reveals New Candidate Genes for Androgenetic Alopecia. *J. Investig. Dermatol.* **2016**, *136*, 1559–1567. [CrossRef]

 agriculture

Article

OTUD7A Regulates Inflammation- and Immune-Related Gene Expression in Goose Fatty Liver

Minmeng Zhao [1], Kang Wen [1], Xiang Fan [1], Qingyun Sun [1], Diego Jauregui [1], Mawahib K. Khogali [1], Long Liu [1], Tuoyu Geng [1,2] and Daoqing Gong [1,2,*]

1 College of Animal Science and Technology, Yangzhou University, Yangzhou 225009, China; mmzhao@yzu.edu.cn (M.Z.); dx120190116@yzu.edu.cn (K.W.); mx120190650@yzu.edu.cn (X.F.); mx120200832@yzu.edu.cn (Q.S.); dh18051@yzu.edu.cn (D.J.); dh16013@yzu.edu.cn (M.K.K.); 005978@yzu.edu.cn (L.L.); tygeng@yzu.edu.cn (T.G.)

2 Joint International Research Laboratory of Agriculture and Agri-Product Safety of the Ministry of Education of China, Yangzhou University, Yangzhou 225009, China

* Correspondence: dqgong@yzu.edu.cn; Tel.: +86-138-0527-5847

Abstract: OTU deubiquitinase 7A (*OTUD7A*) can suppress inflammation signaling pathways, but it is unclear whether the gene can inhibit inflammation in goose fatty liver. In order to investigate the functions of *OTUD7A* and identify the genes and pathways subjected to the regulation of OTUD7A in the formation of goose fatty liver, we conducted transcriptomic analysis of cells, which revealed several genes related to inflammation and immunity that were significantly differentially expressed after *OTUD7A* overexpression. Moreover, the expression of interferon-induced protein with tetratricopeptide repeats 5 (*IFIT5*), tumor necrosis factor ligand superfamily member 8 (*TNFSF8*), sterile alpha motif domain-containing protein 9 (*SAMD9*), radical *S*-adenosyl methionine domain-containing protein 2 (*RSAD2*), interferon-induced GTP-binding protein Mx1 (*MX1*), and interferon-induced guanylate binding protein 1-like (*GBP1*) was inhibited by *OTUD7A* overexpression but induced by *OTUD7A* knockdown with small interfering RNA in goose hepatocytes. Furthermore, the mRNA expression of *IFIT5*, *TNFSF8*, *SAMD9*, *RSAD2*, *MX1*, and *GBP1* was downregulated, whereas *OTUD7A* expression was upregulated in goose fatty liver after 12 days of overfeeding. In contrast, the expression patterns of these genes showed nearly the opposite trend after 24 days of overfeeding. Taken together, these findings indicate that *OTUD7A* regulates the expression of inflammation- and immune-related genes in the development of goose fatty liver.

Keywords: *OTUD7A*; goose; inflammation; immune; nonalcoholic fatty liver disease

Citation: Zhao, M.; Wen, K.; Fan, X.; Sun, Q.; Jauregui, D.; K. Khogali, M.; Liu, L.; Geng, T.; Gong, D. *OTUD7A* Regulates Inflammation- and Immune-Related Gene Expression in Goose Fatty Liver. *Agriculture* **2022**, *12*, 105. https://doi.org/10.3390/agriculture12010105

Academic Editors: Heather Burrow and Michael Goddard

Received: 28 November 2021
Accepted: 11 January 2022
Published: 13 January 2022

1. Introduction

Some fish and birds are able to pre-deposit large amounts of fat in the liver for use during migration, and then the liver can return to its normal state without any obvious pathological symptoms. Geese, as the offspring of migratory birds, also have this characteristic. In agricultural production, this ability of geese is often used for fatty liver production. As a well-known liver-producing species, the fatty liver (typically composed of approximately 60% fat) of Landes geese can reach an 8–10-fold higher weight than the normal liver in a short period through overfeeding [1]. The changes that occur in goose fatty liver are physiological, with no overt injury or pathological symptoms [2,3]. Recent work indicates that the pro-inflammatory factor is suppressed in goose fatty liver vs. normal liver [4]. However, nonalcoholic fatty liver disease (NAFLD) in humans and mammals is frequently accompanied by inflammation [5,6]. Human NAFLD is prone to developing from simple steatosis to nonalcoholic steatohepatitis (NASH), cirrhosis, and even liver cancer, posing a serious threat to human health [7]. In addition, the incidence of fatty liver in livestock and poultry has also increased due to improved feed nutrition levels, reduced animal activity, and increased environmental stress in modern intensive livestock production, thereby bringing economic losses to livestock production [8]. The differences between geese and other animals suggest that the goose liver utilizes a protection mechanism, although the underlying mechanisms remain unclear. Studies of this mechanism may provide information for developing approaches for preventing or treating NAFLD-associated complications in humans and economically important animals.

A previous study demonstrated that inflammation is important in the progression from simple steatosis to NASH [9]. The hepatic inflammatory response plays an important role in insulin resistance, oxidative stress, and endoplasmic reticulum stress. Inflammation can interact with insulin resistance, and inflammatory factors can interfere with insulin signaling pathways and deepen the degree of insulin resistance, leading to a further decrease in insulin sensitivity in liver cells. On the other hand, inflammation causes the activation of Kupffer cells and stellate cells in the liver, and induces the development of liver fibrosis and cirrhosis [10]. Therefore, inflammation has been widely studied as a possible target for NASH therapy. The nuclear factor κB (NF-κB) signaling pathway is activated in human and animal NAFLD models, and the pathway is a key pro-inflammatory signaling pathway leading to the progression of NAFLD to NASH [5,6]. Inhibition of the NF-κB pathway has been reported to reduce the expression of inflammatory factors in the liver and alleviate the progression of NAFLD [11].

The results of our preliminary transcriptomic analysis showed that a number of deubiquitinating enzymes, including ovarian tumor deubiquitinase 7A (*OTUD7A*), were significantly altered at the transcriptional level in the livers of overfed geese compared to controls, suggesting that deubiquitinating modifications are involved in the formation of fatty liver in geese. *OTUD7A*, also known as cellular zinc-finger anti-NF-κB 2, is a member of the ovarian tumor deubiquitinase family [12,13]. OTUD7A promotes deubiquitination of its target proteins, including tumor necrosis factor receptor-associated factor 6 (TRAF6) [13], thus suppressing the NF-κB signaling pathway. Thus, we speculated that *OTUD7A* could inhibit inflammation. However, the functions of *OTUD7A* in the development of NAFLD are unclear. Therefore, we investigated the function of *OTUD7A*, and identified genes and pathways involved in regulating *OTUD7A*, using goose fatty liver as a model.

2. Materials and Methods

2.1. Animal Experiment

Thirty-two healthy 63-day-old male Landes geese were selected and randomly assigned to a control group and an overfeeding group (16 geese per group). Geese in the control group were provided water and feed ad libitum, whereas the geese in the overfeeding treatment were overfed using previously described procedures and diets [14]. In brief, geese in the overfeeding group were subjected to 1 week of pre-overfeeding, followed by 24 days of overfeeding. During the period of pre-overfeeding, the feed intake was gradually increased from 100 g to 300 g per day. For formal overfeeding, the daily feed intake was 500 g for three meals per day in the first 5 days, followed by 1200 g for 5 meals per day in the remaining time. The feed used in this study was cooked maize (maize boiled for 5 min) supplemented with 1% plant oil and 1% salt. All geese were raised in cages. At 81 and 93 days of age, six geese per treatment were randomly selected and fasted overnight with free access to water; in the next morning (at 82 and 94 days of age), the geese were weighed and killed with an electrolethaler. After the geese were exsanguinated, the liver samples were collected and stored at −80 °C. All animal protocols were approved by the Institutional Animal Ethics Committee of Yangzhou University, with permission number 202103309.

2.2. Preparation of Goose Primary Hepatocytes

Hepatocytes were isolated from Landes goose embryos after 23 days of incubation [2]. Specifically, the goose embryo was removed from the egg and placed in a pre-sterilized tray. The abdominal quills were gently removed and the embryo was sterilized with 75% alcohol. The liver was quickly harvested, immersed in PBS, and rinsed 2–3 times. The chopped liver was transferred to a Petri dish and digested with 0.1% type IV collagenase (Worthington Biochemical Corporation, Lakewood, NJ, USA) at 37 °C for 25 min. Subsequently, an equal volume of pre-warmed complete medium that consisted of Dulbecco's modified Eagle's medium (DMEM Gibco, Grand Island, NY, USA), 0.02 mL/L epidermal growth factor (PeproTech, London, UK), 100 IU/mL penicillin (Sigma-Aldrich, St. Louis, MO, USA), 10% fetal bovine serum (Gibco, USA), and 100 mg/mL streptomycin (Sigma-Aldrich, St. Louis, MO, USA) was added to terminate the digestion. The hepatocyte suspension was obtained by filtering through a 220-mesh sterile nylon mesh to remove large tissue clumps and cell clusters. After treating the cells with erythrocyte lysate (Solarbio Co., Ltd., Beijing, China), complete medium was added to the hepatocytes to form a new hepatocyte suspension. The cells were inoculated into 12-well plates at a density of 1×10^6/well and transferred to a 37 °C incubator with 5% CO_2. The medium was renewed after the first 6 h of incubation, and then every 24 h during subsequent incubation.

2.3. Overexpression of Goose OTUD7A

The pcDNA3.1(+) vector containing the goose *OTUD7A* coding sequence (CDS) and the empty vector were designed, isolated, and purified from Shanghai GenePharma Co., Ltd. (Shanghai, China). The CDS of *OTUD7A* was found to be 2808 base pairs. The *OTUD7A* sequence fragment was obtained by PCR; the fragment was digested separately using enzymatic digestion from the pcDNA3.1 vector, and the product was purified and ligated. The ligated products were transformed into bacterial receptor cells, and the clones grown were first identified by enzymatic cleavage to demonstrate that the target gene had been connected to the target vector. The positive clones were then sequenced and analyzed for comparison, and those that were correct were considered to be successful. The recombinant vector was extracted via ultrapure extraction to obtain the pcDNA3.1-*OTUD7A* vector. Goose primary hepatocytes that were transformed with empty vectors were taken as controls (control treatment), and the *OTUD7A* CDS vector was used for overexpression. The vectors were transfected using Lipofectamine 2000 (Biosharp, Hefei, China) according to the instructions of the manufacturer. Six replicates were used for each treatment. After 6 h of transfection, the culture medium was changed from Opti-MEM (Thermo Fisher Scientific, Waltham, MA, USA) to complete medium. Cells were harvested after 24 h of culture.

2.4. Transcriptome Analysis

The samples of the control treatment and *OTUD7A* overexpression were subjected to RNA sequencing analysis. Briefly, mRNA was purified from total RNA using poly-T oligo-attached magnetic beads (Sigma-Aldrich, St. Louis, MO, USA), after which fragmentation was performed. RNA degradation and contamination were monitored on 1% agarose gels. The RNA purity was checked using a spectrophotometer (IMPLEN, Westlake Village, CA, USA), and the integrity was assessed using the RNA Nano 6000 Assay Kit of the Bioanalyzer 2100 system (Agilent Technologies, Santa Clara, CA, USA). A total of 1 μg of RNA per sample was used as input material for cDNA library preparation. The libraries were generated using the NEBNext® Ultra™ RNA Library Prep Kit for Illumina® (NEB, Ipswich, MA, USA), following the manufacturer's recommendations. After library construction, initial quantification was carried out using a Qubit 2.0 Fluorometer (Thermo Fisher Scientific, Waltham, MA, USA). The insert size of the library was then measured using an 2100 Bioanalyzer (Agilent Technologies, Santa Clara, CA, USA). When the insert size met expectations, real-time quantitative PCR (RT-qPCR) was performed to accurately quantify the effective library concentration (effective library concentration should be higher than 2 ng/μL) in order to ensure library quality. Eight cell lanes were used for each cDNA library.

The clustering of the index-coded samples was performed on a cBot Cluster Generation System using the TruSeq PE Cluster Kit v3-cBot-HS (Illumina), according to the manufacturer's instructions. After cluster generation, the library preparations were sequenced on an Illumina NovaSeq platform (San Diego, CA, USA) using four fluorescently labelled dNTP, DNA polymerase, and splice primers, and 150 bp paired-end reads were generated. Raw data (raw reads) of FASTQ format were firstly processed through in-house Perl scripts. In this step, clean data (clean reads) were obtained by removing reads containing adapters, reads containing poly-N, and low-quality reads from raw data. The average clean base (clean reads × 150 bp) was 8.19 G. At the same time, the Q20, Q30, and GC content of the clean data were calculated. All of the downstream analyses were based on the clean data, with high quality; the reference genome was Anser cygnoides domesticus (assembly AnsCyg_PRJNA183603_v1.0, https://www.ncbi.nlm.nih.gov/genome/?term=Anser+cygnoides+domesticus, accessed on 28 November 2021). The reads per kilobase million (FPKM) of each gene were calculated based on the length of the gene and the reads count mapped to the gene. Differential expression analysis of two groups was performed using the DESeq2 R package (Bioconductor, version 1.16.1, http://www.bioconductor.org/about/, accessed on 28 November 2021). The resulting p-values were adjusted using the Benjamini–Hochberg approach for controlling the false discovery rate. The differentially expressed genes (DEGs) were defined as genes with a fold change of treatment over control >2 or <0.5, and p-value < 0.05. In addition, the clusterProfiler R package (Bioconductor, version 3.4.4, http://www.bioconductor.org/about/, accessed on 28 November 2021) was used to test the statistical enrichment of differential expression genes in Kyoto Encyclopedia of Genes and Genomes (KEGG) pathways.

2.5. RNA Interference Assay

The small interfering RNA (siRNA) was designed to target the goose *OTUD7A* CDS, and was generated by Shanghai GenePharma Co., Ltd. (Shanghai, China). The siRNAs were separately transfected into the primary hepatocytes (prepared as described in Section 3.2), and cultured in serum-free and antibiotic-free Opti-MEM (Thermo Fisher Scientific, Waltham, MA, USA) using Lipofectamine 2000 (Biosharp). Briefly, 5 μL of Lipofectamine 2000 was added to 95 μL of Opti-MEM and incubated at room temperature for 5 min to prepare solution A, whereas 5 μL of siRNA or negative control was added to 95 μL of Opti-MEM to prepare solution B. Subsequently, 200 μL of the liquid was added to each well after solutions A and B were mixed and incubated at room temperature for 23 min. The Opti-MEM was replaced with complete culture medium after transfection, and the transfection time was 6 h. The siRNA dose was 100 nM. Scrambled siRNA was taken as a negative control. Six replicates were used for the negative control and RNA interference groups. The best siRNA was selected based on its ability to suppress *OTUD7A* expression (Supplementary Figure S2). After evaluation, the sense strand sequence of the chosen siRNA was 5′-GCGUGUACAGUGAAGAUUUTT-3′, while the antisense strand sequence was 5′-AAAUCUUCACUGUACACGCTT-3′.

2.6. Gene Expression Analysis

Total RNA from liver samples and primary hepatocytes that were transformed with an empty vector or *OTUD7A* CDS vector, or with scrambled siRNA or siRNA targeting *OTUD7A*, was obtained using TRIzol reagent (TaKaRa Biotechnology, Shiga, Japan). The quality and quantity of mRNA were assessed using a NanoDrop 1000 spectrophotometer (Thermo Scientific, Wilmington, DE, USA). A total of 600 ng of RNA per sample was reverse-transcribed into cDNA using PrimeScript RT Master Mix kits (TaKaRa Biotechnology, Shiga, Japan). Real-time PCR was performed using an ABI 7500 real-time quantitative PCR (RT-qPCR) system (Applied Biosystems, Foster City, CA, USA) using SYBR® Premix Ex Taq™ kits (Takara Biotechnology Co., Ltd., Dalian, China). The reactions were as follows: 95 °C for 30 s, followed by 40 cycles of 95 °C for 5 s and 60 °C for 30 s, then 95 °C for 15 s, 60 °C for 60 s, and 95 °C for 15 s. Two technical replicates were made for each sample. The primers used are listed in Table 1. Relative quantification methods were used, and glyceraldehyde-3-phosphate dehydrogenase (*GAPDH*) was chosen as the reference gene. The fold change level of mRNA was analyzed using the $2^{-\Delta\Delta CT}$ method [15].

Table 1. Primer sequences for real-time quantitative PCR analysis.

Gene [1]	GenBank Number	Primer Sequence (5′ to 3′)	Product Size
OTUD7A	XM_013177443.1	F: CAGACTTTGTTCGGTCCACA R: AGCAATGTCATCCTGCCTCT	236
MPAO	XM_013199765.1	F: TCCGACATGCAGTCCTTGTA R: AACCTTGTTGCCGTAGTTGG	201
IFIT5	XM_013194152.1	F: CCTTGAAGAAGTCCCTGCTG R: ATTTTCCAGGGCTTCCTTGT	194
IL23R	XM_013193090.1	F: CTGATGGGCTCCAACATCTC R: CTGTAGGGCAGCCTGAAGTC	161
TNFSF8	XM_013182840.1	F: AGGCTTGGAGCCACTAACAA R: GGTCGGGACATTCTGTAGGA	201
SAMD9	XM_013187038.1	F: CTCTCGCACAAATGGCACTA R: AGCTTTTGCAGCTCCAATGT	157
RSAD2	XM_013172803.1	F: GAGAGCGGTGGTTCAAGAAG R: TTGAGTGCCATGATCTGCTC	238
MX1	XM_013181384.1	F: TGGAGCAAGTAAACGCCTCT R: TTCAGGCACTGGTAGGCTTT	216
GBP1	XM_013170751.1	F: GAGCAGGAGAGAGAGGCTGA R: TGTTCTTCTCGGAGCCACTT	190
GAPDH	XM_013199522.1	F: GCCATCAATGATCCCTTCAT R: CTGGGGTCACGCTCCTG	155

[1] *OTUD7A*: ovarian tumor deubiquitinase 7A; *MPAO*: membrane primary amine oxidase; *IFIT5*: interferon-induced protein with tetratricopeptide repeats 5; *IL23R*: interleukin-23 receptor; *TNFSF8*: tumor necrosis factor ligand superfamily member 8; *SAMD9*: sterile alpha motif domain-containing protein 9; *RSAD2*: radical *S*-adenosyl methionine domain-containing protein 2; *GBP1*: interferon-induced guanylate binding protein 1-like; *MX1*: interferon-induced GTP-binding protein Mx1.

2.7. Statistical Analysis

The data were confirmed for normal distribution using the Shapiro–Wilk test. Significance was determined using Student's *t*-test for pairwise comparisons, and considered significant at $p < 0.05$.

3. Results

3.1. Genes and Pathways Affected by OTUD7A

RNA sequencing analysis of the transcriptomes of goose hepatocytes transfected with the *OTUD7A* overexpression vector vs. empty vector (as control) revealed 34 DEGs (19 upregulated and 15 downregulated) (Supplementary Figure S1). The upregulated and downregulated genes are shown in Tables 2 and 3, respectively. The enriched KEGG pathways were "cytokine–cytokine receptor interaction", "tropane, piperidine, and pyridine alkaloid biosynthesis", "isoquinoline alkaloid biosynthesis", "Janus kinase-signal transducer and activator of transcription (JAK-STAT) signaling pathway", and "PI3K-Akt signaling pathway" (Figure 1). RT-qPCR analysis of nine DEGs was performed to confirm the results of transcriptome analysis. Consistently, the expression of *OTUD7A*, membrane primary amine oxidase (*MPAO*), and interleukin-23 receptor (*IL23R*) was markedly induced by *OTUD7A* overexpression, whereas that of interferon-induced protein with tetratricopeptide repeats 5 (*IFIT5*), TNF ligand superfamily member 8 (*TNFSF8*), sterile alpha motif domain-containing protein 9 (*SAMD9*), radical *S*-adenosyl methionine domain-containing protein 2 (*RSAD2*), interferon-induced GTP-binding protein Mx1 (*MX1*), and interferon-induced guanylate binding protein 1-like (*GBP1*) was significantly inhibited ($p < 0.05$, Figure 2).

These findings were validated in a knockdown assay in goose primary hepatocytes with siRNA against goose *OTUD7A*. Compared with the controls, the mRNA expression of *MPAO*, *IFIT5*, *TNFSF8*, *SAMD9*, *RSAD2*, *MX1*, and *GBP1* was significantly increased by *OTUD7A* siRNA, whereas the *IL23R* expression was decreased ($p < 0.05$, Figure 3).

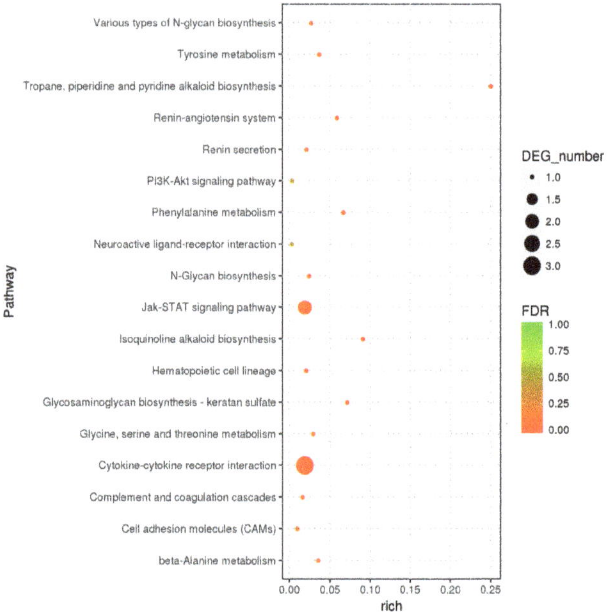

Figure 1. Kyoto Encyclopedia of Genes and Genomes (KEGG) analysis of the control and ovarian tumor deubiquitinase 7A (*OTUD7A*) overexpression groups in goose hepatocytes. Primary hepatocytes were isolated from goose embryos after 23 days of incubation. The cells transfected with the empty pcDNA3.1(+) vector were used as controls, while those transfected with pcDNA3.1(+) containing the *OTUD7A* coding sequence were used for overexpression. Differentially expressed genes (DEGs) were identified using DESeq2 R software. The DEGs were selected based on genes showing a fold change in *OTUD7A* overexpression compared to the control group of >2 or <0.5, and p-value < 0.05. The clusterProfiler R package was used to statistically analyze the enrichment of the DEGs in the KEGG pathways.

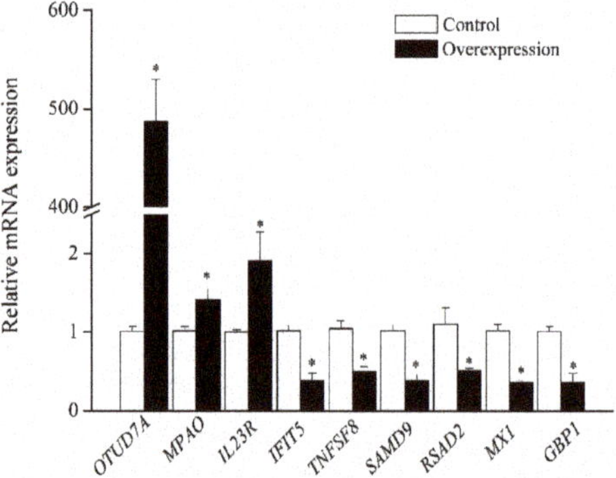

Figure 2. Validation of differentially expressed genes of the control and ovarian tumor deubiquitinase 7A (*OTUD7A*) overexpression groups in goose hepatocytes; * $p < 0.05$. Data are shown as the mean ± SEM of six replicates for each treatment. *MPAO*: membrane primary amine oxidase; *MX1*: interferon-induced GTP-binding protein Mx1; *IL23R*: interleukin-23 receptor; *IFIT5*: interferon-induced protein with tetratricopeptide repeats 5; *TNFSF8*: tumor necrosis factor ligand superfamily member 8; *RSAD2*: radical S-adenosyl methionine domain-containing protein 2; *GBP1*: interferon-induced guanylate binding protein 1-like; *SAMD9*: sterile alpha motif domain-containing protein 9.

Table 2. The upregulated genes in hepatocytes transfected with pcDNA3.1(+) containing goose OTU deubiquitinase 7A CDS and an empty vector.

Name	Description	*p*-Value	Fold Change
SNCG	Gamma-synuclein isoform	0.019	2.11
GPRIN2	G protein-regulated inducer of neurite outgrowth 2 isoform	0.039	2.29
NMU	Neuromedin-U isoform	0.004	7.86
OTUD7A	OTU domain-containing protein 7A	<0.001	134.69
LOC106034831	CMRF35-like molecule 3 isoform	0.018	4.68
ARHGAP28	Rho GTPase-activating protein 28 isoform	0.040	2.10
SH3BP1	SH3 domain-binding protein 1	0.040	2.82
MEST	Mesoderm-specific transcript homolog protein, partial	0.045	2.75
PROCR	Endothelial protein C receptor	0.004	2.61
CLMP	CXADR-like membrane protein	0.012	2.40
KLHDC8A	Kelch domain-containing protein 8A	0.047	3.32
MYL10	Myosin regulatory light chain 2B, cardiac muscle isoform	0.019	2.07
IL23R	Interleukin-23 receptor	0.017	2.14
LOC106045877	Alpha-1,3-mannosyl-glycoprotein 4-beta-N-acetylglucosaminyltransferase C-like	0.042	2.06
ACE	Angiotensin-converting enzyme	0.002	2.05
TMEM114	Transmembrane protein 114	0.004	3.64
PPP1R27	Protein phosphatase 1 regulatory subunit 27	0.05	2.99
MPAO	Membrane primary amine oxidase	0.045	2.03
LOC106048707	Uncharacterized protein LOC106048707 isoform	0.045	4.52

Table 3. The downregulated genes in hepatocytes transfected with pcDNA3.1(+) containing goose OTU deubiquitinase 7A CDS and an empty vector.

Name	Description	*p*-Value	Fold Change
IFIT5	Interferon-induced protein with tetratricopeptide repeats 5	<0.001	0.42
OTOGL	Otogelin-like protein	0.045	0.45
LOC106030637	Interleukin-7-like	0.022	0.17
RSAD2	Radical S-adenosyl methionine domain-containing protein 2	<0.001	0.48
DDX60	Probable ATP-dependent RNA helicase DDX60 isoform	<0.001	0.42
MX1	Interferon-induced GTP-binding protein Mx1	<0.001	0.47
NCAM2	Neural cell adhesion molecule 2 isoform	0.007	0.27
TNFSF8	Tumor necrosis factor ligand superfamily member 8	0.041	0.35
PIH1D3	Protein PIH1D3	0.035	0.39
LOC106038693	Low-affinity vacuolar monovalent cation antiporter	0.043	0.08
SAMD9	Sterile alpha motif domain-containing protein 9	<0.001	0.45
LOC106040189	Arylacetamide deacetylase-like 4	0.010	0.42
LOC106041938	Placenta-specific gene 8 protein-like isoform	<0.001	0.44
B3GNT7	UDP-GlcNAc:betaGal beta-1,3-N-acetylglucosaminyltransferase 7	<0.001	0.36
GBP1	Interferon-induced guanylate-binding protein 1	<0.001	0.45

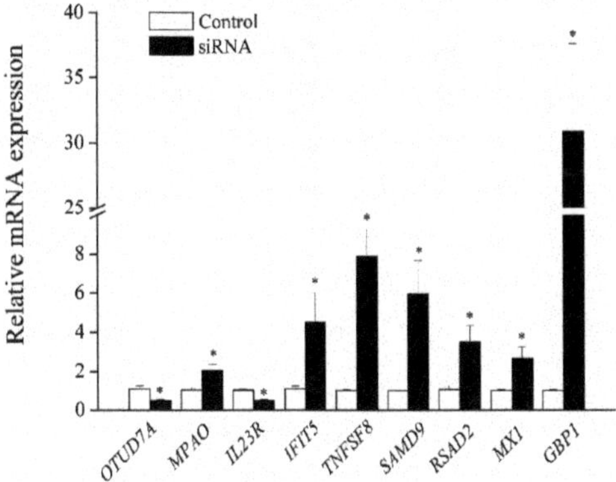

Figure 3. Effects of transfection with siRNA targeting ovarian tumor deubiquitinase 7A (*OTUD7A*) on the expression of downstream genes in goose primary hepatocytes. Scrambled siRNA was used as a control; * $p < 0.05$. Six replicates were evaluated for each treatment. Data are shown as the mean ± SEM.

3.2. Expression of OTUD7A and Its Downstream Genes

The expression of *OTUD7A*, *MPAO*, and *IL23R* was upregulated, whereas that of *IFIT5*, *TNFSF8*, *SAMD9*, *RSAD2*, *MX1*, and *GBP1* was downregulated at day 12 of overfeeding ($p < 0.05$, Figure 4A). However, the *OTUD7A* and *MPAO* expression was downregulated in goose fatty liver, whereas that of *IL23R*, *IFIT5*, *SAMD9*, *MX1*, and *GBP1* was upregulated on day 24 of overfeeding ($p < 0.05$, Figure 4B).

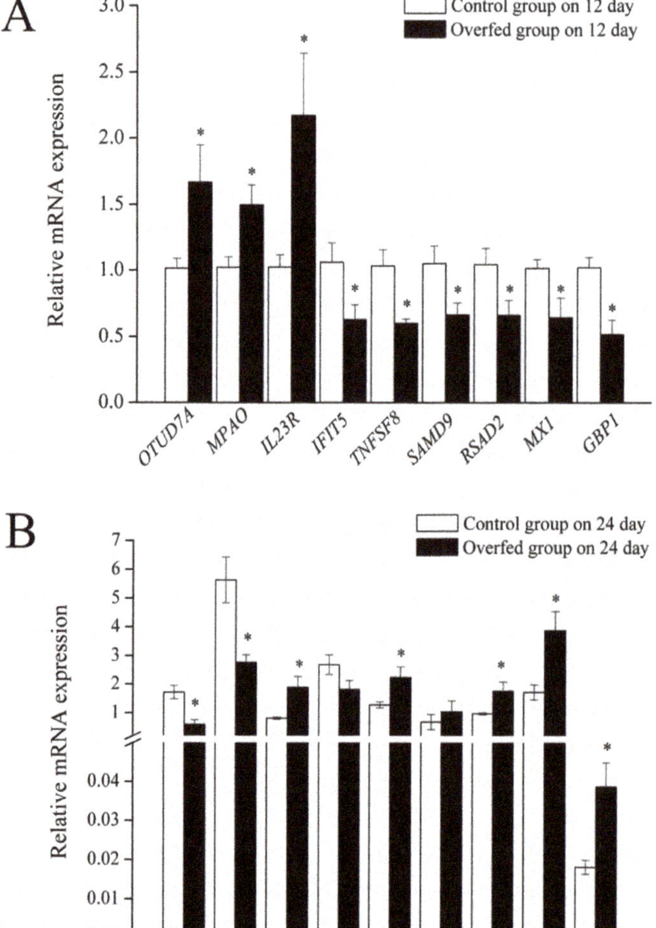

Figure 4. Effects of overfeeding on expression of downstream genes. Control group denotes geese that were fed normally; overfed group denotes geese subjected to 12 (**A**) and 24 days (**B**) of overfeeding; * $p < 0.05$. Samples in the control group on day 12 of overfeeding were used as calibrators for all groups. Six replicates were evaluated for each treatment. Data are shown as the mean ± SEM.

4. Discussion

The pathogenesis of NAFLD is still unclear. Previous studies have indicated that inflammatory responses play important roles in NAFLD, and inflammation is considered to be a marker of progression from simple steatosis to NASH [5,16,17]. Inflammation not only aggravates insulin resistance, but also causes hepatic stellate cell activation, and induces liver fibrosis and cirrhosis [10,18]. The NF-κB pathway—a major inflammatory signaling pathway—is activated in patients with NAFLD and in animal models [5,6]. Activation of the NF-κB pathway can promote transcription of downstream inflammatory response genes, resulting in an increase in inflammatory factor production and release. Inflammatory factors can successively reactivate NF-κB, creating positive feedback regulation that leads to further amplification of the initial inflammatory signal [19]. In addition, NF-κB activation can activate pro-apoptotic proteins on mitochondria, such as BAX, leading to apoptosis of hepatocytes and further exacerbating the inflammatory response [20]. Reducing inflammatory factor expression in the liver can alleviate NAFLD development [11]; therefore, inflammation has received considerable attention as a possible target for NAFLD therapy.

As an excellent waterfowl product, goose fatty liver can be used as both a high-grade food and a unique model for NAFLD research. Our preliminary research has identified some characteristics that are different from mammalian NAFLD, such as elevated fatty acid desaturase, increased expression of lipocalin receptor genes, increased expression of mitochondria-related genes, and downregulated expression of complement genes and pro-inflammatory factors in goose fatty liver [2–4,21]. These results suggest the existence of certain mechanisms in the formation of goose fatty liver that can resist the occurrence and development of inflammation. Some studies suggest that deubiquitinating enzymes are involved in protecting against inflammation in NAFLD. Ubiquitin-specific protease 18—a member of the deubiquitinating enzyme family—is downregulated in the livers of obese mice [22]. Similarly, USP4 protects against the inflammatory response, as USP4 depletion was shown to exacerbate inflammation in mice with high-fat-diet-induced NAFLD, and USP4 may suppress activation of the downstream NF-κB pathway [23]. Mevissen et al. [12] found that *OTUD7A* is a potential tumor suppressor, and can regulate multiple signaling pathways. Another study indicated that *OTUD7A* inhibits the NF-κB pathway through deubiquitination of TRAF6 protein in HepG2 cells [10]. Therefore, *OTUD7A* may be involved in the development of fatty liver.

In this study, transcriptome sequencing analysis showed that a total of 19 genes were upregulated and 15 genes were downregulated in the *OTUD7A* overexpression group compared to controls. The DEGs were mainly enriched in cytokine–cytokine receptor interaction; tropane, piperidine, and pyridine alkaloid biosynthesis; isoquinoline alkaloid biosynthesis, the JAK-STAT signaling pathway; and the PI3K-Akt signaling pathway. Further analysis showed that numerous DEGs affected by *OTUD7A* were related to inflammation, as their mRNA expression was subject to *OTUD7A* overexpression or suppression by siRNA against *OTUD7A*, which is similar to the *OTUD7A* subfamily that can regulate some inflammation-related genes [24,25]. Our findings also support the hypothesis that OTUD7A can regulate the NF-κB pathway via deubiquitination of TRAF6. Pro-inflammatory factors can activate the inflammatory cascade response and increase the expression of genes such as *IL-6* and inducible nitric oxide synthase, causing inflammatory responses and tissue damage in the body. During the formation of mammalian NAFLD, the levels of pro-inflammatory factors are usually significantly elevated [18,26]. These results suggest that *OTUD7A* may be associated with the inhibition of inflammatory responses in goose fatty liver—at least in the mid-overfeeding period.

Interestingly, cytokine–cytokine receptor interactions, which are important for the maintenance and regulatory functions of multicellular organisms [27], were among the enriched KEGG pathways in the *OTUD7A* assay. Tumor necrosis factor superfamily (TNFSF) is a cytokine secreted by immune cells. The interaction between TNFSF and tumor necrosis factor receptor is involved in regulating cell growth, immune response, apoptosis, and inflammatory response [28]. The TNFSF8 protein can activate the NF-κB pathway by binding to its receptor, TNFRSF8, thereby mediating the secretion of IL-2, IL-6, TNF-α, and other cytokines [29]. The expression of *TNFSF8* was downregulated by *OTUD7A* overexpression in primary goose hepatocytes, as well as in goose fatty liver after 12 days of overfeeding, whereas the expression of *TNFSF8* was increased after *OTUD7A* suppression in primary goose hepatocytes, suggesting that *TNFSF8* is regulated by *OTUD7A* in goose hepatocytes through the cytokine receptor interaction pathway.

The JAK-STAT pathway, as a downstream pathway of cytokine receptors, was also enriched according to the transcriptomic analysis of goose primary hepatocytes overexpressing *OTUD7A*. Several cytokines, including interferons, can modulate intracellular signaling by activating the JAK-STAT pathway [30]. Upon the binding of cytokines to their cognate receptors, STATs can modulate the expression of their target genes and participate in inflammation [31,32]. Enrichment of DEGs in the JAK-STAT signaling pathway resulting from *OTUD7A* overexpression suggests that *OTUD7A* regulates inflammation through the JAK-STAT signaling pathway in goose hepatocytes. The IFIT family, as interferon-induced genes, participate in the immune response [33]. Previous studies indicated that *IFIT5* promotes NF-κB activation and synergizes NF-κB-mediated gene expression, whereas knockdown of *IFIT5* inhibits NF-κB pathway activation and downstream gene expression [34,35]. Our data indicate that *IFIT5* expression was significantly decreased by *OTUD7A* overexpression, but was induced by *OTUD7A* knockdown in primary goose hepatocytes. Downregulation of *IFIT5* was accompanied by upregulation of *OTUD7A* on day 12 of overfeeding, whereas upregulation of *IFIT5* was accompanied by downregulation of *OTUD7A* on day 24 of overfeeding, which is consistent with the results of the cell research. Thus, *OTUD7A* may regulate *IFIT5* expression in the goose fatty liver. In addition, previous studies have suggested that *SAMD9* is a downstream target of inflammatory cytokines [36,37], and may function as an anti-inflammatory factor [38]; it is also associated with the immune response [39]. Our data indicate that *SAMD9*, like *IFIT5*, is regulated by *OTUD7A*.

Moreover, many studies have indicated that innate immune response is connected with inflammation in NAFLD/NASH, thereby promoting the development of fibrosis, cirrhosis, and carcinogenesis. The results from our study showed that *OTUD7A* regulates the expression of some immune-related genes—for example, *IL23R*—in goose fatty liver. *IL23R* mediates the stimulation of T cells, natural killer cells, and possibly certain macrophage/myeloid cells—likely through the JAK-STAT pathway [40]. In addition, the mRNA expression of *RSAD2*, *MX1*, and *GBP1* was downregulated by *OTUD7A* overexpression, but upregulated by *OTUD7A* knockdown. These results suggest that *RSAD2*, *MX1*, and *GBP1* are downstream of *OTUD7A*. *RSAD2*, also known as viperin, is an important component of innate immunity [41]. MX1 is also a vital antiviral protein during the innate immune response [42,43]. Some studies have shown that *GBP1* can be induced by IFN-α, IFN-β, IFN-γ, and inflammatory cytokines [44,45]. *GBP1* expression is increased in inflammatory skin diseases, and the gene is a cellular activation marker characterizing the inflammatory cytokine-activated phenotype of cells [46]. As immune-related genes, *IL23R*, *RSAD2*, *MX1*, and *GBP1* were regulated by *OTUD7A*, suggesting that *OTUD7A* can regulate immune response in liver physiology and pathology.

Additionally, KEGG results suggested that the DEGs were related to tropane, piperidine, pyridine, and isoquinoline alkaloid biosynthesis following *OTUD7A* overexpression; for example, *MPAO*, which participates in the metabolism of alkaloid biosynthesis, was upregulated by *OTUD7A* overexpression in goose primary hepatocytes. Consistently, upregulation of *MPAO* was accompanied by upregulation of *OTUD7A* on day 12 of overfeeding, whereas the downregulation of *MPAO* was accompanied by downregulation of *OTUD7A* on day 24 of overfeeding. Therefore, *OTUD7A* may participate in alkaloid biosynthesis in the goose fatty liver. Bour et al. [47] found that *MPAO* expression was increased in the adipose tissue, suggesting that it is involved in adipogenesis. Fat metabolism is very active in the formation of goose fatty liver [3]; thus, *MPAO* may be involved in lipid metabolism in the goose fatty liver. Further research is required in order to confirm this hypothesis.

5. Conclusions

Transcriptome sequencing analysis showed that *OTUD7A* is involved in the development of goose fatty liver, mainly through cytokine–cytokine receptor interaction; tropane, piperidine, and pyridine alkaloid biosynthesis; isoquinoline alkaloid bio-synthesis; the JAK-STAT signaling pathway, and the PI3K-Akt signaling pathway. In addition, *OTUD7A* may regulate the expression of inflammation- and immune-related genes such as *TNFSF8*, *IFIT5*, *IL23R*, *RSAD2*, *MX1*, and *GBP1* in the goose fatty liver.

Supplementary Materials: The following are available online at https://www.mdpi.com/article/10.3390/agriculture12010105/s1: Figure S1: Genes affected by OTU deubiquitinase 7A (*OTUD7A*) overexpression; Figure S2: Screening of small interfering RNAs (siRNAs) for OTU deubiquitinase 7A (*OTUD7A*).

Author Contributions: Conceptualization, T.G. and D.G.; methodology, K.W., D.J. and M.K.K.; software, M.Z.; validation, M.Z. and L.L.; formal analysis, M.Z. and X.F.; investigation, D.G.; resources, M.Z.; data curation, Q.S.; writing—original draft preparation, M.Z.; writing—review and editing, T.G., D.J. and M.K.K.; visualization, D.G.; supervision, D.G.; project administration, D.G.; funding acquisition, M.Z. and D.G. All authors have read and agreed to the published version of the manuscript.

Funding: This research was funded by the National Nature Science Foundation of China, grant numbers 31802052, 31972546, and 32072785; the China Postdoctoral Science Foundation (2017M621840); and the Priority Academic Program Development of Jiangsu Higher Education Institutions (PAPD).

Institutional Review Board Statement: This study was approved by the Institutional Ethics Committee of Yangzhou University (protocol code 202103309, 9 March 2021).

Data Availability Statement: The data presented in this study are available upon request from the corresponding author.

Conflicts of Interest: The authors declare no conflict of interest.

References

1. Hermier, D.; Rousselot-Pailley, D.; Peresson, R.; Sellier, N. Influence of orotic acid and estrogen on hepatic lipid storage and secretion in the goose susceptible to liver steatosis. *Biochim. Biophys. Acta* **1994**, *1211*, 97–106. [CrossRef]
2. Geng, T.Y.; Xia, L.L.; Li, F.Y.; Xia, J.; Zhang, Y.H.; Wang, Q.Q.; Yang, B.; Montgomery, S.; Cui, H.M.; Gong, D.Q. The role of endoplasmic reticulum stress and insulin resistance in the occurrence of goose fatty liver. *Biochem. Biophys. Res. Commun.* **2015**, *465*, 83–87. [CrossRef]
3. Geng, T.Y.; Yang, B.; Li, F.Y.; Xia, L.L.; Wang, Q.Q.; Zhao, X.; Gong, D.Q. Identification of protective components that prevent the exacerbation of goose fatty liver: Characterization, expression and regulation of adiponectin receptors. *Comp. Biochem. Physiol. B Biochem. Mol. Biol.* **2016**, *194–195*, 32–38. [CrossRef]
4. Liu, L.; Zhao, X.; Wang, Q.; Sun, X.X.; Xia, L.L.; Wang, Q.Q.; Yang, B.; Zhang, Y.H.; Montgomery, S.; Meng, H.; et al. Prosteatotic and protective components in a unique model of fatty liver: Gut microbiota and suppressed complement system. *Sci. Rep.* **2016**, *6*, 31763. [CrossRef]
5. Baker, R.G.; Hayden, M.S.; Ghosh, S. NF-κB, inflammation, and metabolic disease. *Cell Metab.* **2011**, *13*, 11–22. [CrossRef] [PubMed]
6. Zhou, Y.; Ding, Y.L.; Zhang, J.L.; Zhang, P.; Wang, J.Q.; Li, Z.H. Alpinetin improved high fat diet-induced non-alcoholic fatty liver disease (NAFLD) through improving oxidative stress, inflammatory response and lipid metabolism. *Biomed. Pharmacother.* **2018**, *97*, 1397–1408. [CrossRef]
7. Hossain, N.; Kanwar, P.; Mohanty, S.R. A comprehensive updated review of pharmaceutical and nonpharmaceutical treatment for NAFLD. Gastroenterol. *Res. Pract.* **2016**, *144*, 7109270.
8. Rahman, A.M.; Mahdavi, A.H.; Rahmani, H.R.; Jahanian, E. Clove bud (*Syzygium aromaticum*) improved blood and hepatic antioxidant indices in laying hens receiving low n-6 to n-3 ratios. *J. Anim. Physiol. Anim. Nutr.* **2017**, *101*, 881–892. [CrossRef] [PubMed]
9. Marra, F. Nuclear factor-κB inhibition and non-alcoholic steatohepatitis: Inflammation as a target for therapy. *Gut* **2008**, *57*, 570–572. [CrossRef] [PubMed]
10. Tilg, H.; Moschen, A.R. Evolution of inflammation in nonalcoholic fatty liver disease: The multiple parallel hits hypothesis. *Hepatology* **2010**, *52*, 1836–1846. [CrossRef] [PubMed]
11. Tian, Y.L.; Ma, J.T.; Wang, W.D.; Zhang, L.J.; Xu, J.; Wang, K.; Li, D.F. Resveratrol supplement inhibited the NF-κB inflammation pathway through activating AMPKα-SIRT1 pathway in mice with fatty liver. *Mol. Cell. Biochem.* **2016**, *422*, 75–84. [CrossRef]
12. Mevissen, T.E.; Hospenthal, M.K.; Geurink, P.P.; Elliott, P.R.; Akutsu, M.; Arnaudo, N.; Ekkebus, R.; Kulathu, Y.; Wauer, T.; Oualid, F.E.I.; et al. OTU deubiquitinases reveal mechanisms of linkage specificity and enable ubiquitin chain restriction analysis. *Cell* **2013**, *154*, 169–184. [CrossRef]
13. Xu, Z.; Pei, L.; Wang, L.; Zhang, F.; Hu, X.; Gui, Y. Snail1-dependent transcriptional repression of Cezanne2 in hepatocellular carcinoma. *Oncogene* **2014**, *33*, 2836–2845. [CrossRef]

14. Zhao, M.M.; Liu, T.J.; Wang, Q.; Zhang, R.; Liu, L.; Gong, D.Q.; Geng, T.Y. Fatty acids modulate the expression of pyruvate kinase and arachidonate-lipoxygenase through PPARγ/CYP2C45 pathway: A link to goose fatty liver. *Poult. Sci.* **2019**, *98*, 4346–4358. [CrossRef] [PubMed]
15. Schmittgen, T.D.; Livak, K.J. Analyzing real-time PCR data by the comparative CT method. *Nat. Protoc.* **2008**, *3*, 1101–1108. [CrossRef] [PubMed]
16. Parola, M.; Vajro, P. Nocturnal hypoxia in obese-related obstructive sleep apnea as a putative trigger of oxidative stress in pediatric NAFLD progression. *J. Hepatol.* **2016**, *65*, 470–472. [CrossRef]
17. Wruck, W.; Graffmann, N.; Kawala, M.A.; Adjaye, J. Concise review: Current status and future directions on research related to nonalcoholic fatty liver disease. *Stem Cells* **2017**, *35*, 89–96. [CrossRef]
18. Pirozzi, C.; Lama, A.; Simeoli, R.; Paciello, O.; Pagano, T.B.; Mollica, M.P.; Di Guida, F.; Russo, R.; Magliocca, S.; Canani, R.B.; et al. Hydroxytyrosol prevents metabolic impairment reducing hepatic inflammation and restoring duodenal integrity in a rat model of NAFLD. *J. Nutr. Biochem.* **2016**, *30*, 108–115. [CrossRef] [PubMed]
19. Kondylis, V.; Kumari, S.; Vlantis, K.; Pasparakis, M. The interplay of IKK, NF-κB and RIPK1 signaling in the regulation of cell death, tissue homeostasis and inflammation. *Immunol. Rev.* **2017**, *277*, 113–127. [CrossRef]
20. John, J.L.; Qian, T.; He, L.H.; Kim, J.S.; Elmore, S.P.; Cascio, W.E.; Brenner, D.A. Role of mitochondrial inner membrane permeabilization in necrotic cell death, apoptosis, and autophagy. *Antioxid. Redox. Signal.* **2002**, *4*, 769–781.
21. Xing, Y.; Xu, C.; Lin, X.; Zhao, M.M.; Gong, D.Q.; Liu, L.; Geng, T.Y. Complement C3 participates in the development of goose fatty liver potentially by regulating the expression of FASN and ETNK1. *Anim. Sci. J.* **2021**, *92*, e13527. [CrossRef] [PubMed]
22. An, S.; Zhao, L.P.; Shen, L.J.; Wang, S.; Zhang, K.; Qi, Y.; Zheng, J.; Zhang, X.J.; Zhu, X.Y.; Bao, R.; et al. USP18 protects against hepatic steatosis and insulin resistance through its deubiquitinating activity. *Hepatology* **2017**, *666*, 1866–1884. [CrossRef]
23. Zhao, Y.C.; Wang, F.; Gao, L.C.; Xu, L.W.; Tong, R.Y.; Lin, N.; Su, Y.Y.; Yan, Y.; Gao, Y.; He, J.; et al. Ubiquitin-specific protease 4 is an endogenous negative regulator of metabolic dysfunctions in nonalcoholic fatty liver disease in mice. *Hepatology* **2018**, *68*, 897–917. [CrossRef] [PubMed]
24. Lin, D.D.; Shen, Y.; Qiao, S.; Liu, W.W.; Zheng, L.S.; Wang, Y.N.; Cui, N.P.; Wang, Y.F.; Zhao, S.L.; Shi, J.H. Upregulation of OTUD7B (cezanne) promotes tumor progression via AKT/VEGF pathway in lung squamous carcinoma and adenocarcinoma. *Front. Oncol.* **2019**, *9*, 862. [CrossRef]
25. Deng, J.J.; Hou, G.X.; Fang, Z.X.; Liu, J.L.; Lv, X.D. Distinct expression and prognostic value of OTU domain-containing proteins in non-small-cell lung cancer. *Oncol. Lett.* **2019**, *5*, 5417–5427. [CrossRef] [PubMed]
26. Ceccarelli, S.; Panera, N.; Mina, M.; Gnani, D.; Stefanis, C.D.; Crudele, A.; Rychlicki, C.; Petrini, S.; Bruscalupi, G.; Agostinelli, L.; et al. LPS-induced TNF-α factor mediates pro-inflammatory and pro-fibrogenic pattern in non-alcoholic fatty liver disease. *Oncotarget* **2015**, *6*, 41434–41452. [CrossRef] [PubMed]
27. Leonard, W.J.; Lin, J.X. Cytokine receptor signaling pathways. *J. Allergy Clin. Immunol.* **2000**, *105*, 877–888. [CrossRef]
28. Croft, M.; Duan, W.; Choi, H.; Eun, S.Y.; Madireddi, S.; Mehta, A. TNF superfamily in inflammatory disease: Translating basic insights. *Trends Immunol.* **2012**, *33*, 144–152. [CrossRef] [PubMed]
29. Horie, R.; Aizawa, S.; Nagai, M.; Ito, K.; Higashihara, M.; Ishida, T.; Inoue, J.; Watanabe, T. A novel domain in the CD30 cytoplasmic tail mediates NFkappaB activation. *Int. Immunol.* **1998**, *10*, 203–210. [CrossRef]
30. McGillicuddy, F.C.; Chiquoine, E.H.; Hinkle, C.C.; Kim, R.J.; Shah, R.; Roche, H.M.; Smyth, E.M.; Reilly, M.P. Interferon γ attenuates insulin signaling, lipid storage, and differentiation in human adipocytes via activation of the JAK/STAT pathway. *J. Biol. Chem.* **2009**, *284*, 31936–31944. [CrossRef]
31. Imada, K.; Leonard, W.J. The Jak-STAT pathway. *Mol. Immunol.* **2000**, *37*, 1–11. [CrossRef]
32. Coskun, M.; Salem, M.; Pedersen, J.; Nielsen, O.H. Involvement of JAK/STAT signaling in the pathogenesis of inflammatory bowel disease. *Pharmacol. Res.* **2013**, *76*, 1–8. [CrossRef] [PubMed]
33. Rohaim, M.A.; Santhakumar, D.; Naggar, R.F.E.; Iqbal, M.; Hussein, H.A.; Munir, M. Chickens expressing IFIT5 ameliorate clinical outcome and pathology of highly pathogenic avian influenza and velogenic newcastle disease viruses. *Front. Immunol.* **2018**, *9*, 2025. [CrossRef] [PubMed]
34. Zhang, B.H.; Liu, X.Y.; Chen, W.; Chen, L. IFIT5 potentiates anti-viral response through enhancing innate immune signaling pathways. *Acta Biochim. Biophys. Sin.* **2013**, *45*, 867–874. [CrossRef]
35. Zheng, C.S.; Zheng, Z.H.; Zhang, Z.H.; Meng, J.; Liu, Y.; Ke, X.L.; Hu, Q.X.; Wang, H.Z. IFIT5 positively regulates NF-κB signaling through synergizing the recruitment of IκB kinase (IKK) to TGF-β-activated kinase 1 (TAK1). *Cell. Signal.* **2015**, *27*, 2343–2354. [CrossRef]
36. Chefetz, I.; Ben, A.D.; Browning, S.; Skorecki, K.; Adir, N.; Thomas, M.G.; Kogleck, L.; Topaz, O.; Indelman, M.; Uitto, J.; et al. Normophosphatemic familial tumoral calcinosis is caused by deleterious mutations in SAMD9, encoding a TNF-α responsive protein. *J. Investig. Dermatol.* **2008**, *128*, 1423–1429. [CrossRef]
37. Hershkovitz, D.; Gross, Y.; Nahum, S.; Yehezkel, S.; Sarig, O.; Uitto, J.; Sprecher, E. Functional characterization of SAMD9, a protein deficient in normophosphatemic familial tumoral calcinosis. *J. Investig. Dermatol.* **2011**, *131*, 662–669. [CrossRef]
38. He, P.; Wu, L.-F.; Bing, P.-F.; Xia, W.; Wang, L.; Xie, F.-F.; Lu, X.; Lei, S.-F.; Deng, F.-Y. SAMD9 is a (epi-) genetically regulated anti-inflammatory factor activated in RA patients. *Mol. Cell Biochem.* **2019**, *456*, 135–144. [CrossRef]
39. Nounamo, B.; Li, Y.B.; Byrne, P.O.; Kearney, A.M.; Khan, A.; Liu, J. An interaction domain in human SAMD9 is essential for myxoma virus host-range determinant M062 antagonism of host anti-viral function. *Virology* **2017**, *503*, 94–102. [CrossRef]

40. Parham, C.; Chirica, M.; Timans, J.; Vaisberg, E.; Travis, M.; Cheung, J.; Pflanz, S.; Zhang, R.; Singh, K.P.; Vega, F.; et al. A receptor for the heterodimeric cytokine IL-23 is composed of IL-12Rβ1 and a novel cytokine receptor subunit, IL-23R. *J. Immunol.* **2002**, *168*, 5699–5708. [CrossRef]
41. Rivieccio, M.A.; Suh, H.S.; Zhao, Y.M.; Zhao, M.L.; Chin, K.C.; Lee, S.C.; Brosnan, C.F. TLR3 ligation activates an antiviral response in human fetal astrocytes: A role for viperin/cig5. *J. Immunol.* **2006**, *177*, 4735–4741. [CrossRef] [PubMed]
42. Wisskirchen, C.; Ludersdorfer, T.H.; Müller, D.A.; Moritz, E.; Pavlovic, J. Interferon-induced antiviral protein MxA interacts with the cellular RNA helicases UAP56 and URH49. *J. Biol. Chem.* **2011**, *286*, 34743–34751. [CrossRef] [PubMed]
43. Xiao, J.; Yan, J.; Chen, H.; Li, J.; Tian, Y.; Tang, L.S.; Feng, H. Mx1 of black carp functions importantly in the antiviral innate immune response. *Fish Shellfish Immunol.* **2016**, *58*, 584–592. [CrossRef]
44. Tripal, P.; Bauer, M.; Naschberger, E.; Mörtinger, T.; Hohenadl, C.; Cornali, E.; Thurau, M.; Stürzl, M. Unique features of different members of the human guanylate-binding protein family. *J. Interferon Cytokine Res.* **2007**, *27*, 44–52. [CrossRef]
45. Britzen-Laurent, N.; Bauer, M.; Berton, V.; Fischer, N.; Syguda, A.; Reipschläger, S.; Naschberger, E.; Herrmann, C.; Stürzl, M. Intracellular trafficking of guanylate-binding proteins is regulated by heterodimerization in a hierarchical manner. *PLoS ONE* **2010**, *12*, e14246. [CrossRef] [PubMed]
46. Lubeseder-Martellato, C.; Guenzi, E.; Jörg, A.; Töpolt, K.; Naschberger, E.; Kremmer, E.; Zietz, C.; Tschachler, E.; Hutzler, P.; Schwemmle, M.; et al. Guanylate-binding protein-1 expression is selectively induced by inflammatory cytokines and is an activation marker of endothelial cells during inflammatory diseases. *Am. J. Pathol.* **2002**, *161*, 1749–1759. [CrossRef]
47. Bour, S.; Daviaud, D.; Gres, S.; Lefort, C.; Prévot, D.; Zorzano, A.; Wabitsch, M.; Saulnier-Blache, J.-S.; Valet, P.; Carpéné, C. Adipogenesis-related increase of semicarbazide-sensitive amine oxidase and monoamine oxidase in human adipocytes. *Biochimie* **2007**, *89*, 916–925. [CrossRef] [PubMed]

Article

Positive Selection and Adaptive Introgression of Haplotypes from *Bos indicus* Improve the Modern *Bos taurus* Cattle

Qianqian Zhang [1,2,*], Anna Amanda Schönherz [2,3], Mogens Sandø Lund [2] and Bernt Guldbrandtsen [2,4]

[1] Institute of Biotechnology Research, Beijing Academy of Agriculture and Forestry Sciences, Beijing 100097, China
[2] Center for Quantitative Genetics and Genomics, Aarhus University, 8000 Aarhus, Denmark; anna.schonherz@anis.au.dk (A.A.S.); mogens.lund@qgg.au.dk (M.S.L.); bernt.guldbrandtsen@sund.ku.dk (B.G.)
[3] Department of Animal Science, Aarhus University, 8000 Aarhus, Denmark
[4] Department of Veterinary and Animal Sciences, Copenhagen University, 1165 Copenhagen, Denmark
* Correspondence: zhangqianqian@baafs.net.cn

Abstract: Complex evolutionary processes, such as positive selection and introgression can be characterized by in-depth assessment of sequence variation on a whole-genome scale. Here, we demonstrate the combined effects of positive selection and adaptive introgression on genomes, resulting in observed hotspots of runs of homozygosity (ROH) haplotypes on the modern bovine (*Bos taurus*) genome. We first confirm that these observed ROH hotspot haplotypes are results of positive selection. The haplotypes under selection, including genes of biological interest, such as *PLAG1, KIT, CYP19A1* and *TSHB*, were known to be associated with productive traits in modern *Bos taurus* cattle breeds. Among the haplotypes under selection, we demonstrate that the *CYP19A1* haplotype under selection was associated with milk yield, a trait under strong recent selection, demonstrating a likely cause of the selective sweep. We further deduce that selection on haplotypes containing *KIT* variants affecting coat color occurred approximately 250 generations ago. The study on the genealogies and phylogenies of these haplotypes identifies that the introgression events of the *RERE* and *REG3G* haplotypes happened from *Bos indicus* to *Bos taurus*. With the aid of sequencing data and evolutionary analyses, we here report introgression events in the formation of the current bovine genome.

Keywords: positive selection; adaptive introgression; runs of homozygosity; haplotype; cattle

Citation: Zhang, Q.; Schönherz, A.A.; Lund, M.S.; Guldbrandtsen, B. Positive Selection and Adaptive Introgression of Haplotypes from *Bos indicus* Improve the Modern *Bos taurus* Cattle. *Agriculture* **2022**, *12*, 844. https://doi.org/10.3390/agriculture12060844

Academic Editors: Heather Burrow and Michael Goddard

Received: 7 April 2022
Accepted: 12 May 2022
Published: 11 June 2022

1. Introduction

Whole-genome sequencing technology and genomic tools provide opportunities to investigate and understand the interplay between complex evolutionary processes, such as positive selection, introgression and inbreeding [1–3]. Using genome-wide or genomic region-specific analyses with single base pair resolution, an in-depth understanding of the selective processes shaping patterns of genetic variation in a population can be achieved [4–6]. Among these complex processes, positive selection has played a very important role in changing genomes through adaptation driven by frequency changes of favorable alleles in the modern population. Strong selection on favorable alleles often happens in response to environmental changes or diseases. An example of the former is the selection at the lactase locus in humans for the ability to digest milk [7]. In farm animals, strong drivers of adaptation include domestication and recent strong artificial selection for desired phenotypes. Signatures of selection in the genome can be detected through high levels of linkage disequilibrium resulting in extended haplotypes, deviations of allele frequencies from the neutral model and reduced local heterozygosity [8,9]. However, the process of positive selection has its own complexity as it interacts with other evolutionary processes, resulting in distinct patterns of genetic variation. Thus far, there is a poor understanding of the interaction between different complex evolutionary processes on the observed genomic phenomenon for haplotypes on a genome-wide scale, such as the distribution of homozygosity [10]. Here, we study the impact of selection resulting in genomic homozygous haplotypes, known as runs of homozygosity (ROH), in connection to the candidate target regions for positive selection and adaptive introgression.

Domestic cattle is an appealing model to study the effects of demography and positive selection on their genomes resulting in extreme patterns of ROH distribution, because of the known selection history, detailed records and controlled environments. Since domestication from wild aurochs ~10,000 years ago [11–13], the bovine genome has been heavily shaped by intensive artificial selection. Selection has been especially strong during the most recent 60 years [14,15], resulting in a significant improvement in milk and meat production as well as changes in disease resistance [1,16,17]. The extreme use of the genetically best sires has left many and clear signals of selection in the cattle genome. It provides opportunities for detecting haplotypes not only under strong positive selection, but also resulting from introgression from other species. These important haplotypes have largely contributed to the improvement of the modern taurine cattle breeds. Evidence already shows that there is pervasive introgression aiding in domestication and adaptation in the *Bos* species [18]. Hence, these unique conditions make the genome of domestic cattle an excellent model to study how long-term intensive positive selection together with introgression shapes their genomes.

This study used large-scale genome sequencing data from domestic cattle as a model to demonstrate that positive selection can cause uniform ROH haplotype patterns at the population level. We first characterized and compared the distribution of polymorphisms with single-base resolution to quantify genetic diversity in different breeds. We next examined the effects of positive selection on the distribution of ROH in the current domestic cattle population. Based on the length distribution of ROH in the population, we inferred the time-scale of positive selection on specific ROH haplotypes in the population. Finally, we identified introgression events from *Bos indicus* to *Bos taurus* by examining the phylogenetics of ROH hotspot haplotypes under positive selection. Overall, we utilized the bovine genome as an excellent and unique model system to demonstrate the effects of demography and positive selection in shaping the occurrence and distribution of ROH in cattle genomes, and provide important and novel insights in inferring genomic footprints from demographic history and positive selection using ROH as a tool from large-scale genomes.

We confirmed these candidate regions under selection using other statistics, such as the integrated haplotype score (iHS), and its interplay with demography, shaping the genomes of modern species.

2. Materials and Methods

2.1. Sequencing and SNP Discovery

A total of 684 bulls with high genetic contributions to the current cattle populations, including 267 Holstein, 102 Fleckvieh, 89 Angus, 34 Jersey, 30 Simmental, 29 Brown Swiss, 26 Charolais, 25 Hereford, 25 Gelbvieh, 17 Finnish Ayrshire, 16 Swedish Red, 15 Danish Red and 9 Belgium Blue, were obtained from the 1000 Bull Genomes Project's Run4 [19]. Among these, the Danish Red cattle are from the Old Danish Red cattle population [20], and Principal Component analysis (PCA) showed that Simmental and Fleckvieh are separated as different breeds (unpublished results). Animals were selected following the same criteria as Boitard et al. [21]. Sequences of *Bos indicus*, Gaur, Bison, Wisent, Banteng and Gayal individuals were downloaded from NCBI (for accession numbers, see Data Availability) [5,18].

Sequence reads from sequenced individuals were aligned to the cattle reference genome assembly UMD3.1 [22] using bwa [23]. Duplicate reads were marked using samtools [24]. Local realignment and quality recalibration was performed using the Genome Analysis Toolkit (GATK) [25] following the Human 1000 Genome guidelines, incorporating information from dbSNP [26]. Subsequently, variants were called using HaplotypeCaller from GATK [25] and annotated using information from dbSNP [26]. Only variants with

PHRED scores above 100 were kept and indels were excluded from further analyses. Nucleotide diversity was calculated using a sliding window of 10 kbp over the whole genome in all sequenced individuals, following Bosse et al. [27]. We corrected the SNP counts per 10 kbp bin for the number of bases within each 10 kbp bin proportionally to 10,000 covered bases, i.e., bases with coverage between half and twice the average genome coverage. The correction factor was calculated as DP/bin size, where DP = coverage in bp/bin.

2.2. *Runs of Homozygosity*

ROH were identified on all autosomes of the sequenced individuals using the method developed by Bosse et al. [27]. We set the threshold to declare an ROH as an SNP count maximum of 0.25 times the genome coverage in a window of 10 kbp. Detected ROH were classified into three size categories: (1) short ROH smaller than 100 kbp (size class S); (2) medium ROH between 0.1 and 5 Mbp (size class M); and (3) long ROH longer than 5 Mbp (size class L). For each breed analyzed, the number of ROH in each individual was plotted against the total sum of lengths of the ROH detected in that individual. The number of individuals within each breed and across all breeds sharing an ROH in a sliding window of 100 kb were counted across the whole genome (i.e., ROH hotspot score). Genomic regions where the fraction of individuals sharing an ROH exceeded the 99th percentile of the empirical distribution were defined as ROH hotspots.

2.3. *QTL Enrichment*

Quantitative trait loci (QTLs) in cattle were extracted from the Animal QTL Database [28]. QTLs on the X chromosome or without locations were excluded. QTLs without references in the Animal QTL Database were excluded [28]. The remaining QTLs were then classified from the Animal QTL Database into six groups by the type of associated trait, i.e., milk, reproduction, production, health, meat and carcass, and exterior traits. The QTLs located within detected ROH hotspots were identified. When two QTLs were found to have the same exact genomic interval or to be in the same associated trait group, they were counted as one QTL. To test whether the enrichment of QTLs in the candidate ROH hotspots was random or not, a permutation test was applied where ROH hotspot regions are simulated in each permutation. Briefly, candidate ROH hotspot regions were randomly distributed across the whole genome 10,000 times. Relative proportion and length of ROH hotspot regions were kept constant to preserve their correlation structure. Next, we repeated trait group assignment of QTLs located within the permuted ROH hotspot regions and computed the number of QTLs in each of the six trait groups. The distribution of numbers of QTLs observed in the permutated ROH hotspot regions was treated as the null distribution, from which we computed the significance levels of the number of QTLs observed in the real data. Moreover, genes located in the candidate ROH hotspots were annotated to the *Ensembl* Genes 89 Database using BioMart [29] and GO enrichment analysis was performed. The PANTHER classification system [30] was used to identify over-represented biological process-related GO terms. We used the human one-to-one orthologues for all cattle genes, because human genes are annotated more comprehensively. Significance levels were adjusted based on the Benjamini and Hochberg correction [31] for multiple comparisons, implemented in the PANTHER classification system ($FDR < 0.05$). Finally, the haplotype structure of ROH hotspots was examined among different breeds and species using haplostrips (version 1.1) [32], and phylogenetic trees of genomic sequences in ROH hotspots were constructed using Neighbor-Joining and bootstrapping methods, implemented in MEGA (version X) [33].

2.4. *Detection of Selection Signatures*

The Integrated Extended Haplotype Homozygosity (EHH) and Integrated Haplotype Score (iHS) statistics [8], as well as the posterior probabilities from the hidden Markov model (HMM) [21], were calculated and obtained within Holstein, Fleckvieh and Angus populations with relatively large sample size. The integrated EHH is a metric to identify genomic regions of excess haplotype homozygosity. It measures the excess of homozygosity due to identity by descent around an ancestral or derived allele of interest [8]. Consequently, an SNP at a very high allele frequency with strong and long-range LD and thus excessive integrated EHH scores indicates recent positive selection that has rapidly brought the haplotype close to fixation in a population. The iHS test identifies chromosome segments where the derived allele occurs at unusually high frequencies, indicating hitchhiking with a selectively favored variant. Unlike the EHH, it requires the definition of ancestral alleles. Integrated EHH statistics, hence, identify genomic regions under selection which have been fixed or are close to fixation, while iHS has high power to identify haplotypes under selection which have not yet been fixed [8]. The posterior probabilities from HMM are a measure of hard-sweep within breed and reported in [21], and were transformed in the following way: $\log_{10}(p/(1\text{-}p))$, where p is the posterior probability following [21].

Since the ancestral state of SNPs is usually used for detecting selection, for the dbSNPs from the sequence data, we inferred the ancestral alleles in *Bos taurus* using the method of Rocha et al. [34], in which the variants in dbSNPs were compared with sheep, water buffalo and yak. In this study, the allele was assigned as ancestral if it was observed at least twice in either sheep, water buffalo or yak. In total, there were 4,839,909 dbSNPs with inferred ancestral states and this information was used to estimate the integrated EHH and iHS scores. Finally, we calculated both the correlations between the integrated EHH scores and the number of individuals sharing an ROH, and the correlations between the iHS scores and the number of individuals sharing an ROH in a window of 100 kb for Holstein, Fleckvieh and Angus populations. The p values of correlations were calculated to determine whether they were significantly different from 0 using the R (http://www.r-project.org/, accessed 20 May 2019) *cor* and *cor.test* functions.

2.5. Association Mapping of Haplotypes from the ROH Hotspot Containing the CYP19A1 Gene

Due to its role in mammary gland development [35], we hypothesized that there is a phenotypic effect of the ROH hotspot containing *CYP19A1*. To test the phenotypic effect, we implemented a haplotype-based mixed linear model test for the effect of haplotypes on milk yield in ROH hotspot haplotypes containing *CYP19A1* variants. A total number of 5,199 Holstein individuals with HD genotypes were used to test the haplotype association with de-regressed proofs (DRP) of milk yield. The genotypes from gene *CYP19A1* were extracted and phased using Beagle [36]. The following haplotype-based mixed linear model was used: $y = 1\mu + \mathbf{Za} + \mathbf{h}_1 + \mathbf{h}_2 + \mathbf{e}$, where **y** was a vector of phenotypes (milk yield); **1** was a vector of ones; μ was the intercept; a was a vector of random polygenic effects following a multivariate normal distributed as $\mathbf{a} \sim N(\mathbf{0},\mathbf{A}\sigma_a^2)$; **A** was the pedigree-based additive relationship matrix; σ_a^2 was the polygenic variance; $\mathbf{h_1}$ and $\mathbf{h_2}$ were vectors of random haplotype effects, assumed to follow $\mathbf{h_i} \sim N(\mathbf{0},\mathbf{I}\sigma_h^2)$; **Z** was an incidence matrix, relating phenotypes to the corresponding random polygenic effects; e was the vector of random individual error terms, where $\mathbf{e} \sim N(0, \mathbf{I}\sigma_e^2)$; **I** was an identity matrix; and σ_h^2 and σ_e^2 were the variance of haplotype effects and error variance, respectively. We quantified the significance of the haplotype substitution effect by using the likelihood ratio test, comparing the full haplotype-based association mixed linear model with a null model with mean, polygenic effect and random error term, but without haplotype effects.

3. Results

3.1. The Distribution of ROH on Genomes Shaped by Demography

Runs of homozygosity on autosomes were determined for the sequenced cattle individuals from different breeds with a high genetic contribution to the current domestic cattle populations. The samples were grouped based on their breed origin, with Holstein, Jersey, Brown Swiss, Ayrshire Finnish Red, Swedish Red and Danish Red being dairy breeds, Angus, Charolais, Hereford, Gelbvieh and Belgium Blue being beef breeds and Fleckvieh and Simmental being dual-purpose breeds. There was an average number of 1221 ROH per genome, with an average size of 202 kbp across all individuals. The mean number and size of ROH varied from breed to breed, which reflects the different population histories and levels of inbreeding in these populations (Figure 1). The highest mean ROH size of 460 kbp was observed in the Danish Red population, while the lowest mean ROH size of 106 kbp was observed in the Swedish Red population. The highest average number of 2199 ROH was found in Jersey and the lowest average number of 641 ROH was observed in Charolais. On average, across all the populations, 9.39% of the genome was contained in ROH, ranging from 3.59% in Swedish Red cattle to 22.4% in Danish Red cattle, in which the proportion of ROH in the genome is defined as the ROH length divided by the total length of the cattle genome. The proportion of ROH is relatively moderate for Hereford, Jersey and Angus compared with Swedish Red and Danish Red. ROH segments were grouped into three classes (S, M, L) by length. S ROH were the most abundant in number, followed by M ROH and L ROH segments. Clusters of S ROH, i.e., many individuals with ROH at a site, indicate sites of low haplotype diversity, which may result from past or ongoing selection. However, the average total length of S ROH segments across the genome was small compared to the total length of M ROH segments. M ROH were fewer in number, but their average total length across the genome was longest among S, M and L ROH.

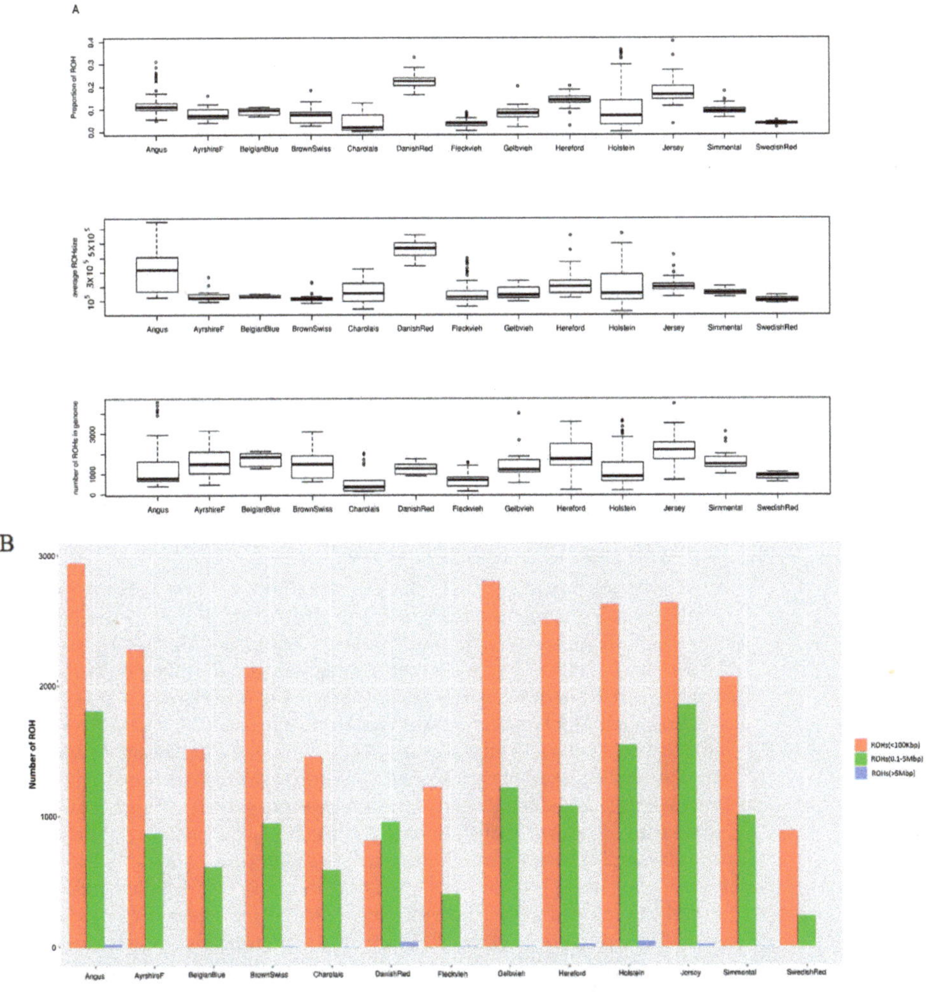

Figure 1. *Cont.*

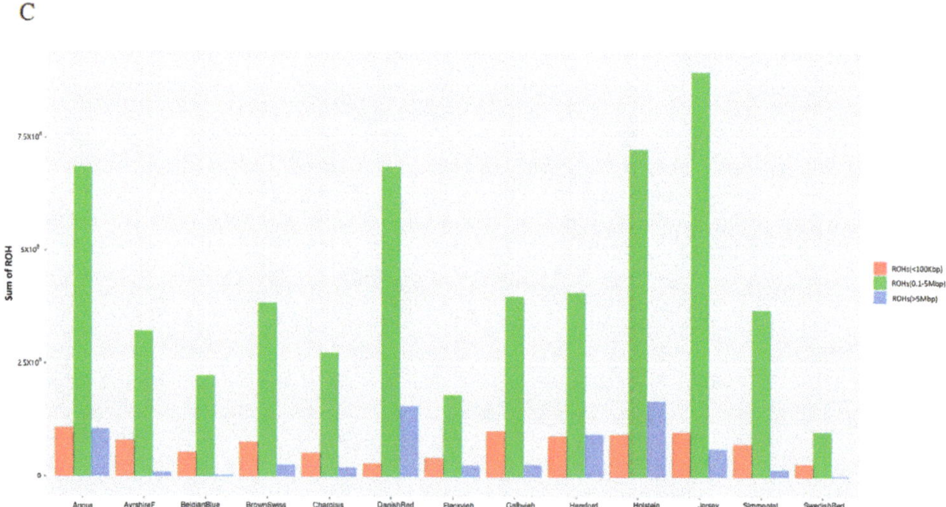

Figure 1. General statistics of ROH distribution in sequenced populations. (**A**) The proportion of the genome in ROH, the average ROH size and the number of ROH segments in the genome. (**B**) The number of ROH segments classified as short (S; red), medium (M; green) and long (L; blue) ROH. (**C**) The sum length of ROH segments classified as short (S), medium (M) and long (L) ROH.

To reveal the demographic history of the sequenced populations, we plotted the numbers of ROH against the sum lengths of ROH (Figure 2). Out of 13 cattle breeds, Angus, Jersey, Ayrshire Finnish Red and Simmental showed a medium number of ROH and medium total length in ROH, with most points locating in the middle of the plot (Figure 2). Fleckvieh, Charolais, Brow Swiss and Swedish Red had most points in the lower left corner of the plot, indicating a small number of ROH with a small sum of total ROH length. Danish Red was an extreme case, with a small number of ROH and large total ROH length. For the Holstein population, no clear patterns were observed. Instead, the ROH distribution between Holstein individuals was characterized by large variation, with a few Holstein individuals from the Netherlands showing extreme levels of inbreeding.

3.2. *Effect of Positive Selection on ROH Occurrence*

The correlations between ROH occurrence and selection signatures were firstly calculated by using integrated EHH, iHS and HMM tests, and ROH hotspot scores for the breeds with large sample size (i.e., Holstein, Angus, and Fleckvieh). Generally, we observed significantly high, positive correlations between the integrated EHH scores for the ancestral and the derived alleles and ROH hotspot scores (Figure S1) (0.54, 0.56 and 0.34 for ancestral alleles for Holstein, Angus and Fleckvieh; 0.47, 0.42 and 0.25 for derived alleles for Holstein, Angus and Fleckvieh, $p < 0.01$). Similarly, a significantly positive correlation was observed between the transformed posterior probabilities from HMM tests detecting hard sweeps [21] and the ROH hotspot scores on the genome (Figure S2) (0.20 and 0.23 for Holstein and Angus, $p < 0.01$). In contrast, much smaller, but still significant, correlations were observed between the proportion of SNPs with $|iHS| > 2$ and ROH hotspot scores in a window of 100 kbp (Figure S3) (0.07, 0.05 and 0.14 for Holstein, Angus and Fleckvieh, $p < 0.01$). Compared with selective sweeps detected from integrated EHH, HMM and iHS tests, a large proportion of the ROH hotspots were validated as candidate regions under positive selection in either Holstein, Angus or Fleckvieh (Tables S1–S3). For example, the integrated EHH test identified selective sweeps around ROH hotspots including the genes

TSHB, *RERE* and *CTNNA1*. We confirmed the hypothesis that ROH hotspot scores can be used to detect candidate regions under positive selection.

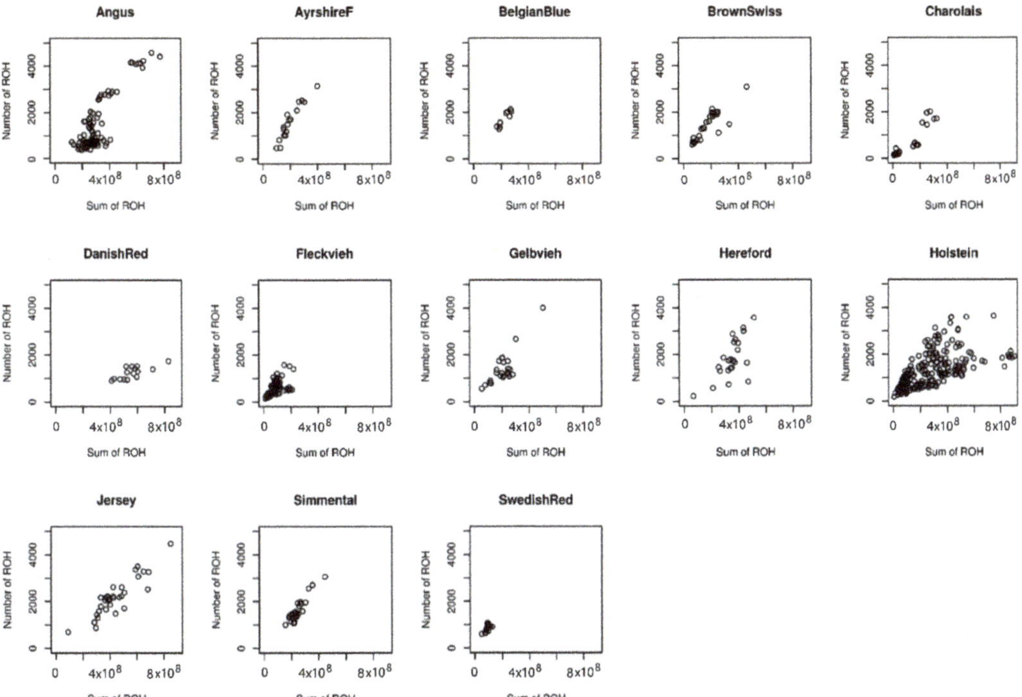

Figure 2. The number of ROH plotted against the sum of ROH in sequenced populations. The x axis shows the total length of ROH in bp. The y axis shows the total number of ROH in the genome. Each dot represents one individual.

We next examined the effect of positive selection on ROH occurrence in a genome-wide scale. Several ROH hotspots were observed in the genome (Figure 3). In total, 31 ROH hotspots were identified across the whole genome and a number of annotated genes were located in these ROH hotspots (Table S4). The most pronounced ROH hotspots were observed on chromosomes 7 and 16. They contained the genes *RERE* and *CTNNA1*, which were previously found to be under positive selection [37]. The genes *CAV1* and *TSHB*, which are related to mammary gland development, were also located in ROH hotspots, while *CAV1* and *TSHB* have not previously been found to be associated with selective sweeps. Other genes in ROH hotspots, such as *PLAG1* and *KIT*, were associated with well-known signatures of selection [38,39]. Genes such as *CYP19A1*, *CHCHD7*, *CLSTN1* and *SLC25A33* in ROH hotspots were previously identified as candidate genes in hard selective sweep regions in cattle using sequencing data [21]. The GO enrichment analysis of genes located in ROH hotspots revealed a significant over-representation of GO terms related to cellular component organization, including the cellular process and cellular component organization or biogenesis ($FDR < 3.35 \times 10^{-2}$). Moreover, enriched QTLs were identified in these ROH hotspots. However, only QTLs associated with health-related traits were significantly enriched in the ROH hotspot regions ($p < 0.01$).

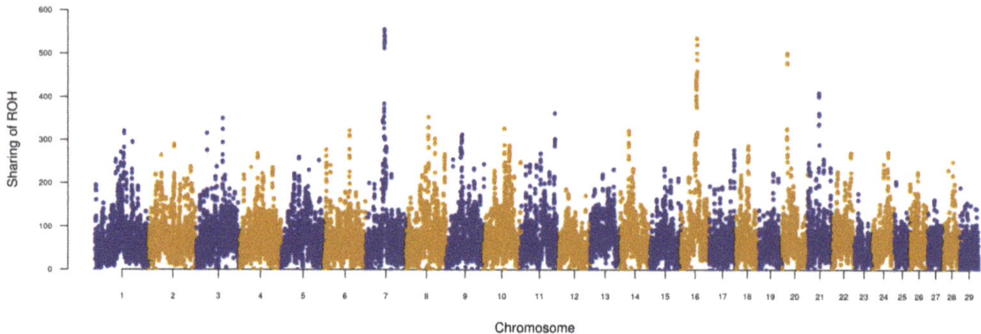

Figure 3. The ROH hotspot scores across all the sequenced individuals. The x axis shows the chromosome location on the 29 bovine autosomes. The y axis shows the number of animals with an ROH at this position (i.e., ROH hotspot scores). The total number of animals examined was 684. Each dot represents the count in windows of 100 kb.

We further deduced the time-scale of selection of the ROH hotspot candidates under positive selection based on the length of the ROH hotspots (Figure 4). The ROH hotspot around the *KIT* gene was selected to infer the time-scale of selection due to the role of the *KIT* gene in the coloring pattern in cattle [40]. The mean length of ROH around *KIT* was 988 kbp (N = 362). The length (in unit of 100 kbp) of ROH around the *KIT* gene fitted a chi-squared distribution with 8.2 degrees of freedom corresponding to a mean of 820 kbp. The expected length of shared ROH in the population is $2/T_c$, where T_c is the length of ROH haplotypes in Morgan. Therefore, this haplotype seems to have become a target of selection on the order of 250 generations ago.

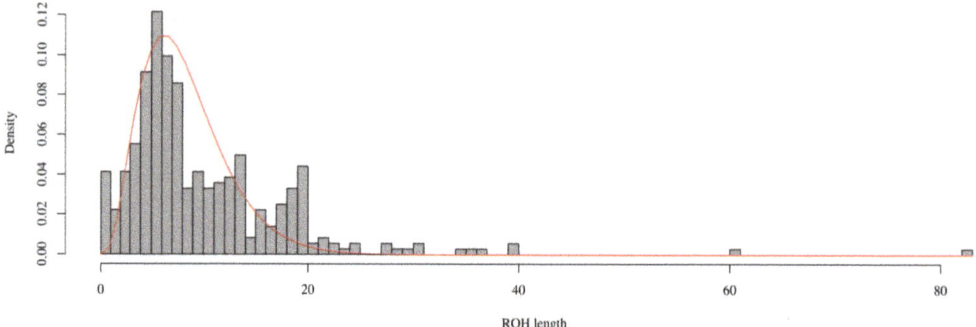

Figure 4. The distribution of ROH hotspot haplotype length in the *KIT* gene. The x axis is the length of ROH in units of 100 kbp. The red curve indicates the density function for a chi-squared distribution with parameter of degrees of freedom of 8.2 fitted to the distribution of lengths of ROH.

3.3. Phylogenies and Genealogies in ROH Hotspots

We examined the phylogenies and genealogies of haplotype structures in ROH hotspot regions, comparing between different cattle breeds and species including Zebu (*Bos indicus*), Gaur (*Bos gaurus*), Bison (*Bison bison*), Wisent (*Bison bonasus*), Banteng (*Bos javanicus*) and Gayal (*Bos frontalis*). Comparison of *Bos taurus* genealogies revealed the striking difference between a tree topology with shallow branches for a random haplotype and a tree topology with a very deep branch for a haplotype under selection (Figure 5). Patterns were especially pronounced for the ROH hotspot containing the *RERE* gene, a selection signature in most *Bos taurus* breeds (Figures 5A and S5). The genealogy in a non-ROH hotspot is shown for comparison (Figure 5B). The difference in genealogies around *RERE* compared to the non-ROH region is quite striking. In the ROH hotspot region, the majority of animals, independent of breed origin, clustered within one group of very closely related haplotypes. A few distantly related and rare haplotypes segregate in some *Bos taurus* breeds. In order to trace the origin of the haplotypes under selection, we examined the haplotype structure across species close to *Bos taurus* (Figure 6). Some of the haplotypes in the group dominant in *Bos taurus RERE* were found to be identical to haplotypes observed in *Bos indicus*, but very different from the haplotypes of Gaur, Bison, Wisent, Banteng and Gayal, as well as the alternative haplotypes in *Bos taurus*. This suggests that an introgression event happened in the *RERE* haplotype from *Bos indicus* to *Bos taurus*.

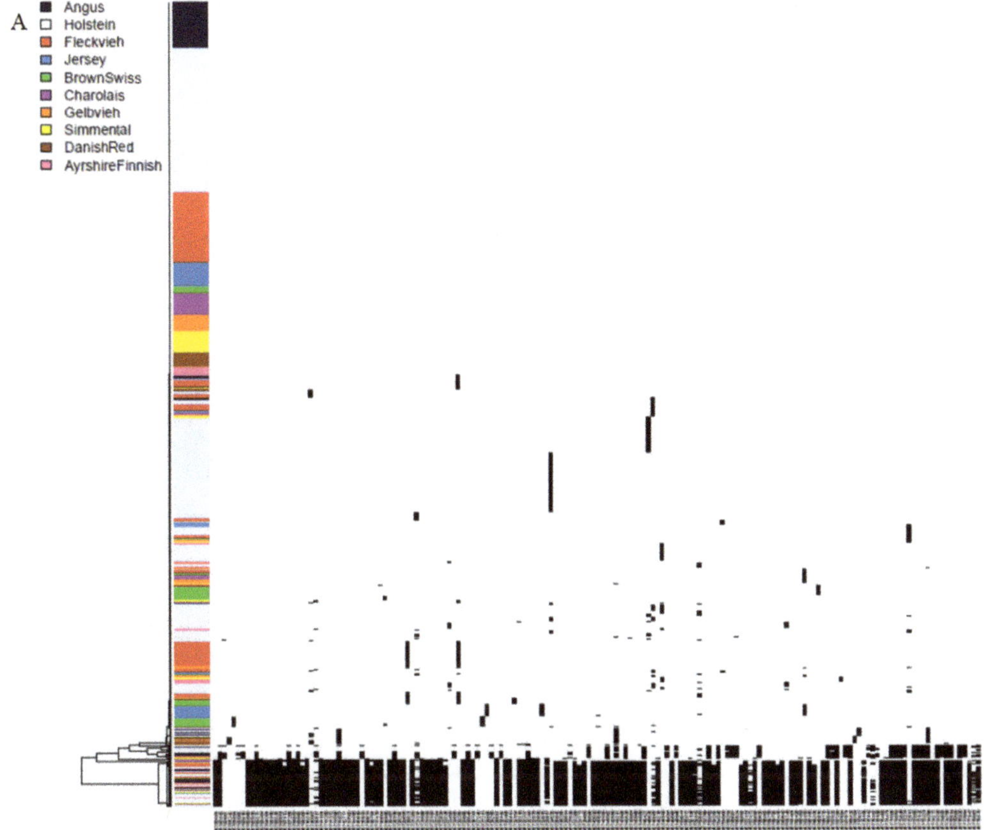

Figure 5. *Cont.*

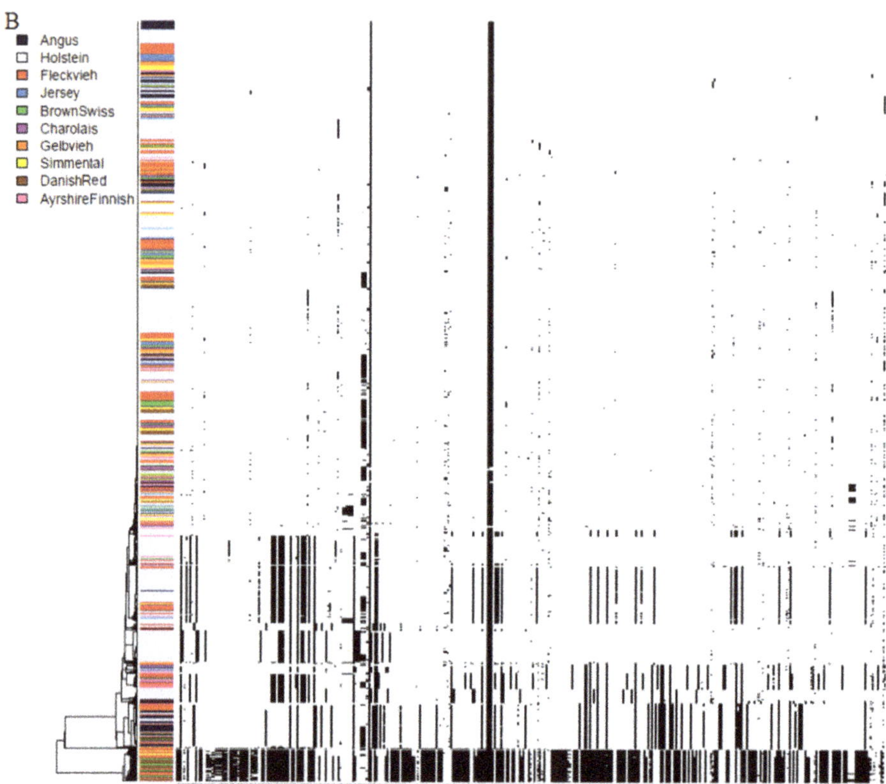

Figure 5. Haplotype structure and genealogies of haplotypes containing *RERE* variants located in ROH hotspots in different *Bos taurus* populations. (**A**) The structure and genealogies of haplotypes containing *RERE* gene. (**B**) The haplotype structure and genealogies in a non-ROH hotspot (chromosome 7:21,100,000–21,180,000 bp). The two alleles at each bi-allelic SNP are shown as black or white lines. Haplotypes are clustered based on based on the Manhattan distance, bringing together similar haplotypes and ordered by decreasing similarity. Breed association is indicated by the dendrogram on the left side.

3.4. Effect of Haplotype in the ROH Hotspot around the CYP19A1 Gene on Milk Yield

The phenotypic effect of an ROH hotspot potentially under positive selection was examined. We tested the effects of haplotypes in an ROH hotspot containing the *CYP19A1* gene on milk yield using 5,199 Holstein individuals under the hypothesis of the important biological role of *CYP19A1* in mammary gland development [35]. Four different haplotypes were observed. Haplotypes had significant substitution effects for milk yield in Holstein ($p < 0.05$) (Table 1). Interestingly, we observed a frequency of 94% of the selectively favored haplotype, with an effect of 0.629 in the Holstein population, while the frequency of the alternative homozygous haplotype with an effect of −0.846 was 5%. This suggested that this selectively favored haplotype with a positive effect on milk yield is under positive selection and nearly fixed in the Holstein population.

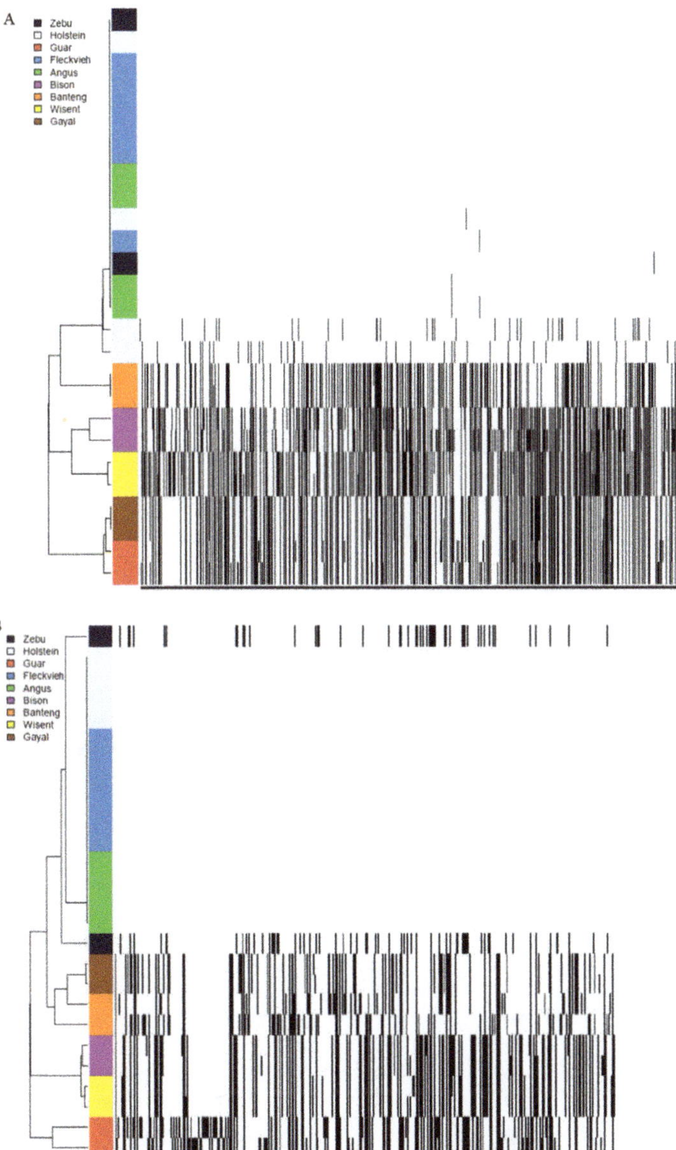

Figure 6. Haplotype structure and phylogenies of haplotypes from an ROH hotspot region containing the *RERE* gene for *Bos taurus* (Holstein, Angus and Fleckvieh), Zebu (*Bos indicus)*, Gaur, Bison, Banteng, Wisent and Gayal. (**A**) Left panel: the phylogeny of *RERE* haplotypes. Haplotypes are clustered based on based on the Manhattan distance, bringing together similar haplotypes and ordered by decreasing similarity. Right panel: the haplotype structure around the *RERE* gene. Colored blocks indicate the origin of haplotypes. The two alleles at each bi-allelic SNP are shown as black or white lines. (**B**) Left panel: the genealogy of *SLC25A51* haplotypes. Right panel: the haplotype structure around the *SLC25A51* gene. Colored blocks indicate the origin of haplotypes. The two alleles at each bi-allelic SNP are shown as black or white lines.

Table 1. Haplotype effects from *CYP19A1* locus on milk yield in Holstein population. * refers to p value < 0.05.

Haplotype	Effect Size	Standard Error	Significance (p Value) of Substitution Effects between Haplotypes
11,111	0.629	0.730	
12,111	0.135	0.984	0.010 *
22,211	0.081	0.971	
22,222	−0.846	0.740	

4. Discussion

Signatures of positive selection and demographic history in domestic cattle can be studied by examining ROH hotspot scores in their genomes. This makes domestic cattle an excellent model to demonstrate the interplay between positive selection and demography. Studies have shown that the distribution and burden of ROH is highly related to the current or previous population sizes [41,42]. It is expected that there are more and longer ROH segments distributed in populations with non-random mating, while in admixed populations, the number of ROH is reduced and ROH remain short due to the introgression of different haplotypes. Bottlenecks result from increased numbers of short ROH [43,44]. On the other hand, mating of close relatives increases the number of long ROH, while the variance of sum of length of ROH increases [45,46].

Different demographic histories result in the diverse locations of points when plotting the number of ROH and the sum lengths of ROH (Figure 2). In most breeds, we see that most of the animals roughly lie on a line. The slope of this line reflects the average length of ROH. A steep slope corresponds to a short average ROH length, while a shallow slope reflects a long average ROH length. The shorter the average ROH, the longer ago we find their origin. Out of 13 cattle breeds in this study, Angus and Jersey have relatively small populations and have experienced bottlenecks, as indicated by the high fraction of the genome in ROH. Nonetheless, even in Jersey, we find individuals with a very low level of ROH. Fleckvieh, Charolais, Brown Swiss and Swedish Red show very low amounts of ROH. This is consistent with recent admixture in the sequenced individuals [1]. Danish Red exhibits a very shallow line corresponding to very long ROH combined with a very large fraction of the genome in ROH. This reflects a pattern of strong recent inbreeding in Danish Red. In most populations, we see individuals with a low amount of ROH in terms of both total sum and number of ROH, except in Danish Red and Belgian Blue. This probably reflects an absence of admixture in these two breeds. Brown Swiss, Jersey and Hereford contain individuals to the right of the slope, which is evidence of consanguinity among their parents. The Holstein population looks heterogeneous. Points cluster on two lines, one steep and one shallow, reflecting the population structure, with subpopulations being characterized by different ROH patterns and different amounts of inbreeding. However, there are many individuals with points in between the two slopes. Finnish Ayrshire and Simmental probably had larger population sizes in the past, as shown by a steep slope and moderate total length of ROH. The distribution of ROH numbers and lengths illustrates the demographic diversity among cattle breeds.

We observed an abundance of short and medium-sized ROH in cattle genomes in different breeds (Figures 1 and 2). The high occurrence of ROH sites among individuals may be the result of intensive artificial selection, nowadays performed by animal breeders. Selection enacted in the population results in a less diverse haplotype distribution, and thereby more non-randomly distributed ROH in cattle populations are observed. Genomic regions located in an ROH region might be a result of close inbreeding and skewed haplotype spectra [47]. We confirmed the effect of positive selection on most of the ROH hotspots by examining the integrated EHH scores, iHS scores and the posterior probabilities from HMM tests for hard selective sweeps, and correlating them with ROH hotspot scores in these genomic regions. As a whole, these tests confirmed genes *PLAG1, KIT, RERE, CAV1, TSHB, CYO19A1, CHCHD7, CLSTN1* and *SLC25A33* as likely targets of selection in ROH hotspots. Among these genes, *RERE* plays an important role in development and in cell survival [48], and histone methyl transferases in regulating gene expression [49], so the different haplotypes in *Bos taurus* populations in gene *RERE* might cause different expression levels associated with production or disease. *CTNNA1* may play a role in disease susceptibility [50]. *CAV1* could regulate the release of milk from the mammary gland during lactation and progress the mammary gland to a mature structure [35], while *TSHB* was found to be associated with milk fat percentage [51,52]. *PLAG1* is associated with calving ease and stature, while *KIT* is associated with coat color patterning and pigmentation in cattle [40,53].

These results suggest that the occurrence of ROH hotspots is highly positively associated with selective sweeps close to fixation due to the high prevalence of the ROH hotspot haplotypes, but less so with ongoing selection. The relatively lower correlation between integrated EHH scores and ROH hotspot scores in the Fleckvieh population compared with the Holstein population suggests a difference in selection, such as selection intensity, in Holstein compared with the Fleckvieh population. To correlate the ROH hotspots with phenotypes, we performed a QTL enrichment analysis and we observed a significant enrichment of health-related QTLs in ROH hotspots in bovine genomes. This suggests that ROH hotspot regions in cattle are more associated with health-related traits. Furthermore, an overrepresentation of GO terms related to cellular component organization was observed. It implies that the different haplotypes in genes located in ROH hotspots might result in an abnormality within a specific cell component, thereby causing a disease [54,55]. However, it is noticeable that the ROH hotspot regions are probably more related to selective sweeps close to fixation through positive selection. Hence, SNPs located in ROH hotspots are no longer segregating in the populations and are therefore difficult to detect in a QTL mapping study. *CYP19A1* plays a biological role in female gonad development, mammary gland development and the development of male sexual characteristics [35]. The widespread occurrence of ROH around *CYP19A1* agrees with strong artificial selection on the milk- or fertility-related traits in dairy cattle.

We examined and reported the time-scale of positive selection, phylogenies and genealogies and phenotypic effects in one ROH hotspot. The time-scale of selection in ROH hotspots can be inferred by examining the length distribution of ROH hotspots across individuals. The mean length of ROH from the common ancestor is inversely proportional to the number of generations since the most recent common ancestor giving rise to the ROH [56]. Thereby, we are able to deduce an approximate selection time-scale based on the length distribution of the ROH across individuals. The haplotype containing *KIT* variants was used as an example to demonstrate the time-scale in an ROH hotspot. The selection acting on the haplotype containing *KIT* variants seems to have started around 250 generations or 1250 years ago, assuming a generation interval of 5 years. Haplotypes containing *KIT* variants are associated with the color patterns. This suggests an onset of selection around the 7th century CE.

Our observed result suggests a possible introgression from *Bos indicus* to *Bos taurus* for a haplotype of the ROH hotspot region containing the *RERE* gene (Figure 6), where we identified that the *RERE* haplotype of *Bos indicus* is identical to a haplotype in *Bos taurus*. A second potential introgression event from *Bos indicus* to *Bos taurus* was detected for the ROH hotspot containing the *REG3G* gene (Figure S4). Gene *REG3G* is associated with the immune response to pathogens and bacteria by stimulating toll-like receptors (TLRs) [57], suggesting that this possible introgression, followed by subsequent selection, improved the fitness and disease resistance in *Bos taurus*. The widespread distribution of nearly identical haplotypes in *Bos taurus* demonstrates that this haplotype has been under recent intensive selection in *Bos taurus*. The introgressed haplotype is highly represented in several cattle breeds; thus, it is a strong indication of adaptive introgression. However, very different haplotypes are still present in some *Bos taurus* breeds. This supports that the direction of introgression is from *Bos indicus* to *Bos taurus*—and not vice versa. Finally, we observed that an ROH hotspot containing *CYP19A1* was associated with a signal of positive selection.

Haplotypes in this region were associated with effects on milk yield. This provides a likely explanation for the selective force creating the selection signature.

5. Conclusions

Our study demonstrates that the formation and distribution of ROH in bovine populations is highly influenced by demography and positive selection. We illustrate that strong positive selection strongly shapes the occurrence of ROH and, for the first time, show the phylogenies in ROH hotspots, timing of selection on ROH hotspots and the phenotypic haplotype effects of ROH hotspots under strong selection in bovine populations. These ROH hotspots are very likely to significantly influence the fitness and economic traits of individuals in the population. We demonstrate that ROH hotspots are positively correlated with selective sweeps close to fixation. This highlights the importance of positive selection on shaping ROH distributions across individuals. Moreover, it sheds light on the importance of including effects of positive selection when estimating inbreeding from ROH using whole-genome sequence data. We provide an example with strong evidence of a significant association between ROH haplotypes under positive selection and milk yield in cattle populations, strongly supporting our findings. Furthermore, we show ROH as a tool to study the effects of demography, introgression and positive selection in the bovine population; however, it is generally applicable for any species. We highlight the importance of effects of positive selection and demography on shaping ROH localization on genomes in a domestic population under strong artificial selection for long time.

Supplementary Materials: The following supporting information can be downloaded at: https://www.mdpi.com/article/10.3390/agriculture12060844/s1. Figure S1: The ROH hotspots scores plot against integrative extended haplotype homozygosity (EHH) scores for ancestral and derived alleles in Holstein population. a. ancestral alleles; b. derived alleles; Figure S2: The ROH hotspots scores plot against the posterior probabilities from the hidden Markov model (HMM) measuring hard sweeps for holstein and angus populations. a. Holstein population; b. Angus population; Figure S3: The ROH hotspots scores plot against integrated haplotype scores (iHS) for Holstein and angus populations. a. Holstein population; b. Angus population; Figure S4: Haplotype structure and phylogenies of haplotypes from an ROH hotspot region containing the REG3G gene for *Bos taurus* (Holstein, Angus and Fleckvieh), *Bos indicus* (Zebu), Gaur, Bison, Banteng, Wisent and Gayal. Left panel: the genealogy of REG3G haplotypes. Right panel: the haplotype structure around the REG3G gene. Colored blocks indicate the origin of haplotypes. The two alleles at each bi-allelic SNP are shown as black or white lines. Figure S5: The genealogies tree of RERE haplotypes from MEGA. Table S1: Candidate selective sweep regions comparing between the ROH hotspots scores and integrative extended haplotype homozygosity (EHH) scores for Holstein and angus and fleckvieh populations; Table S2: Candidate selective sweep regions comparing between the ROH hotspots scores and integrated haplotype scores (iHS) for Holstein and angus and fleckvieh populations; Table S3: Candidate selective sweep regions comparing between the ROH hotspots scores and the posterior probabilities from the hidden Markov model (HMM) measuring hard sweeps for Holstein and angus populations; Table S4: ROH hotspots candidate regions and annotated genes in the regions.

Author Contributions: Q.Z. developed and planned the design of the study, coordinated the study, performed data analyses and drafted the manuscript. A.A.S. and B.G. participated in the design of the study and drafting of the manuscript. M.S.L. participated in the design of the study. All authors have read and agreed to the published version of the manuscript.

Funding: We are grateful to the Nordic Cattle Genetic Evaluation (NAV, Aarhus, Denmark) for providing the phenotypic data used in this study and 1000 Bull Genome Project for providing sequence data. Qianqian Zhang benefited from a joint grant from the European Commission within the framework of the Erasmus-Mundus joint doctorate "EGS-ABG". This research was supported by the Center for Genomic Selection in Animals and Plants (GenSAP) funded by Innovation Fund Denmark (grant 0603-00519B) and Beijing Nova program from Beijing Academy of Science and Technology, Beijing, China (grant Z201100006820910).

Institutional Review Board Statement: Not applicable.

Informed Consent Statement: Not applicable.

Data Availability Statement: The data used in this study originated from the 1000 Bull Genome Project (Daetwyler et al. 2014 Nature Genet. 46:858-865). Whole-genome sequence data of individual bulls of the 1000 Bull Genomes Project are available at NCBI using SRA no. SRP039339 (http://www.ncbi.nlm.nih.gov/bioproject/PRJNA238491, access 10 January 2022). The whole-genome sequence data from *Bos indicus*, Gaur, Bison, Wisent, Banteng and Gayal individuals were download from NCBI with SRA number: SRR6423855, SRR6448720, SRR6448721, SRR6448737, SRR6448738, SRR6448739, SRR6448740, SRR6448732, SRR6448733, SRR6448734, SRR6448735, SRR6448580, SRR6448581, SRR6448670, SRR6448682, SRR6448683, SRR6448684.

Acknowledgments: We thank Simon Biotard and Marlies Dolezal for the help in sample data selection and helpful discussions, Thomas Bataillon and Doug Speed for helpful discussions and Amanda Chamberlain for the help in running the script to process the bam files.

Conflicts of Interest: The authors declare that they have no competing interests.

References

1. Zhang, Q.; Calus, M.P.L.; Bosse, M.; Sahana, G.; Lund, M.S.; Guldbrandtsen, B. Human-Mediated Introgression of Haplotypes in a Modern Dairy Cattle Breed. *Genetics* **2018**, *209*, 1305–1317. [CrossRef] [PubMed]
2. Figueiró, H.V.; Li, G.; Trindade, F.J.; Assis, J.; Pais, F.; Fernandes, G.; Santos, S.H.D.; Hughes, G.M.; Komissarov, A.; Antunes, A.; et al. Genome-wide signatures of complex introgression and adaptive evolution in the big cats. *Sci. Adv.* **2017**, *3*, e1700299. [CrossRef] [PubMed]
3. Jones, E.R.; Gonzalez-Fortes, G.; Connell, S.; Siska, V.; Eriksson, A.; Martiniano, R.; McLaughlin, R.; Llorente, M.G.; Cassidy, L.M.; Gamba, C.; et al. Upper Palaeolithic genomes reveal deep roots of modern Eurasians. *Nat. Commun.* **2015**, *6*, 8912. [CrossRef] [PubMed]
4. Pearson, R.D.; Amato, R.; Auburn, S.; Miotto, O.; Almagro-Garcia, J.; Amaratunga, C.; Suon, S.; Mao, S.; Noviyanti, R.; Trimarsanto, H.; et al. Genomic analysis of local variation and recent evolution in *Plasmodium vivax*. *Nat. Genet.* **2016**, *48*, 959–964. [CrossRef]
5. Chen, N.; Cai, Y.; Chen, Q.; Li, R.; Wang, K.; Huang, Y.; Hu, S.; Huang, S.; Zhang, H.; Zheng, Z.; et al. Whole-genome resequencing reveals world-wide ancestry and adaptive introgression events of domesticated cattle in East Asia. *Nat. Commun.* **2018**, *9*, 1–13. [CrossRef]
6. Pool, J.E.; Hellmann, I.; Jensen, J.D.; Nielsen, R. Population genetic inference from genomic sequence variation. *Genome Res.* **2010**, *20*, 291–300. [CrossRef]
7. Bersaglieri, T.; Sabeti, P.C.; Patterson, N.; Vanderploeg, T.; Schaffner, S.F.; Drake, J.A.; Rhodes, M.; Reich, D.E.; Hirschhorn, J.N. Genetic signatures of strong recent positive selection at the lactase gene. *Am. J. Hum. Genet.* **2004**, *74*, 1111–1120. [CrossRef]
8. Voight, B.F.; Kudaravalli, S.; Wen, X.Q.; Pritchard, J.K. A map of recent positive selection in the human genome. *PLoS Biol.* **2006**, *4*, 446–458.
9. Sabeti, P.C.; Varilly, P.; Fry, B.; Lohmueller, J.; Hostetter, E.; Cotsapas, C.; Xie, X.; Byrne, E.H.; McCarroll, S.A.; Gaudet, R. Genome-wide detection and characterization of positive selection in human populations. *Nature* **2007**, *449*, 913–918. [CrossRef]
10. Ellegren, H.; Galtier, N. Determinants of genetic diversity. *Nat. Rev. Genet.* **2016**, *17*, 422–433. [CrossRef]
11. Loftus, R.T.; MacHugh, D.E.; Bradley, D.G.; Sharp, P.M.; Cunningham, P. Evidence for two independent domestications of cattle. *Proc. Natl. Acad. Sci. USA* **1994**, *91*, 2757–2761. [CrossRef] [PubMed]
12. Bollongino, R.; Burger, J.; Powell, A.; Mashkour, M.; Vigne, J.-D.; Thomas, M. Modern Taurine Cattle Descended from Small Number of Near-Eastern Founders. *Mol. Biol. Evol.* **2012**, *29*, 2101–2104. [CrossRef] [PubMed]
13. Götherström, A.; Anderung, C.; Hellborg, L.; Galil, R.; Smith, E.C.; Bradley, D.G.; Ellegren, H. Cattle domestication in the Near East was followed by hybridization with aurochs bulls in Europe. *Proc. R. Soc. B Boil. Sci.* **2005**, *272*, 2660. [CrossRef]
14. Flori, L.; Fritz, S.; Jaffrézic, F.; Boussaha, M.; Gut, I.; Heath, S.; Foulley, J.-L.; Gautier, M. The Genome Response to Artificial Selection: A Case Study in Dairy Cattle. *PLoS ONE* **2009**, *4*, e6595. [CrossRef]
15. Pryce, J.E.; Daetwyler, H.D. Designing dairy cattle breeding schemes under genomic selection: A review of international research. *Anim. Prod. Sci.* **2012**, *52*, 107–114. [CrossRef]
16. Hayes, B.; Bowman, P.; Chamberlain, A.; Goddard, M. Invited review: Genomic selection in dairy cattle: Progress and challenges. *J. Dairy Sci.* **2009**, *92*, 433–443. [CrossRef]
17. Berglund, B. Genetic improvement of dairy cow reproductive performance. *Reprod. Domest. Anim.* **2008**, *43*, 89–95. [CrossRef]
18. Wu, D.-D.; Ding, X.-D.; Wang, S.; Wójcik, J.; Zhang, Y.; Tokarska, M.; Li, Y.; Wang, M.-S.; Faruque, O.; Nielsen, R.; et al. Pervasive introgression facilitated domestication and adaptation in the *Bos* species complex. *Nat. Ecol. Evol.* **2018**, *2*, 1139–1145. [CrossRef]
19. Daetwyler, H.D.; Capitan, A.; Pausch, H.; Stothard, P.; van Binsbergen, R.; Brøndum, R.F.; Liao, X.; Djari, A.; Rodriguez, S.C.; Grohs, C.; et al. Whole-genome sequencing of 234 bulls facilitates mapping of monogenic and complex traits in cattle. *Nat. Genet.* **2014**, *46*, 858–865. [CrossRef]
20. Zhang, Q.; Guldbrandtsen, B.; Bosse, M.; Lund, M.S.; Sahana, G. Runs of homozygosity and distribution of functional variants in the cattle genome. *BMC Genom.* **2015**, *16*, 542. [CrossRef]

21. Boitard, S.; Boussaha, M.; Capitan, A.; Rocha, D.; Servin, B. Uncovering Adaptation from Sequence Data: Lessons from Genome Resequencing of Four Cattle Breeds. *Genetics* **2016**, *203*, 433–450. [CrossRef] [PubMed]
22. Zimin, A.V.; Delcher, A.L.; Florea, L.; Kelley, D.R.; Schatz, M.C.; Puiu, D.; Hanrahan, F.; Pertea, G.; van Tassell, C.P.; Sonstegard, T.S.; et al. A whole-genome assembly of the domestic cow, *Bos taurus*. *Genome Biol.* **2009**, *10*, R42. [CrossRef] [PubMed]
23. Li, H.; Durbin, R. Fast and accurate short read alignment with Burrows–Wheeler transform. *Bioinformatics* **2009**, *25*, 1754–1760. [CrossRef] [PubMed]
24. Li, H.; Handsaker, B.; Wysoker, A.; Fennell, T.; Ruan, J.; Homer, N.; Marth, G.; Abecasis, G.; Durbin, R.; Proc GPD. The Sequence Alignment/Map format and SAMtools. *Bioinformatics* **2009**, *25*, 2078–2079. [CrossRef] [PubMed]
25. McKenna, A.; Hanna, M.; Banks, E.; Sivachenko, A.; Cibulskis, K.; Kernytsky, A.; Garimella, K.; Altshuler, D.; Gabriel, S.; Daly, M.; et al. The Genome Analysis Toolkit: A MapReduce framework for analyzing next-generation DNA sequencing data. *Genome Res.* **2010**, *20*, 1297–1303. [CrossRef] [PubMed]
26. Sherry, S.T.; Ward, M.-H.; Kholodov, M.; Baker, J.; Phan, L.; Smigielski, E.M.; Sirotkin, K. dbSNP: The NCBI database of genetic variation. *Nucleic Acids Res.* **2001**, *29*, 308–311. [CrossRef] [PubMed]
27. Bosse, M.; Megens, H.-J.; Madsen, O.; Paudel, Y.; Frantz, L.A.F.; Schook, L.B.; Crooijmans, R.P.M.A.; Groenen, M.A.M. Regions of Homozygosity in the Porcine Genome: Consequence of Demography and the Recombination Landscape. *PLoS Genet.* **2012**, *8*, e1003100. [CrossRef]
28. Hu, Z.-L.; Park, C.A.; Reecy, J.M. Developmental progress and current status of the Animal QTLdb. *Nucleic Acids Res.* **2016**, *44*, D827–D833. [CrossRef]
29. Kinsella, R.J.; Kähäri, A.; Haider, S.; Zamora, J.; Proctor, G.; Spudich, G.; Almeida-King, J.; Staines, D.; Derwent, P.; Kerhornou, A.; et al. Ensembl BioMarts: A hub for data retrieval across taxonomic space. *Database* **2011**, *2011*, bar030. [CrossRef]
30. Mi, H.; Huang, X.; Muruganujan, A.; Tang, H.; Mills, C.; Kang, D.; Thomas, P.D. PANTHER version 11: Expanded annotation data from Gene Ontology and Reactome pathways, and data analysis tool enhancements. *Nucleic Acids Res.* **2017**, *45*, D183–D189. [CrossRef]
31. Benjamini, Y.; Yekutieli, D. The control of the false discovery rate in multiple testing under dependency. *Ann. Stat.* **2001**, *29*, 1165–1188. [CrossRef]
32. Marnetto, D.; Huerta-Sánchez, E. Haplostrips: Revealing population structure through haplotype visualization. *Methods Ecol. Evol.* **2017**, *8*, 1389–1392. [CrossRef]
33. Tamura, K.; Dudley, J.; Nei, M.; Kumar, S. MEGA4: Molecular evolutionary genetics analysis (MEGA) software version 4.0. *Mol. Biol. Evol.* **2007**, *24*, 1596–1599. [CrossRef] [PubMed]
34. Rocha, D.; Billerey, C.; Samson, F.; Boichard, D.; Boussaha, M. Identification of the putative ancestral allele of bovine single-nucleotide polymorphisms. *J. Anim. Breed. Genet.* **2014**, *131*, 483–486. [CrossRef] [PubMed]
35. Uyttendaele, H.; Soriano, J.V.; Montesano, R.; Kitajewski, J. Notch4 and Wnt-1 proteins function to regulate branching morphogenesis of mammary epithelial cells in an opposing fashion. *Dev. Biol.* **1998**, *196*, 204–217. [CrossRef]
36. Browning, S.R.; Browning, B.L. Rapid and accurate haplotype phasing and missing-data inference for whole-genome association studies by use of localized haplotype clustering. *Am. J. Hum. Genet.* **2007**, *81*, 1084–1097. [CrossRef]
37. Gutiérrez-Gil, B.; Arranz, J.J.; Wiener, P. An interpretive review of selective sweep studies in *Bos taurus* cattle populations: Identification of unique and shared selection signals across breeds. *Front. Genet.* **2015**, *6*, 167. [CrossRef]
38. Pausch, H.; Flisikowski, K.; Jung, S.; Emmerling, R.; Edel, C.; Götz, K.-U.; Fries, R. Genome-Wide Association Study Identifies Two Major Loci Affecting Calving Ease and Growth-Related Traits in Cattle. *Genetics* **2011**, *187*, 289–297. [CrossRef]
39. Karim, L.; Takeda, H.; Lin, L.; Druet, T.; Arias, J.A.C.; Baurain, D.; Cambisano, N.; Davis, S.R.; Farnir, F.; Grisart, B.; et al. Variants modulating the expression of a chromosome domain encompassing PLAG1 influence bovine stature. *Nat. Genet.* **2011**, *43*, 405–413. [CrossRef]
40. Hou, L.; Panthier, J.-J.; Arnheiter, H. Signaling and transcriptional regulation in the neural crest-derived melanocyte lineage: Interactions between KIT and MITF. *Development* **2000**, *127*, 5379–5389. [CrossRef]
41. Kirin, M.; McQuillan, R.; Franklin, C.S.; Campbell, H.; McKeigue, P.M.; Wilson, J.F. Genomic Runs of Homozygosity Record Population History and Consanguinity. *PLoS ONE* **2010**, *5*, e13996. [CrossRef] [PubMed]
42. Purfield, D.C.; Berry, D.P.; McParland, S.; Bradley, D.G. Runs of homozygosity and population history in cattle. *BMC Genet.* **2012**, *13*, 70. [CrossRef] [PubMed]
43. Pemberton, T.J.; Absher, D.; Feldman, M.W.; Myers, R.M.; Rosenberg, N.A.; Li, J.Z. Genomic Patterns of Homozygosity in Worldwide Human Populations. *Am. J. Hum. Genet.* **2012**, *91*, 275–292. [CrossRef] [PubMed]
44. Von Holdt, B.M.; Pollinger, J.P.; Earl, D.A.; Knowles, J.C.; Boyko, A.R.; Parker, H.; Geffen, E.; Pilot, M.; Jedrzejewski, W.; Jedrzejewska, B.; et al. A genome-wide perspective on the evolutionary history of enigmatic wolf-like canids. *Genome Res.* **2011**, *21*, 1294–1305. [CrossRef] [PubMed]
45. Ferenčaković, M.; Hamzic, E.; Gredler, B.; Solberg, T.R.; Klemetsdal, G.; Curik, I.; Sölkner, J. Estimates of autozygosity derived from runs of homozygosity: Empirical evidence from selected cattle populations. *J. Anim. Breed. Genet.* **2013**, *130*, 286–293. [CrossRef] [PubMed]
46. Curik, I.; Ferenčaković, M.; Sölkner, J. Inbreeding and runs of homozygosity: A possible solution to an old problem. *Livest. Sci.* **2014**, *166*, 26–34. [CrossRef]

47. Qanbari, S.; Simianer, H. Mapping signatures of positive selection in the genome of livestock. *Livest. Sci.* **2014**, *166*, 133–143. [CrossRef]
48. Fregeau, B.; Kim, B.J.; Hernandez-Garcia, A.; Jordan, V.K.; Cho, M.T.; Schnur, R.E.; Monaghan, K.G.; Juusola, J.; Rosenfeld, J.A.; Bhoj, E.; et al. De Novo Mutations of RERE Cause a Genetic Syndrome with Features that Overlap Those Associated with Proximal 1p36 Deletions. *Am. J. Hum. Genet.* **2016**, *98*, 963–970. [CrossRef]
49. Kim, B.J.; Scott, D.A. Mouse Model Reveals the Role of RERE in Cerebellar Foliation and the Migration and Maturation of Purkinje Cells. *PLoS ONE* **2014**, *9*, e87518. [CrossRef]
50. Sheikh, F.; Chen, Y.; Liang, X.; Hirschy, A.; Stenbit, A.E.; Gu, Y.; Dalton, N.D.; Yajima, T.; Lu, Y.C.; Knowlton, K.U.; et al. Alpha-E-Catenin inactivation disrupts the cardiomyocyte adherens junction, resulting in cardiomyopathy and susceptibility to wall rupture. *Circulation* **2006**, *114*, 1046–1055. [CrossRef]
51. Cochran, S.D.; Cole, J.B.; Null, D.J.; Hansen, P.J. Discovery of single nucleotide polymorphisms in candidate genes associated with fertility and production traits in Holstein cattle. *BMC Genet.* **2013**, *14*, 49. [CrossRef] [PubMed]
52. Ortega, M.S.; Denicol, A.C.; Cole, J.B.; Null, D.J.; Taylor, J.F.; Schnabel, R.D.; Hansen, P.J. Association of single nucleotide polymorphisms in candidate genes previously related to genetic variation in fertility with phenotypic measurements of reproductive function in Holstein cows. *J. Dairy Sci.* **2017**, *100*, 3725–3734. [CrossRef] [PubMed]
53. Hayes, B.J.; Pryce, J.; Chamberlain, A.J.; Bowman, P.J.; Goddard, M.E. Genetic Architecture of Complex Traits and Accuracy of Genomic Prediction: Coat Colour, Milk-Fat Percentage, and Type in Holstein Cattle as Contrasting Model Traits. *PLoS Genet.* **2010**, *6*, e1001139. [CrossRef] [PubMed]
54. Christofidou, P.; Nelson, C.; Nikpay, M.; Qu, L.; Li, M.; Loley, C.; Debiec, R.; Braund, P.S.; Denniff, M.; Charchar, F.; et al. Runs of Homozygosity: Association with Coronary Artery Disease and Gene Expression in Monocytes and Macrophages. *Am. J. Hum. Genet.* **2015**, *97*, 228–237. [CrossRef]
55. Ghani, M.; Reitz, C.; Cheng, R.; Vardarajan, B.N.; Jun, G.; Sato, C.; Naj, A.C.; Rajbhandary, R.; Wang, L.-S.; Valladares, O.; et al. Association of Long Runs of Homozygosity with Alzheimer Disease Among African American Individuals. *JAMA Neurol.* **2015**, *72*, 1313–1323. [CrossRef]
56. Fisher, R.A. A Fuller Theory of Junctions in Inbreeding. *Heredity* **1954**, *8*, 187–197. [CrossRef]
57. Abreu, M.T. Toll-like receptor signalling in the intestinal epithelium: How bacterial recognition shapes intestinal function. *Nat. Rev. Immunol.* **2010**, *10*, 131–143. [CrossRef]

Article

Using Genomics to Measure Phenomics: Repeatability of Bull Prolificacy in Multiple-Bull Pastures

Gary L. Bennett, John W. Keele *, Larry A. Kuehn, Warren M. Snelling, Aaron M. Dickey, Darrell Light, Robert A. Cushman and Tara G. McDaneld

U.S. Department of Agriculture, Agricultural Research Service, U.S. Meat Animal Research Center, Clay Center, NE 68933, USA; gary.bennett@usda.gov (G.L.B.); larry.kuehn@usda.gov (L.A.K.); warren.snelling@usda.gov (W.M.S.); aaron.dickey@usda.gov (A.M.D.); darrell.light@usda.gov (D.L.); bob.cushman@usda.gov (R.A.C.); tara.mcdaneld@usda.gov (T.G.M.)
* Correspondence: john.keele@usda.gov

Abstract: Phenotypes are necessary for genomic evaluations and management. Sometimes genomics can be used to measure phenotypes when other methods are difficult or expensive. Prolificacy of bulls used in multiple-bull pastures for commercial beef production is an example. A retrospective study of 79 bulls aged 2 and older used 141 times in 4–5 pastures across 4 years was used to estimate repeatability from variance components. Traits available before each season's use were tested for predictive ability. Sires were matched to calves using individual genotypes and evaluating exclusions. A lower-cost method of measuring prolificacy was simulated for five pastures using the bulls' genotypes and pooled genotypes to estimate average allele frequencies of calves and of cows. Repeatability of prolificacy was 0.62 ± 0.09. A combination of age-class and scrotal circumference accounted for less than 5% of variation. Simulated estimation of prolificacy by pooling DNA of calves was accurate. Adding pooling of cow DNA or actual genotypes both increased accuracy about the same. Knowing a bull's prior prolificacy would help predict future prolificacy for management purposes and could be used in genomic evaluations and research with coordination of breeders and commercial beef producers.

Keywords: DNA pooling; parentage; reproduction

Citation: Bennett, G.L.; Keele, J.W.; Kuehn, L.A.; Snelling, W.M.; Dickey, A.M.; Light, D.; Cushman, R.A.; McDaneld, T.G. Using Genomics to Measure Phenomics: Repeatability of Bull Prolificacy in Multiple-Bull Pastures. *Agriculture* **2021**, *11*, 603. https://doi.org/10.3390/agriculture11070603

Academic Editors: Heather Burrow and Michael Goddard

Received: 3 June 2021
Accepted: 25 June 2021
Published: 28 June 2021

1. Introduction

The conceptual linking of genome to phenome is fundamental to animal improvement. Current genetic evaluation systems utilize many genotypes and associated phenotypes to accelerate improvement. Beyond the genome-to-phenome paradigm, some phenotypes used for management or inputs to genetic evaluation, particularly those notoriously difficult to be measured, may be best estimated by genomics. Bull prolificacy in multiple-bull pastures is one of those phenotypes.

Efficient use of bulls for mating cows in commercial beef production involves both the number of bulls maintained and pregnancy rates of the cows. Maintaining fewer bulls reduces costs, and greater pregnancy rates increase calves born. However, maintaining and using too few bulls can lower pregnancy rates. Using multiple bulls in breeding pastures is one technique used to simplify some aspects of management and may increase pregnancy rates without increasing the number of bulls [1]. Microsatellite and SNP markers have allowed calves resulting from multiple-bull breeding pastures to be matched to their sire [2–5] and fill in the sire-side of their pedigrees. Knowing calves' sires allows evaluating bulls for prolificacy—their ability to sire calves in a multiple-bull pasture situation. Behavioral, physical, or other fertility factors may cause some bulls to sire more or fewer calves [2,6,7]. Cattle breeders could use a bull's predicted prolificacy to help decide whether to keep, cull, or use the bull in a different way if predictions were accurate.

Genomic markers allow matching sires to calves, but the cost of genotyping all commercial calves may not be economical if prolificacy is the only use of the genotypes. The method of pooling DNA to estimate genomic differences between groups with different phenotypes is being studied for use in genomic association and prediction [8–11]. Another potential use is to estimate the number of calves for sires by pooling DNA from all calves, i.e., using DNA pooling to estimate phenotypic prolificacy differences.

One objective of this study is to estimate the repeatability of Bos taurus bull prolificacy across years in a temperate climate when used in multiple-bull breeding pastures and identify whether some bull traits could predict prolificacy before use. Another objective is to evaluate the potential for using the DNA pooling method to estimate prolificacy of bulls used in multiple-bull pastures at a lower cost.

2. Materials and Methods

2.1. Real Data

This retrospective study used 4 years of multiple-bull pasture breeding (2012 to 2015) and subsequent calving data (2013 to 2016). Cattle populations were purebred Angus and advanced generation composites MARC I, MARC II, and MARC III [12]. Only bulls and cows aged 2 and older at the time of breeding were included in data analyzed. In addition, only bulls passing breeding soundness exams for physical and semen traits were used in breeding pastures. A few bulls were removed for injury or other reasons and were replaced by different bulls unless it was late in the breeding season. Any bull not in the breeding pasture for at least 90% of the breeding season was eliminated from analyses. In 2012, some older cows in each of the 3 composite populations were synchronized and bred by AI 15 d after younger cows entered breeding pastures. The synchronized cows entered the same breeding pastures younger cows were in 9 d after AI and 39 d before the end of pasture breeding. Calves sired by AI were eliminated from data, but those sired by the pasture bulls were retained for analysis. Pastures and cow and bull assignments are described in Table 1. Assuming half of AI cows were open after timed AI, the average assigned open cows per bull was 23.6.

Table 1. Natural service assignments of bulls and cows and resulting calves.

Calf Birth Year	Population	Pasture	Bulls Assigned, N	Bulls Analyzed [1], N	Cows Assigned, N	Cow Age, Year	Natural Service, d	Calves Born, N	Sire Known, N
	MARC I	A	8	6	157	3.92	63	140	138
					66 [2]	7.45	39	57	57
	MARC II	B	7	5	130	3.24	63	115	110
2013					86 [2]	7.20	39	75	69
	MARC III	C	6	6	65	3.06	63	43	42
					165 [2]	6.39	39	150	148
	Angus	D	10	10	246	4.77	63	227	227
	MARC I	E	6	6	156	4.08	63	146	144
		F	3	2	62	8.42	63	49	48
2014	MARC II	G	9	9	214	5.05	63	180	178
	MARC III	H	10	9	245	5.37	63	201	201
	Angus	I	11	11	267	4.97	63	226	225
	MARC I	J	9	9	222	5.52	63	182	182
2015	MARC II	K	9	7	212	5.13	61	179	161
	MARC III	L	11	11	251	4.97	62	195	194
	Angus	M	11	10	256	5.21	60	235	210
	MARC I	N	9	9	221	5.31	49	198	198
	MARC II	O	9	9	203	5.04	49	168	165
2016	MARC III	P	6	6	126	3.07	49	110	109
		Q	5	5	102	6.32	49	90	90
	Angus	R	11	11	251	5.14	49	214	213

[1] Bulls that were removed, or their replacements, were not used in analyses unless they were in the breeding pasture for at least 90% of the natural service period. [2] Older cows were synchronized and artificially inseminated 15 d after younger cows began natural service mating and entered the breeding pastures 9 d later for the remaining 39 d of natural service.

The patterns of bull usage in the edited data are shown in Table 2. Included were 79 unique bulls with 141 breeding opportunities, ranging from 38 bulls with one opportunity each to 4 bulls used each of the 4 years. Across years there were 27, 37, 37, and 40 breeding opportunities for calf years 2013, 2014, 2015, and 2016, respectively. Forty-one bulls had an average of 2.5 breeding opportunities, and 38 bulls had a single opportunity.

Table 2. Patterns of edited bull usage by calf birth year.

No. Bulls	2013	2014	2015	2016
4	1	1	1	1
7	1	1	1	0
1	1	0	1	1
5	0	1	1	1
7	1	1	0	0
6	0	1	1	0
11	0	0	1	1
8	1	0	0	0
8	0	1	0	0
3	0	0	1	0
19	0	0	0	1

Sires and calves were genotyped with 4 genotyping panels across 4 years. Two were low-density panels based on parentage markers [13] using the Bovine Parentage Panel (Eureka Genomics, Hercules, CA, USA) or implemented with TruSeq DNA technology (Illumina Inc., San Diego, CA, USA). Two were higher density panels with more than 50,000 SNP consisting of parentage, linkage, and functional SNP (BovineSNP50, Illumina Inc., San Diego, CA; GeneSeek Genomic Profiler F250, Neogen, Lansing, MI, USA). A higher proportion of calves born in early years were genotyped with only parentage panels. Most calves born in 2016 and all but 3 bulls (1 Angus, 1 MARC I, and 1 MARC II) were genotyped with GeneSeek Genomic Profiler F250.

A set of parentage SNP [13] in common across the 4 genotyping panels was identified. These SNP were used to identify sires based on exclusions [4]. Additional steps were taken to try to resolve some ambiguous sire identifications, including expanding exclusions to additional SNP if available, calculating the genotypic correlations between a calf and potential sires [14], and re-genotyping some animals. Calves genotyped with higher density panels (48.8%) were matched with sires genotyped with high density (33.8%) or only parentage (15.0%) genotyped with parentage panels were matched with sires genotyped with high density (15.6%) or only parentage (35.6%). Most genotyped calves were matched to a single sire, but several MARC II calves born in 2013 and 2015 and some Angus calves born in 2015 were not genotyped. Sires of 3 calves with <100 genotypes also were not identified.

The distribution of number-of-calves per bull opportunity (bull within pasture and year) was skewed (Table 3) as has been observed in other studies of sire prolificacy. The median and mode of the distribution is 18 with values from 0 to 57. A square root transformation was applied to the data before analysis.

Table 3. Distribution of number of calves (CalfN) by bull within pasture and year.

CalfN	No. Observations
0	3
1 to 7	18
8 to 14	31
15 to 21	35
22 to 28	21
29 to 35	12
36 to 42	13
43 to 49	4
50 to 56	3
57	1

Repeatability was estimated from the variance components for random effects for between bull (bull) and within bull across years and pastures (e) as $\sigma^2_{bull} \times \left(\sigma^2_{bull} + \sigma^2_e\right)^{-1}$. Variance components were estimated using PROC GLIMMIX (SAS Institute Inc., Cary, NC, USA) from the model $CalfN^{0.5}_{i,j,k} = \mu + Y_i + P_j(Y_i) + bull_k + e_{i,j,k}$, where $CalfN^{0.5}_{i,j,k}$ is number of calves sired by bull k in pasture j (P_j) nested within year i (Y_i). Additional fixed categorical or continuous factors were individually added to this base model to test for possible explanatory variables for bull prolificacy. Based on results of individual variables, bull age category and scrotal circumference were added to the base model for testing jointly. All reported statistical probabilities are based on data transformed by square root, but reported means and regression coefficients of explanatory variables are from analyses of untransformed data.

2.2. Simulating Errors in Pooling Allele Frequency

The concept of using pooled calf DNA to estimate bulls' prolificacies was tested by simulating DNA pooling of actual genotypes of calves born in 2016 and genotyped with the Genomic Profiler F250 and sired by bulls genotyped with the same panel. Five pastures (N, O, P, Q, and R; Table 1) with 189,165, 76, 89, and 198 calves and 9, 9, 5, 5, and 10 bulls, respectively, were used after removing bulls and calves without Genomic Profiler F250 genotypes. Simulations used 14,190 autosomal SNP common to both BovineSNP50 and GeneSeek Genomic Profiler F250 panels.

Simulation of pooling was a function of the actual allele frequencies for the calves in a pasture as well as pool construction and technical errors. Pool construction error or random unintended differences in the contribution of individual calves to the pool can result from incorrect DNA measurement or quantification; pipetting error; or cross contamination between pools, or between pools and individual animals. Technical error is the result of variation in the ratio of X (red dye intensity) to Y (green dye intensity) for samples with the same allele frequency (replicated pools) or the same genotype (replicated individuals or individuals of the same genotype). Standard deviations for pool construction error and technical error were estimated from replicated pools in earlier studies that have the same real or underlying allele frequency as if the animals in the pool had been individually genotyped [15,16]. Pooling allele frequency was estimated as the average of genotypes (copies of B allele) weighted by the random calf contribution divided by 2. Pool construction error was simulated as a Dirichlet distribution with SD = 0.0024 equivalent to using symmetric shape (alpha) parameter of 20 when pool size is 92. The Dirichlet distribution is parameterized by a shape parameter, alpha, for each calf in the pool. For example, for a pool size of 150 calves the parameterization included a vector of 150 elements all having the same value of 20. The magnitude of alpha determines how peaked the distribution of calf contributions is. Alpha = 10 would have less peaked and more variable animal contributions than alpha = 20. Simulated technical error was drawn from a normal distribution with a mean of zero and SD of 0.07.

Estimating the number of calves sired by a bull can be improved by knowing average genotype frequencies of the dams. Three levels of average dam allele frequency information were compared: (1) none, (2) a simulated, pooled estimate of allele frequencies from the 94.5% of dams with genotypes, and (3) average allele frequencies of the dams with genotypes. Quadratic programming [17] to compute sire contributions while both not adjusting and adjusting for dam pooling allele frequency is included in an R script that is part of the Supplemental Files.

When dam information was not included in the quadratic programming analysis, dam allele frequencies were proportional to the residual after subtracting predicted sire allele frequencies from calf pooling allele frequency (r^2 ~0.8; data not shown). Adding dam frequencies would be expected to improve the accuracy of sire solutions at an additional cost of sampling cows and genotyping pools. Three levels of average dam allele frequency information were compared: (1) none, (2) pooled estimate of allele frequencies from the 94.5% of dams with genotypes with simulated pool construction and technical error incorporated, and (3) allele frequencies of the dams with genotypes.

3. Results

3.1. Repeatability Estimates

Estimated repeatability of bull prolificacy was 0.62 ± 0.09, and the SD of untransformed calf N was 13.6 (Table 4). The data used to estimate repeatability are in supplementary materials (Table S1) Previous numbers of calves from bulls used in multiple-bull pastures as 2-year-olds and older are good predictors of future numbers of calves. This result is consistent with or greater than other studies based on data from various pasture situations with different numbers of bulls and cows assigned to pastures, usually for fewer years and including yearling bulls [2,5]. It is also consistent with several studies observing bulls' rankings across years [18,19].

Table 4. Estimated repeatability of bull prolificacy and component variances.

Estimate	Value [1]	SE
Between bull	1.72	0.45
Within bull	1.05	0.21
Repeatability	0.62	0.09

[1] Variances of square roots of number of calves per bull within pasture and year. Estimated variances of untransformed numbers are 114.72 (between) and 70.43 (within).

3.2. Explanatory Variables

A bull's life history and measurements made before entering the breeding pasture may explain some difference in bull prolificacy. Several retrospective variables were individually added to the statistical model for repeatability. Life history variables for bulls were (1) dam was 2 years old, (2) previously used for breeding as a yearling, (3) used for breeding any prior season, and (4) breeding age was >2. Bull measurements made during breeding soundness exams before each breeding season were included, such as bull weight and scrotal circumference.

None of the life history traits or bull measurements were significant when added individually to the model (Table 5). However, some measurements and life-history traits were partially confounded, especially bull measurements and breeding age. A model including breeding age category (2 years old vs. older than 2) and scrotal circumference measurement (at breeding soundness exam) resulted in significant effects for both. At the same scrotal circumference, bulls older than 2 years sired 6.33 more calves. At the same age classification, a 1.0 cm increase in scrotal circumference was associated with a decrease of 1.56 calves, possibly because only bulls that had scrotal circumferences greater than breeding soundness standards were used. The addition of these two variables accounted for less than 5% of variation in prolificacy and did not change estimated repeatability. The ability of individual bull measurements and life-history traits to predict prolificacy is limited compared to knowing a bull's earlier prolificacy results, when available.

Table 5. Patterns of edited bull usage by calf birth year.

Contrasts and Regression Coefficients	Value, N [1]	SE [1]	P [2]
Dam was 2-year-old	2.46	3.93	0.32
Previously used as yearling	5.25	3.12	0.11
Previously used, any age	3.64	2.51	0.18
Breeding age older than 2 years	3.79	2.66	0.16
Breeding exam weight, N/kg	−0.005	0.015	0.79
Breeding scrotal circumference, N/cm	−1.04	0.59	0.11
Values from adding 2 variates simultaneously			
Scrotal circumference, N/cm	−1.56	0.63	0.03
Breeding age older than 2 years	6.33	2.80	0.02

[1] Values reported are calf number per bull per season. [2] Probabilities were based on analysis of calf $N^{0.5}$.

3.3. Estimated Prolificacy

Genotypes of calves, dams and sires used to simulate pooling and estimate sire contributions are in supplementary materials (Tables S2–S6 for the 5 pastures N, O, P, Q, and R, respectively). Results of simulated pooling of calf DNA samples (Table 6; Figure 1) suggest that the concept of pooling to estimate prolificacy is valid. The accuracy of predicting actual proportions of calves is moderate to high regardless of whether cow allele frequencies are unknown, estimated by pooling, or known. Allele frequencies of dams estimated by pooling did increase accuracy and consistency across different pastures. There was little difference in either accuracy or consistency when allele frequencies of dams were estimated by individual cow or pooled genotyping.

Table 6. Intercept, slope and r2 from regressing pooling estimates of sire progeny proportions on known proportions 100 simulated replicates using 14,190 autosomal.

Dam Genotypes	Pasture	Intercept	SE	Slope	SE	r^2	SD
None	N	−0.006	0.009	0.553	0.077	0.879	0.011
	O	−0.003	0.015	0.529	0.133	0.692	0.012
	P	−0.007	0.011	0.536	0.046	0.977	0.014
	Q	−0.007	0.013	0.537	0.063	0.956	0.009
	R	0.005	0.008	0.452	0.069	0.842	0.013
Pooled	N	0.002	0.003	0.486	0.025	0.980	0.004
	O	−0.002	0.005	0.516	0.043	0.950	0.004
	P	−0.000	0.005	0.502	0.021	0.994	0.006
	Q	0.006	0.008	0.471	0.043	0.971	0.006
	R	−0.000	0.002	0.501	0.020	0.987	0.004
Individual	N	0.002	0.003	0.487	0.023	0.983	0.003
	O	−0.003	0.005	0.522	0.041	0.956	0.004
	P	0.001	0.005	0.494	0.022	0.993	0.007
	Q	.0005	0.009	0.476	0.043	0.973	0.006
	R	0.000	0.002	0.499	0.017	0.990	0.003

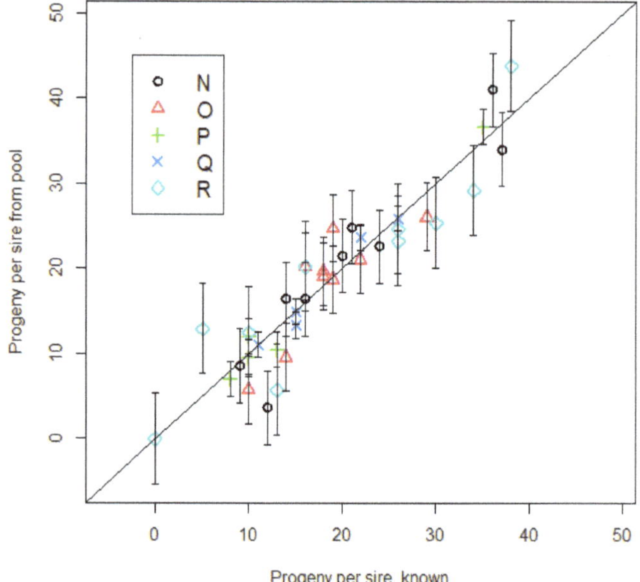

Figure 1. Pooling estimates of sire progeny numbers for 2016 calf crop by known progeny numbers using 14,190 SNP. Dam allele frequencies were assumed unknown for this graph. Error bars are lack of fit SD between pooling sire progeny number and known progeny number. Symbols in the legend identify different pastures.

4. Discussion

A predicted number of calves should be useful for making management decisions about the number of bulls maintained and used. However, we are not aware of experimental evidence for making those management decisions. It seems likely that bulls that previously sired no or few calves should not subsequently be used in multiple-bull pastures. It is possible that these bulls would sire more calves in single-bull pastures or by AI, but culling them would also be a reasonable management strategy. Whether fewer bulls could be used in multiple-bull pastures if one-quarter of them are predicted to be above average (based on prior usage) and three-quarters are average (based on no prior information) is unknown. Any management strategy using predicted prolificacy would need to account for risks of injury, becoming unsound, or death before or during breeding. The estimated repeatability only applies to bulls 2 and older that passed breeding soundness exams prior to the breeding season and completed at least 90% of the breeding season.

Although not significant, prolificacy tended to increase from 2 through about 5 years of age in pastures with mixed-age bulls [5]. Our results showed a significant increase in prolificacy of bulls older than 2 compared with 2-year-olds when adjusted for scrotal circumference.

Scrotal circumference of older bulls measured in breeding soundness exams before breeding has shown little to moderate positive correlation with bull prolificacy in multiple-bull pastures. Results tended to be greater when breeding soundness exams were not rigorously applied [3,18]. Scrotal circumference breeding value was found to have some predictive ability across 2-year-old and older bulls [5].

Scrotal circumference combined with age had significant but small effects on prolificacy in this study. Life-history traits related to previous breeding experience had positive values, but a larger experiment would be needed to conclusively determine any trait effects of that size. Breeding exam live weights showed no indication of effects on prolificacy.

Pooling DNA can potentially reduce costs of accurately estimating bull prolificacy, essentially using genomics to measure phenotype if bull prolificacy is the only information desired by a cattle producer. A basic implementation of this method begins by individually sampling DNA from all bulls used in a multiple-sire pasture and maintaining correspondence between each sample and each bull. Blood from all calves would be individually sampled. Calf blood samples would need to be maintained individually but samples would not need to be connected to individual calves. For higher accuracy, cows assigned to a breeding pasture would need to be sampled sometime from pre-breeding through weaning if the cows were maintained as a group through that period. Like the calf samples, cow blood samples do not need to maintain connection with the individual cows. To make decisions on bull management in time for the current breeding season, samples would need to be collected well before breeding, likely at birth. Research has found that bulls siring large numbers of calves early in the calving period tended to sire more calves throughout [5,19]. This seems to be one likely approach for obtaining a prolificacy estimate early enough to be useful for the breeding season immediately following calving. This study has a few limitations with regards to relevance to commercial production. First, F_1 crossbred dams are important in the commercial beef cattle sector. F_1 dams might reduce precision of estimating sire contributions because alleles that are not inherited in calves would be part of dam allele frequencies but not part of calves. This is more complicated if dams are F_1. We do not know how important this is currently. Second, pool construction and technical errors were simulated in the current study. The distribution of these errors may be different than real pooling data. We are planning future studies to rectify this limitation.

Once samples are collected, a genomics service provider would need to extract and separately pool calf DNA and cow DNA. Other ways of constructing pools are possible [16]. Then, the individual bull samples and each pool would need to be genotyped with a moderate-to-high density panel. A service provider would need to estimate prolificacy from the genotypes. This approach to estimating prolificacy may facilitate additional research on factors affecting bull prolificacy using industry cooperators.

Other benefits from a pooled sample of calves are possible but would require additional technical support and industry coordination. It should be possible to estimate average breed composition from pooled alleles, predicted heterosis from differences between weighted sire allele frequencies and pooled cow allele frequencies, and average performance levels from those two inputs. Estimated performance might be enhanced through genomic prediction of commercial bulls based on their individual genotypes. This could be useful for marketing calves or managing them through the feedlot or as breeding heifers. Estimated bull prolificacy could be used as phenotypic data for genomic evaluation with information transfer to breeding value prediction organizations based on careful documentation of commercial bulls, their use in pastures, and their genotypes. Claims-based marketing programs could be verified for things such as breed composition or genetic potential based on markers associated with large effects for carcass traits, e.g., beef tenderness [20] or carcass leanness [21]. A pool including heifers could be genetically evaluated for reproductive traits, e.g., pubertal age, antral follicle count, and heifer pregnancy rate [22,23], and for common genetic disorders [24] or disease susceptibility to avoid inappropriate matings.

5. Conclusions

The repeatability of bull prolificacy in multi-bull breeding pastures is high. Bull prolificacy is accurately estimated using DNA pools of calves and genotyping the calves at a lower cost compared to individually genotyping. Genotyping pools of dams improves the accuracy of estimating bull prolificacy which requires samples of dams.

Supplementary Materials: The following are available online at https://www.mdpi.com/article/10.3390/agriculture11070603/s1, Table S1: Data used for repeatability estimates, Table S2: Genotypes of calves, bulls and dams in Pasture N, Table S3: Genotypes of calves, bulls and dams in Pasture O, Table S4: Genotypes of calves, bulls and dams in Pasture P, Table S5: Genotypes of calves, sires and dams in Pasture Q, Table S6: Genotypes of calves, sires and dams in Pasture R, quadratic ProgramSireProportions. R: An R script to estimate sire proportions.

Author Contributions: Conceptualization, G.L.B. and J.W.K.; methodology, G.L.B., J.W.K., and L.A.K.; validation, G.L.B., J.W.K., L.A.K., and W.M.S.; formal analysis, G.L.B. and J.W.K.; resources, G.L.B., W.M.S., and T.G.M.; data curation, D.L., A.M.D., and T.G.M.; writing—original draft preparation, G.L.B. and J.W.K.; writing—review and editing, L.A.K., W.M.S., T.G.M., A.M.D., D.L., and R.A.C. All authors have read and agreed to the published version of the manuscript.

Funding: This research received no external funding.

Institutional Review Board Statement: Archival pedigrees, genotypes, animal data, and blood samples from research protocols approved and monitored by the USDA, Agricultural Research Service, U.S. Meat Animal Research Center Institutional and Animal Care Committee following the Guide for the Care and Use of Agricultural Animals in Agricultural Research and Teaching [25] were used in this study.

Informed Consent Statement: Not applicable.

Data Availability Statement: Data is contained within the supplementary tables.

Acknowledgments: Mention of the trade name, proprietary product, or specific equipment does not constitute a guarantee or warranty by the U.S. Department of Agriculture (USDA) and does not imply approval to the exclusion of other products that may be suitable. The USDA is an equal opportunity provider and employer. We acknowledge Richard G. Tait, Jr., for contributions to determining pedigree in the early stages of this research while employed by the U.S. Meat Animal Research Center.

Conflicts of Interest: The authors declare no conflict of interest.

References

1. Lunstra, D.D.; Laster, D.B. Influence of single-sire and multiple-sire natural mating on pregnancy rate of beef heifers. *Theriogenology* **1982**, *18*, 373–382. [CrossRef]
2. Lehrer, A.R.; Brown, M.B.; Schindler, H.; Holzer, Z.; Larsen, B. Paternity tests in multi-sired beef herds by blood grouping. *Acta Vet. Scand.* **1977**, *18*, 433–441. [CrossRef]
3. Holroyd, R.G.; Doogan, V.J.; De Faveri, J.; Fordyce, G.; McGowan, M.R.; Bertram, J.D.; Vankan, D.M.; Fitzpatrick, L.A.; Jayawardhana, G.A.; Miller, R.G. Bull selection and use in northern Australia: 4. Calf output and predictors of fertility of bulls in multiple-sire herds. *Anim. Reprod. Sci.* **2002**, *71*, 67–79. [CrossRef]
4. Van Eenennaam, A.L.; Weaber, R.L.; Drake, D.J.; Penedo, M.C.T.; Quaas, R.L.; Garrick, D.J.; Pollak, E.J. DNA-based paternity analysis and genetic evaluation in a large, commercial cattle ranch setting. *J. Anim. Sci.* **2007**, *85*, 3159–3169. [CrossRef] [PubMed]
5. Van Eenennaam, A.L.; Weber, K.L.; Drake, D.J. Evaluation of bull prolificacy on commercial beef cattle ranches using DNA paternity analysis. *J. Anim. Sci.* **2014**, *92*, 2693–2701. [CrossRef] [PubMed]
6. Ologun, A.G.; Chenoweth, P.J.; Brinks, J.S. Relationships among production traits and estimates of sex drive and dominance value in yearling beef bulls. *Theriogenology* **1981**, *15*, 379–388. [CrossRef]
7. Whitworth, W.A.; Forrest, D.W.; Sprott, L.R.; Holloway, J.W.; Warrington, B.G. Physical, and mating behavior traits of bulls on number of calves sired per bull in a multisire herd. *Prof. Anim. Sci.* **2008**, *24*, 184–188. [CrossRef]
8. Keele, J.W.; Kuehn, L.A.; McDaneld, T.G.; Tait, R.G., Jr.; Jones, S.A.; Smith, T.P.L.; Shackelford, S.D.; King, D.A.; Wheeler, T.L.; Lindholm-Perry, A.K.; et al. Genomewide association study of lung lesions in cattle using sample pooling. *J. Anim. Sci.* **2015**, *93*, 956–964. [CrossRef] [PubMed]
9. Baller, J.L.; Kachman, S.D.; Kuehn, L.A.; Spangler, M.L. Genomic prediction using pooled data in a single-step genomic best linear unbiased prediction framework. *J. Anim. Sci.* **2020**, *98*, skaa184:1–skaa184:12. [CrossRef] [PubMed]
10. Reverter, A.; Porto-Neto, L.R.; Fortes, M.R.S.; McCulloch, R.; Lyons, R.E.; Moore, S.; Nicol, D.; Henshall, J.; Lehnert, S.A. Genomic analyses of tropical beef cattle fertility based on genotyping pools of Brahman cows with unknown pedigree. *J. Anim. Sci.* **2016**, *94*, 4096–4108. [CrossRef] [PubMed]
11. Alexandre, P.A.; Porto-Neto, L.R.; Karaman, E.; Lehnert, S.A.; Reverter, A. Pooled genotyping strategies for rapid construction of genomic reference populations. *J. Anim. Sci.* **2019**, *97*, 4761–4769. [CrossRef] [PubMed]
12. Gregory, K.E.; Cundiff, L.V.; Koch, R.M. Breed effects and heterosis in advanced generations of composite populations for preweaning traits of beef cattle. *J. Anim. Sci.* **1991**, *69*, 947–960. [CrossRef] [PubMed]
13. Heaton, M.P.; Harhay, G.P.; Bennett, G.L.; Stone, R.T.; Grosse, W.M.; Casas, E.; Keele, J.W.; Smith, T.P.; Chitko-McKown, C.G.; Laegreid, W.W. Selection and use of SNP markers for animal identification and paternity analysis in U.S. beef cattle. *Mamm. Genome* **2002**, *13*, 272–281. [CrossRef] [PubMed]
14. amap: Another Multidimensional Analysis Package. Available online: https://cran.r-project.org/web/packages/amap/index.html (accessed on 6 May 2021).
15. Kuehn, L.A.; McDaneld, T.G.; Keele, J.W. Quantification of genomic relationship from DNA pooled samples. In Proceedings of the World Congress on Genetics Applied to Livestock Production, Auckland, New Zealand, 11–16 February 2018; Volume 11, p. 631.
16. Abrams, A.N.; McDaneld, T.G.; Keele, J.W.; Chitko-McKown, C.G.; Kuehn, L.A.; Gonda, M.G. Evaluating accuracy of DNA pool construction based on white blood cell counts. *Front. Genet.* **2021**, *12*, 635846:1–635846:6. [CrossRef] [PubMed]
17. quadprog: Functions to Solve Quadratic Programming Problems. Available online: https://CRAN.R-project.org/package=quadprog (accessed on 11 May 2021).
18. McCosker, T.H.; Turner, A.F.; McCool, C.J.; Post, T.B.; Bell, K. Brahman bull fertility in a north Australian rangeland herd. *Theriogenology* **1989**, *27*, 285–300. [CrossRef]
19. Abell, K.M.; Theurer, M.E.; Larson, R.L.; White, B.J.; Hardin, D.K.; Randle, R.F.; Cushman, R.A. Calving distributions of individual bulls in multiple-sire pastures. *Theriogenology* **2017**, *93*, 7–11. [CrossRef]
20. Tait, R.G., Jr.; Shackelford, S.D.; Wheeler, T.L.; King, D.A.; Keele, J.W.; Casas, E.; Smith, T.P.; Bennett, G.L. CAPN1, CAST, and DGAT1 genetic effects on preweaning performance, carcass quality traits, and residual variance of tenderness in a beef cattle population selected for haplotype and allele equalization. *J. Anim. Sci.* **2014**, *92*, 5382–5393. [CrossRef] [PubMed]
21. Bennett, G.L.; Tait, R.G., Jr.; Shackelford, S.D.; Wheeler, T.L.; King, D.A.; Casas, E.; Smith, T.P.L. Enhanced estimates of carcass and meat quality effects for polymorphisms in myostatin and mu-calpain genes. *J. Anim. Sci.* **2019**, *97*, 569–577. [CrossRef]
22. Snelling, W.M.; Cushman, R.A.; Fortes, M.R.S.; Reverter, A.; Bennett, G.L.; Keele, J.W.; Kuehn, L.A.; McDaneld, T.G.; Thallman, R.M.; Thomas, M.G. How single nucleotide polymorphism chips will advance our knowledge of factors controlling puberty and aid in selecting replacement beef females. *J. Anim. Sci.* **2012**, *90*, 1152–1165. [CrossRef] [PubMed]
23. Cushman, R.A.; Tait, R.G., Jr.; McNeel, A.K.; Forbes, E.D.; Amundson, O.L.; Lents, C.A.; Lindholm-Perry, A.K.; Perry, G.A.; Wood, J.A.; Cupp, A.S.; et al. A polymorphism in myostatin influences puberty but not fertility in beef heifers, whereas μ-calpain affects first calf birth weight. *J. Anim. Sci.* **2015**, *93*, 117–126. [CrossRef] [PubMed]
24. OMIA-Online Mendelian Inheritance in Animals. Available online: http://omia.org/home/ (accessed on 4 May 2021).
25. FASS. *Guide for the Care and Use of Agricultural Animals in Agricultural Research and Teaching*, 1st ed.; Federation of Animal Science Societies: Savoy, IL, USA, 1999.

 *agriculture* 

Article

Genetic Correlations between Days to Calving across Joinings and Lactation Status in a Tropically Adapted Composite Beef Herd

Madeliene L. Facy [1,*], Michelle L. Hebart [1], Helena Oakey [2], Rudi A. McEwin [1] and Wayne S. Pitchford [1]

1 Davies Livestock Research Centre, School of Animal and Veterinary Sciences, University of Adelaide, Roseworthy, SA 5371, Australia
2 Robinson Research Institute, Adelaide Medical School, University of Adelaide, Adelaide, SA 5006, Australia
* Correspondence: maddie.facy@adelaide.edu.au

Abstract: Female fertility is essential to any beef breeding program. However, little genetic gain has been made due to long generation intervals and low levels of phenotyping. Days to calving (DC) is a fertility trait that may provide genetic gain and lead to an increased weaning rate. Genetic parameters and correlations were estimated and compared for DC across multiple joinings (first, second and third+) and lactation status (lactating and non-lactating) for a tropical composite cattle population where cattle were first mated as yearlings. The genetic correlation between first joining DC and mature joining DC (third+) was moderate–high (0.55–0.83). DC was uncorrelated between multiparous lactating and non-lactating cows ($r_G = -0.10$). Mature joining DC was more strongly correlated with second joining lactating DC (0.41–0.69) than with second joining non-lactating DC (−0.14 to −0.16). Thus, first joining DC, second joining DC and mature joining DC should be treated as different traits to maximise genetic gain. Further, for multi-parous cows, lactating and non-lactating DC should be treated as different traits. Three traits were developed to report back to the breeding programs to maximise genetic gain: the first joining days to calving, the second joining days to calving lactating and mature days to calving lactating.

Keywords: cattle; fertility; heritability; genetic evaluation; variance components; index selection

Citation: Facy, M.L.; Hebart, M.L.; Oakey, H.; McEwin, R.A.; Pitchford, W.S. Genetic Correlations between Days to Calving across Joinings and Lactation Status in a Tropically Adapted Composite Beef Herd. *Agriculture* **2023**, *13*, 37. https://doi.org/10.3390/agriculture13010037

Academic Editors: Heather Burrow and Michael Goddard

Received: 1 December 2022
Revised: 19 December 2022
Accepted: 20 December 2022
Published: 22 December 2022

1. Introduction

Fertility traits are often treated with low priority in beef breeding programs worldwide, as many female fertility traits have a low heritability, are difficult to measure and the long generation interval of cattle dissuade phenotype recording [1]. Despite this, in Northern Australia female reproductive performance has been identified as a major economic issue in beef production [1,2]. An Australian beef industry study reported that an increase in weaning rate of 5% for tropically adapted cattle would lead to a 20% increase in average annual net profit, identifying fertility as a key performance driver in northern production systems [2].

In Northern Australia, low reproductive performance tends to be a consequence of late puberty attainment in heifers and extended post-partum anoestrous periods in cows [3,4]; these issues are particularly predominant in lactating first-calf heifers [5]. Cattle are required to adapt to the extreme heat, disease, pests, varying nutrition qualities and quantities as well as be reproductively successful [3–5]. These factors highlight the difficultly in improving reproductive performance in Northern Australia; however, ignoring fertility traits is extremely detrimental to any breeding program [6–8].

Days to calving is defined as the number of days from the start of joining (the day the bull goes into the same paddock) to subsequent calving, as described by Meyer et al. [9]. Days to calving is a trait that can be measured multiple times over a cow's lifetime (each joining) and evaluated as a repeated measure trait [10]. A joining includes any cow given an opportunity to conceive over a period from the bull-in date to the bull-out date. Days to calving is often treated as a repeated trait; however, due to the complexity of the trait, earlier in life joining's should possibly be evaluated separately to mature joinings due to different biological mechanisms.

Days to calving is a potential female fertility trait that can be measured to account for the attainment of puberty and extended post-partum anoestrous periods. Days to calving is a composite trait that is associated with age of puberty in heifers, measuring dam effects such as post-partum anoestrus, conception rate, gestation length and how early they conceive [11,12]. Days to calving has previously been reported to have a high correlation with lifetime weaning rate in both *Bos taurus taurus* and *Bos taurus indicus* breeds, an essential economic driver in breeding programs [1,13]. Incorporating days to calving in a breeding program has the potential to improve female fertility while maintaining adaption to the harsh environment of Northern Australia.

The present study investigates days to calving to enable the greatest genetic gain in breeding programs. This was done through testing different penalty values placed on non-pregnant cows, investigating the effect of joining and lactation status (lactating and non-lactating) and finally using selection index theory was to evaluate the best way to report days to calving to breeding programs. The study models the days to calving traits to estimate variance components, breeding values, response to selection and genetic and phenotypic correlations.

2. Materials and Methods

2.1. Breeding Program

Popplewell Composites, a bull breeding program focused on breeding composite bulls adapted to the semi-arid tropics in Northern Australia, provided the dataset. This breeding program is predominantly situated in southeast Queensland, with bulls sold across Northern Australia. The breeding program started in 2008 and involved four main breeds: Angus, Senepol, Africander and Brahman, to produce four composite products, Transition, Pathfinder, Eureka and Indiplus. Popplewell composites use the four composite products to change existing cow bases in commercial herds from their original breeds (often high content Brahmans) into composite herds of a consistent breed type. The main objective of Popplewell composites is to increase weaning rate through retained heterosis and intense selection pressure for reduced age at puberty and post-partum anoestrus interval. In addition, there is an emphasis on good growth, carcass traits and naturally polled red cattle that are well adapted for tropical environments such as Northern Australia.

2.2. Genotype Data

Genotype data were provided from 2613 composite cattle. Genotyping was conducted over the breeding program's duration, resulting in multiple SNP chips being used (Illumina 777k, Illumina GGPLD V3 30K, Illumina GGPLD V4 30K, Illumina ICB 50K and Weatherby's Scientific Versa50K). A total of 10,830 SNPs overlapped on all SNP chips. All genotypes were imputed from the common SNPs to GGPLD30K (27,755 SNPs) using FImpute 2.1 [14]. The genotype data were then filtered, and all SNPs with $<1\%$ minor allele frequency and duplicate animals were removed. After filtering, there were 2609 animals and 27,638 SNPs for constructing the genomic relationship matrix (GRM; see below). The GRM was built using VanRaden's first method [15].

The heterozygosity fraction (Het) was calculated on imputed data, as the proportion (%) of heterozygous SNPs for each individual animal. Heterozygosity fraction is an indicator of heterosis, which is especially important in a composite population. All models included the heterozygosity fraction as a co-variate to account for heterozygosity when estimating breeding values [16].

2.3. Phenotype Data

Days to calving is the time from the day the bull goes in with the cow to the day the cow gives birth [9]. Days to calving includes every joining for a cow over its lifetime, thus providing multiple records. Days to calving was estimated using the bull-in date, bull-out date (i.e., in the same paddock as cow), pregnancy test results (fetal age) and calf date of birth. Typically, days to calving is treated as one trait, including all joinings; however, the dataset used herein routinely evaluates heifer (first joining) days to calving separately from all other joinings.

There were 2242 days to calving records from 903 unique animals. Of the 903 unique animals, there was 366 animals that had a first, second and third joining. The number of records and summary statistics at each joining are presented (Table 1). The cattle are first joined at 12–15 months of age (12 months earlier than most tropically adapted cattle programs), with subsequent joinings every year. Pregnancy results represent if the cow got pregnant during the joining period; however, lactation status represents if the cow weaned a calf during the mating period.

Table 1. Summary (min, max, mean, SD) of adjusted [1] days to calving over joinings.

Joining Number	Total No. of Cows	No. of Lactating Cows	No of Non-Lactating Cows	No. of Empty [2] Cows	No. of Pregnant Cows	[3] Min days	[4] Max days	Mean Days	[5] SD Days
1	649	-	-	214	435	269	409	340.0	38.0
2 [6]	636	326	287	472	164	271	406	335.0	38.5
3	494	406	88	325	169	260	409	343.6	38.0
4	249	225	24	166	83	272	400	340.8	36.9
5+	214	197	17	166	48	283	400	332.2	34.4

[1] Adjusted phenotypes have a penalty of 31 days added for empty cows. [2] Empty cows refers to non-pregnant cows. [3] Min is the minimum. [4] Max is the maximum. [5] SD is the standard deviation. [6] There are 23 animals with no lactation status for second joining.

Correcting days to calving values by adding a constant penalty (21 days = 1 oestrus cycle) to cows that have not calved is standard practice [10]. However, a longer period was used herein to allow for low-quality sub-tropical pasture as the breeder felt it was too "generous" to assume open heifers and cows would have conceived in a single additional cycle. Those animals that did not conceive during the joining period received an adjusted value of 32 days (1.5 oestrus cycles) added to the maximum value of the animals joining management group. This penalty assumes they would have conceived in the next cycle and a half if given the opportunity. The impact of the penalty was tested herein, with several different penalty values investigated. These values were no-penalty, 21, 32, 43, 63 and 252 days. The breeding values, variance components, heritability and heterozygosity coefficient slope were estimated and compared.

Contemporary groups were defined as management groups plus birth groups. Throughout their lifetime, management groups were recorded. The birth group was defined by year and time of birth, early (1 June to the 4 September, 75% of cattle), mid (5 September to the 30 November, 6% of cattle) and late (1 December to the end of calving season, 19% of cattle). Breed group was partially confounded with contemporary group.

2.4. Genomic Model

A linear mixed model to estimate breeding values can be written as,

$$y = X\beta + Z_a u_a + Z_r u_r + e$$

where y is the vector of phenotypic values for days to calving; β represents fixed effects with the design matrix X; the fixed effects included in all models are contemporary group, joining number, lactation status, days to calving management group and heterozygosity fraction (fit as a linear co-variate); $\mathbf{Z}_a$ is the design matrix for additive effects $\boldsymbol{u}_a$, with $\boldsymbol{u}_a \sim N(\mathbf{0}, \sigma_A^2 \mathbf{G}_A)$; $\mathbf{Z}_r$ is the design matrix for the repeated environmental effects $\boldsymbol{u}_r$ and $\boldsymbol{e}$ is the residual effects.

The genomic relationship matrix (GRM) is calculated as follows [15]:

$$\mathbf{G}_A = \frac{\mathbf{ZZ}\prime}{2\sum_{i=1}^{m} p_i q_i}$$

where the matrix $\mathbf{Z}$ has dimensions of nxm, n is the number of individuals and m is the number of markers. Matrix $\mathbf{Z}$ has the elements z_{ij} for the jth individual at the ith marker is,

$$z_{ij} = \begin{cases} (2-2p_i) \\ (1-2p_i) \\ -2p_i \end{cases} for\ Genotypes \begin{cases} AA \\ AB \\ BB \end{cases}$$

and p_i is the allele frequency of the most frequent (major) allele at Marker i, and $q_i = 1 - p_i$.

2.5. Days to Calving Models

Days to calving was modelled to estimate the variance components, heritability, repeatability and best linear unbiased predictions (BLUPs) or estimated breeding values (EBVs). Overall days to calving (DC) was run as a repeatability model to adjust for multiple records on the same cow over different years. Two models (Table 2) were fitted in RStudio version 4.1.1 [17] using the package ASReml-R v4 [18]. A description of each model are listed in Table 2. Days post-partum (DPP) was calculated as the number of days between bull-in date and last calf date of birth, and it was included in the model to be compared within the contemporary group.

Table 2. Description of models run for days to calving.

Model	Description	Fixed	Random [1]
Model_1	Overall days to calving, with a separate variance component for five joining categories	Join management group, heterozygosity fraction and Contemporary group:DPP	diag(Joining Number):vm(ID, GRM)
Model_2	Overall days to calving, fitted models with a separate variance component for each joining but estimation of the correlation between two joining at a time (due to computational limitations) with a total of 15 components	Join management group, heterozygosity fraction and Contemporary group:DPP	us(Joining Number):vm(ID, GRM)

[1] describes ASReml-R model terms, diag() = diagonal variance model, vm() = known variance structure allows the use of either a relationship or inverse relationship matrix, us() = unstructured general covariance matrix.

The differences and similarities between the joining numbers was explored. Two models were run, Model_1 to estimate separate variance components for each joining, assuming joining numbers were uncorrelated with each other and Model_2 to estimate separate variance components between each joining number and estimate the correlation between each joining number. In Model_2 the correlation between joining numbers were estimated two at a time due to computational limitations. These models were used to calculate the genetic correlation between joining numbers.

As, by definition, maiden joinings can only have a non-lactating lactation status, a further analysis was run with second joining onwards (DC2+), splitting up lactating (DC2+_Wet) and non-lactating (DC2+_Dry) cattle to estimate the genetic correlation between lactating status. The fixed and random effects for these models are listed in Table 3.

Table 3. Description of the fixed effects included first joining days to calving (DC1), second joining days to calving (DC2) and mature joining days to calving (DC3+).

Trait	Description	Number of Records	Number of Contemporary Groups	Fixed Effects	Random Effects
DC1	First joining days to calving	649	45	Birth group, Join group, Dam age, Het, contemporary group	GRM
DC2	Second joining days to calving with all records (wet [1] and dry [2])	636	54	Birth group, Dam age, Het, contemporary group:DPP:Lactation Status	GRM
DC2_Dry	Second joining days to calving with dry [2] records only	287	44	Birth group, Dam age, Het, contemporary group:DPP	GRM
DC2_Wet	Second joining days to calving with wet [1] records only	326	45	Birth group, Dam age, Het, contemporary group:DPP	GRM
DC2+	Days to calving that include records from the second joining, third joining, fourth joining and joining five + with both wet [1] and dry [2] records	1534	213	Joining number, Het, contemporary group:DPP:Lactation Status	GRM, ID
DC2+_Dry	Days to calving that include records from the second joining, third joining, fourth joining and joining five + with dry [2] records only	431	82	Joining number, Het, contemporary group:DPP	GRM, ID
DC2+_Wet	Days to calving that include records from the second joining, third joining, fourth joining and joining five + with wet [1] records only	1103	189	Joining number, Het, contemporary group:DPP	GRM, ID
DC3+	Days to calving that include records from the third joining, fourth joining and joining five + with both wet [1] and dry [2] records	898	162	Joining number, Het, contemporary group:DPP:Lactation Status	GRM, ID
DC3+_Wet	Days to calving that include records from the third joining, fourth joining and joining five + with wet [1] records only	796	143	Joining number, Het, contemporary group:DPP	GRM, ID
DC	All days to calving records, including both wet [1] and dry [2] records	2242	270	Joining number, join group, Hey, contemporary group:DPP	GRM, ID

ID is the unique animal tag, GRM is the genomic relationship matrix. [1] Wet refers to lactating cows. [2] Dry refers to non-lactating cows.

2.6. Univariate Models

Further univariate models were run based on the results from the above Days to calving models; these included first joining days to calving (DC1), second joining days to calving (DC2) and mature joinings (J3+) days to calving (DC3+). These models were further split into lactating (wet) and non-lactating (dry) for DC2, DC2+ and DC3+. Variance components, heritability and estimated breeding values were estimated for all traits. The fixed and random effects are listed in Table 3, and the GRM was included as a random effect for all models. For repeatability models, a unique animal tag (ID) was fit alongside the genomic relationship matrix as a random term to account for multiple records on the same animals, estimating the repeatable environmental variance in addition to the residual.

2.7. Response to Selection

A simple breeding objective of five consecutive pregnancies was assumed and selection index theory used to test what trait combination would maximise response. The index followed standard selection index theory [19,20], and three matrices were calculated (***P***, ***G***, ***C***). The ***P*** matrix is a square matrix of phenotypic (co)variances among the traits used as selection criteria, in this instance each individual joining was used. The ***G*** matrix consists of genetic (co)variances between the selection criteria and breeding objective. The ***C*** matrix is a square matrix with genetic (co)variance among the traits in the breeding objective. The selection index for three different scenarios were calculated (Table 4). The (co)variance estimates for each index were estimated as a univariate model (index one), bivariate model (index two) and tri-variate model (index three).

Table 4. Description of the different scenarios used in the index (description in Table 3).

Index One	Index Two	Index Three
DC	DC1 DC2+_Wet	DC1 DC2_Wet DC3+_Wet

The standard deviation of the breeding objective was calculated by,

$$\sigma_H = \sqrt{\boldsymbol{a}\prime \boldsymbol{C}\boldsymbol{a}}$$

where $\boldsymbol{a}$ is a vector of five economic weights, in these calculations the weighting used was dependent on weaning rate over five joinings. For example, in Index three the DC1 value was 0.2, the DC2_Wet value was 0.2, and the DC3+ value was 0.6 for $\boldsymbol{a}$, representing how many calves can be produced in each trait, with the aim for five calves from five joinings.

A vector of index weights ($\boldsymbol{b}$) was calculated by,

$$\boldsymbol{b} = \boldsymbol{P}^{-1}\boldsymbol{G}\boldsymbol{a}$$

The standard deviation of the index was calculated by,

$$\sigma_I = \sqrt{\boldsymbol{b}\prime \boldsymbol{P}\boldsymbol{b}}$$

The accuracy of selection was calculated by,

$$r_{I,H} = \frac{\sigma_I}{\sigma_H}$$

The selection response was calculated for each trait reported. The response was calculated in two ways, first for the individual traits by the additive genetic standard deviation multiplied by accuracy ($\sigma_A\sqrt{h^2}$), which is equivalent to the accuracy based on a single measure. Second for the index values, the standard deviation of the breeding objective multiplied by the accuracy of selection ($\sigma_H r_{I,H}$). The response calculations are over one generation with an assumed selection intensity of one and with units of days/generation.

2.8. Multi-Variate Models

Multi-variate models were run between the different days to calving traits to estimate genetic and phenotypic correlation. DC2 and DC3+ were further partitioned into lactating and non-lactating. An average DC3+ (DC3+_Ave) was used to estimate phenotypic correlations between DC3+ and other days to calving traits. The mixed model for bivariate is now written as,

$$y = \boldsymbol{X}^{*}\boldsymbol{b} + \boldsymbol{Z}_a^{*}\boldsymbol{u}_a + \boldsymbol{Z}_r\boldsymbol{u}_r + \boldsymbol{e}$$

where $y = (y_1', y_2')'$, is the combined vector of data between two traits; $b = (b_1', b_2')'$ is the $2m$ x 1 vector of fixed effect with $X^* = I_2 \otimes X$ the associated design matrix; $u_a = (u_1', u_2')'$ is the $2n$ x 1 vector of random effects with $Z_a^* = I_2 \otimes Z$ the associated design matrix, Z_r is the design matrix for the repeated environmental effects u_r and $e = (e_1', e_2')'$ the vector of residual variance, where $'$ is the transpose. The variance components estimates from this multi-variate an additive genetic variance $(\sigma_{A1}^2, \sigma_{A2}^2)$ for each trait, an additive covariance between two traits (σ_{A12}^2), one repeatability variance (σ_R^2), a separate residual variances $(\sigma_{\varepsilon 1}^2, \sigma_{\varepsilon 2}^2)$ for each trait and the correlation between residual variances $(\sigma_{\varepsilon 12}^2)$. All bivariate models were fitted for all traits utilising a genomic relationship matrix with heritability, genetic and phenotypic correlations estimated.

3. Results

3.1. Effect of Changing the Penalty Value of Days to Calving

Varying penalty values were assessed for DC1 and DC2+ to determine how the penalty value affected the variance components, breeding values and heritability. For DC1, the heritability increased by including the non-pregnant animals across all the penalty values used. The variance components and the absolute heterozygosity coefficients increased as the penalty increased; however, there was little change in the heritability, which is a ratio of variances (Table 5). A one percent increase in heterozygosity results in fewer days to calving, therefore a negative heterozygosity fraction is advantageous.

Table 5. Variance Components, heritability and heterozygosity (het) coefficient of DC1 with different penalty values with standard error in parenthesis.

Model	Number of Records	Additive Variance	Residual Variance	Heritability	Het Coefficient (Day/%)
No penalty	435	25.9 (44)	394 (47)	0.06 (0.10)	−0.29
21 Days	634	195 (95)	827 (86)	0.19 (0.09)	−1.59
32 Days	634	257 (121)	1049 (110)	0.20 (0.09)	−1.90
43 Days	634	309 (150)	1290 (136)	0.19 (0.09)	−2.16
63 Days	634	446 (221)	1903 (201)	0.19 (0.09)	−2.73
252 Days	634	2693 (1552)	141,213 (1457)	0.16 (0.09)	−7.93

Similar results were observed for DC2+; as the penalty value increased, total variance and the absolute heterozygosity coefficient increased. When no penalty was applied, the heritability was 0.17, compared with 0.09 when a 21-day penalty was applied. The heritability for DC2+ seems to have a direct scale effect, resulting in slight decreases as the penalty value increases (Table 6). The correlation between the BLUPs was high (>0.99; data not shown) for both DC1 and DC2+ for the different penalty values, indicating no re-ranking of the animals.

Table 6. Variance Components, heritability and heterozygosity (Het) coefficient of DC2+ (including J2) with different penalty values with standard error in parenthesis.

Model	Number Records	Additive Variance	Repeatability Variance	Residual Variance	Heritability	Het Coefficient (Day/%)
No penalty	1122	78 (31)	53 (30)	318 (23)	0.17 (0.06)	−0.62
21 Days	1550	81 (30)	0 [1]	783 (37)	0.09 (0.03)	−1.81
32 Days	1550	86 (36)	0 [1]	998 (46)	0.08 (0.03)	−2.05
43 Days	1550	92 (48)	0 [1]	1227 (68)	0.07 (0.04)	−2.26
63 Days	1550	106 (65)	0 [1]	1810 (99)	0.06 (0.03)	−2.71
252 Days	1550	371 (333)	0 [1]	13,267 (584)	0.03 (0.02)	−6.87

[1] Converged to zero.

3.2. *Variance Components and Genetic Correlation of the Different Joining Number*

When estimating the variance for each joining separately (Model 1), the first joining had the largest variance of 563 days squared and joining 5+ had the lowest variance of 104 days squared (Table 7). The estimated genetic variance decreased with joining number.

Table 7. Genetic variance component of each joining for Model_1 model with standard error in parenthesis.

Joining Number	Variance
1	563 (107)
2	280 (79)
3	215 (77)
4	125 (98)
5+	104 (104)

Model_2 (Table 2) was run to estimate the genetic correlation and covariances between joinings with an unstructured covariance matrix. There were many variance components (15), estimated with large standard errors. The correlation between J1 and J2 was −0.12, which is extremely low (Table 8). However, correlations between J1 and J3, J4 and J5+ were higher at 0.73, 0.36 and 0.64, respectively. The correlation between J2 and J3 was low (0.22), and the correlation between J2 and J4 and J5+ was negative, with values of −0.11 and −0.64, respectively. The correlations between J3, J4 and J5+ were all high, except for J3 and J5, which was −0.03 correlation. The correlation matrix was not positive definite, possibly due to the lower number of records for J4 and J5+. The large standard errors on the estimates of correlations were due to the lower number of records in the older ages (4 and 5+). Based on the high correlation between later joining, the decision was made to combine and treat them as a single trait (DC3+), reducing standard error and maximise response to selection.

Table 8. Correlation matrix for days to calving with genetic correlation below for each joining number with standard error in parenthesis.

	1	2	3	4	5+
1	1				
2	−0.12 (0.19)	1			
3	0.73 (0.17)	0.22 (0.25)	1		
4	0.36 (0.35)	−0.11 (0.39)	0.79 (0.37)	1	
5+	0.64 (0.39)	−0.64 (0.54)	−0.03 (0.61)	0.99 (0.62)	1

3.3. *Days to Calving Separated by Lactation Status*

A further analysis was run, splitting days to calving into individual traits based on lactation status for cows on their second and subsequent lactations. There were 1103 records for DC2+_Wet and 431 records for DC2+_Dry. The heritability of DC2+_Dry and DC2+_Wet which are mature joining from joining 2 onward and non-lactating and lactating individuals respectively had similar heritability estimates of 0.22 and 0.17 (Table 9). The residual variance was smaller for DC2+_Wet at (537 days squared) then DC2_Dry (775 days squared). The repeatability of DC2+_Dry could not be estimated as non-lactating cows were not retained for more than two joinings, whereas DC2+_Wet had a repeatability estimate of 0.37. The genetic correlation between DC2+_Dry and DC2+_Wet was −0.10. The correlation between BLUPs was 0.32 (Figure 1). As expected, the animals with phenotypes re-ranked more than those animals without phenotypes (Figure 1).

Table 9. Variance components, heritability and heterozygosity coefficient of days to calving (J2+) split into lactation status (lactating/non-lactating) with standard errors in parenthesis.

Models	Number of Records	Additive Variance	Repeatability Variance	Residual Variance	Heritability	Repeatability [1]	Het Coefficient	Response to Selection
DC2+_Dry	431	217 (90)	NA	775 (88)	0.22 (0.09)	0	−2.38	6.92
DC2+_Wet	1103	144 (53)	166 (58)	538 (42)	0.17 (0.06)	0.37 (0.07)	−2.80	4.92

[1] Repeatability is the additive and repeatable variance divided by the total variance.

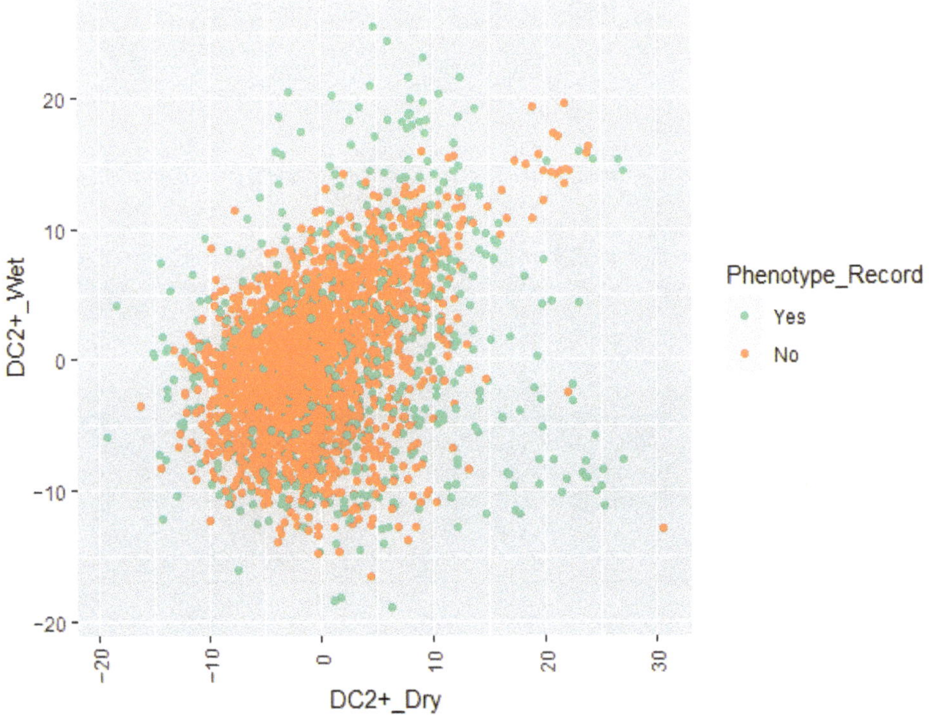

Figure 1. Plot of DC2+_Dry breeding values and DC2+_Wet breeding values.

3.4. Univariate Models of Days to Calving

The heritability for the seven different days to calving traits is presented (Table 10). DC1 had a heritability of 0.20, DC2 had a heritability of 0.18, and DC3+ had a heritability of 0.25. The repeatability for DC3+ is 0.34. DC1 and DC2 do not have repeated records, so they cannot have a repeatability estimate. The heterozygosity coefficient was the lowest for DC2 (−1.07 days/%), with DC3+ having the highest (−4.13 days/%) and the DC1 heterozygosity coefficient being −1.90 days/%. DC1 had the highest residual variance of 1049.22 days^2 and a total variance of 1306.74 days^2. Further models were run with lactating and non-lactating cattle being separated. DC2_Dry heritability 0.22, which is similar to DC2. However, DC2_Wet had a higher heritability (0.39). Conversely, the heterozygosity coefficient in DC2_Dry was higher at −2.40 days/%, whereas DC2_Wet was −0.97 days/%.

Table 10. Variance components, repeatability, heritability and heterozygosity (Het) coefficient of days to calving traits with standard error in parenthesis.

Traits	Number	Repeatability Variance	Additive Variance	Residual Variance	Repeatability	Heritability	Het Coefficient	Response to Selection (Days/Gen)
DC	2184	87 (43)	124 (43)	811 (38)	0.20 (0.05)	0.12 (0.04)	−2.45	4.0
DC1	634	-	257 (122)	1049 (110)	-	0.20 (0.09)	−1.90	7.2
DC2	621	-	192 (90)	886 (89)	-	0.18 (0.08)	−1.07	5.8
DC2_Wet	326	-	353 (180)	551 (143)	-	0.39 (0.12)	−0.97	11.8
DC2_Dry	287	-	211 (124)	736 (117)	-	0.22 (0.18)	−2.40	6.8
DC3+	929	101 (73)	264 (78)	714 (56)	0.34 (0.08)	0.25 (0.06)	−4.13	8.1
DC3+_Wet	841	159 (68)	150 (64)	539 (47)	0.36 (0.08)	0.17 (0.08)	−3.68	5.0

3.5. Response to Selection

The response to selection was calculated for each day to calving trait. The response was highest for second joining days to calving lactating (DC2_wet) at 11.84 days/generation (Table 10), indicating that selection for DC2_Wet in a breeding program will have the biggest impact compared with every other trait. The second highest response was for joining one days to calving (DC1) of 7.21 days/generation (Table 10). The lowest response to selection was for days to calving including all joinings (DC) of 4.00 days/generation. The response to selection for DC3+ was 8.13 days/generation, and DC3+_Wet was 5.03 days/generation (Table 10). The response for DC2+_Wet was 4.92 days/generation, which was lower than DC3+_Wet (Table 9).

Three selection indexes were calculated for three different scenarios (Table 4) using index selection theory. The variance components used for Index one are the DC variance components in Table 10; index two used variance components in Table 11 and index three used variance components in Table 12. Index three had the highest accuracy and response to selection indicating this would be the best scenario to use (Table 13).

Table 11. Variance components estimates used to calculate Index two.

	DC1	DC2+_Wet
DC1	279	-
DC2+_Wet	115	156

Table 12. Variance components estimates used to calculate Index three.

	DC1	DC2_Wet	DC3+_Wet
DC1	332	-	-
DC2_Wet	27	366	-
DC3+_Wet	244	169	250

Table 13. Index values were calculated for three different scenarios using index theory, including breeding objective, the variance of the index, accuracy and response to selection. The response units are days per generation.

	Index One	Index Two	Index Three
Standard deviation of the breeding objective	11.7	12.2	14.8
Standard deviation of the index	5.9	6.4	9.5
Accuracy	0.50	0.52	0.64
Response to selection (days/generation)	2.97	3.32	6.08

Index scenarios are described in Table 4. Variance components used for index calculations can be found in Table 10 for index one, Table 11 for index two and Table 12 for index three.

The effect of additional records was investigated (Figure 2), demonstrating that more records increases accuracy. Using the indexes, showed that treating days to calving as three traits (DC1, DC2_Wet and DC3_Wet) results in the greatest response of 6.08 (index three), with little difference between response for index one and index two (Table 13).

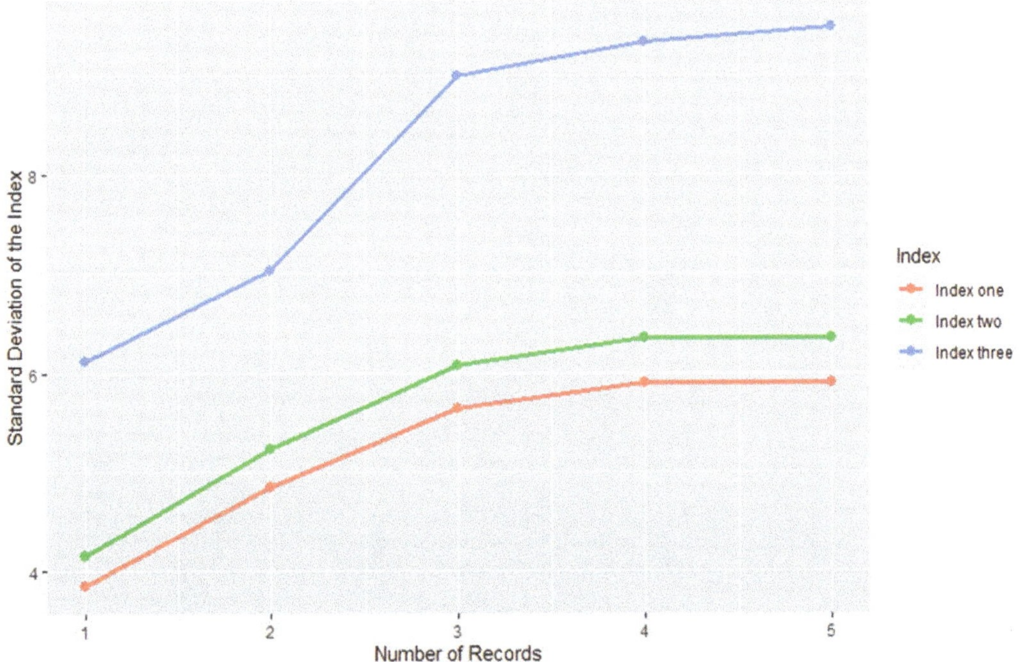

Figure 2. Effects of addition records on the standard deviation of the index. See Table 4 for a description of the index scenarios.

3.6. Bivariate Models for Days to Calving

The genetic correlation between DC1 and DC2 was only −0.06, whereas DC1 had a much higher genetic correlation with DC3+ of 0.83 (Table 14). However, when DC2 was split based on lactation status (lactating or non-lactating), there was a stronger positive correlation of 0.13 between DC1 and DC2_Dry. Conversely, DC2_Wet and DC1 had a large negative genetic correlation of −0.42. Similar changes in genetic correlation occurred when DC3+ was spilt based on lactation status, with the genetic correlation being reduced for DC1 and DC3+_Wet to 0.65. There were insufficient DC3+_Dry records to be able to estimate any genetic correlations. The genetic correlation between DC2 and DC3+ was −0.17. However, when DC2 was split by lactation status, the genetic correlation between DC2_Dry and DC3+ was −0.16, but the genetic correlation was much higher for DC2_Wet at 0.41. The genetic correlation was further increased for DC2_Wet and DC3+_Wet at 0.69. An average of DC3+ was used to calculate phenotypic correlations; the trends were similar to genetic correlations (Table 14). The standard errors for the correlation in the bivariate analysis ranged from 0.20 to 0.48 (Table 14), despite that large standard errors for some DC traits particular between DC1 and DC2_Dry, of the important traits which are DC1, DC2_Wet and DC3+_Wet the standard errors are comparable to previous studies estimates [1,21,22].

Table 14. Correlations between days to calving traits. Genetic correlations below and phenotypic correlations above the diagonal with standard errors in parenthesis.

	DC1	DC2	DC2_Dry	DC2_Wet	DC3+	DC3+_Wet	DC3+_Ave	DC3+_Ave_Wet
DC1	-	−0.22 (0.07)	−0.33 (0.11)	−0.04 (0.10)	NA	NA	0.34 (0.06)	0.31 (0.06)
DC2	−0.06 (0.33)	-	NA	NA	NA	NA	−0.17 (0.06)	−0.03 (0.07)
DC2_Dry	0.13 (0.48)	NA	-	NA	NA	NA	−0.28 (0.07)	−0.23 (0.08)
DC2_Wet	−0.42 (0.37)	NA	NA	-	NA	NA	0.07 (0.10)	0.19 (0.10)
DC3+	0.83 (0.20)	−0.04 (0.24)	−0.16 (0.25)	0.41 (0.29)	-	NA	NA	NA
DC3+_Wet	0.66 (0.26)	−0.16 (0.27)	−0.14 (0.26)	0.69 (0.33)	NA	-	NA	NA
DC3+_Ave	0.57 (0.29)	0.27 (0.32)	−0.10 (0.37)	0.70 (0.34)	NA	NA	-	NA
DC3+_Ave_Wet	0.32 (0.36)	0.46 (0.32)	0.32 (0.40)	0.55 (0.35)	NA	NA	NA	-

3.7. Days to Calving Correlations between Final Traits (Tri-Variate Model)

A final model was run between the traits recommended for breeding programs as a tri-variate model between the final three traits: DC1, DC2_Wet and DC3+_Wet. The genetic correlation between DC1 and DC3+_Wet was 0.85 (Table 15), which is higher than when it was modelled as a bivariate (0.66, Table 14). Similarly, the genetic correlation between DC1 and DC2_Wet in the multi-variate was 0.08 (Table 15), in contrast to the bivariate model (−0.42, Table 14). The genetic correlation between DC2_Wet and DC3+_Wet was 0.56 (Table 15), similar to the bivariate analysis (Table 14). The heritability estimates for the multi-variate analysis were 0.25, 0.40 and 0.30 for DC1, DC2_Wet and DC3+, respectively (Table 15). All heritability estimates from the multi-variate analysis were higher than the univariate analysis (Table 10). A comparison between BLUPs was made between the three traits (DC1, DC2_Wet and DC3+_Wet) and DC (days to calving, including all joinings) (Figure 3). The correlation between DC1 and DC2_Wet BLUPs, was 0.04, indicating that these two traits are uncorrelated and do not have any impact on each other. The correlation between DC1 and DC3+_Wet BLUPs was 0.18, indicating significant re-ranking among the two traits. The correlation between BLUPs was higher between DC1 and DC at 0.55, with the plot showing much less re-ranking than DC3+_Wet BLUPs. The correlation between DC2_Wet and DC3+_Wet was 0.70, with minimal re-ranking of the BLUPs. DC2_Wet had a lower correlation with DC of 0.55, indicating more re-ranking of animals than DC3+_Wet. Finally, the highest correlation was between DC3+_Wet and DC BLUPs at 0.81, implying the lowest amount of re-ranking between traits.

Table 15. Correlation matrix of days to calving traits to be utilised in a breeding program run as a tri-variate. Genetic correlations (below) and heritability estimates (diagonal) with standard error in parenthesis.

	DC1	DC2_Wet	DC3+_Wet
DC1	0.25 (0.09)		
DC2_Wet	0.08 (0.27)	0.40 (0.17)	
DC3+_Wet	0.85 (0.18)	0.56 (0.25)	0.30 (0.06)

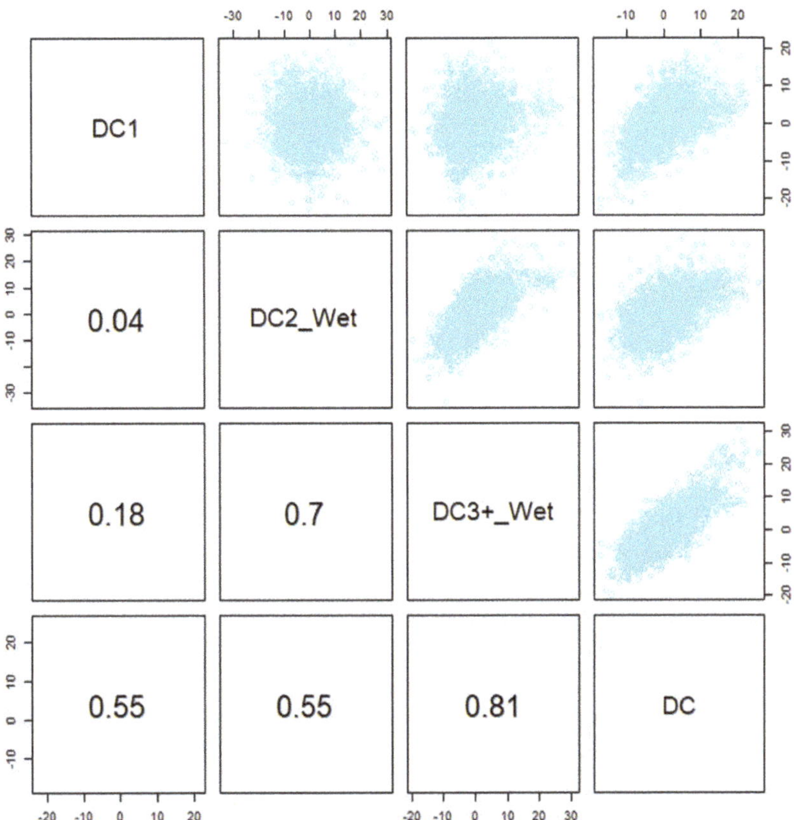

Figure 3. Correlation and plot matrix of breeding values of the final three traits of days to calving that were included in index three and the overall days to calving (DC). Above the diagonal is the plot of breeding values and below is the correlation of those breeding values.

4. Discussion

4.1. Heritability of Days to Calving

A primary breeding objective of Popplewell Composites is to increase lifetime weaning rate. However, it is a complex trait to measure because, by definition, it is only obtained at the end of a cow's breeding life. Days to calving is a good indicator of lifetime weaning rate as it is genetically correlated with lifetime weaning rate and is heritable [1,9,13,23]. Days to calving is a female fertility measure that combines effects such as the age of puberty for first joining, conception rate, how early they conceive in the joining period and gestation length. Female fertility traits are expensive and difficult to measure, and challenging to include in breeding programs due to bias caused by culling cattle that do not conceive. In the breeding program described herein, the strategy is that cattle are expected to wean two calves within the first three matings; those failing to do so are culled from the herd, any further subsequent matings cows are culled if they fail to wean a calf. These culling methods result in missing data as there are many cows without lifetime weaning rate records.

Previous studies in the use of days to calving as a fertility trait proxy, commonly treat days to calving as a single trait including all joinings as a repeated measure. This is the first publication separating both the effect of joining number and lactation status on days to calving. Separating days to calving based on joining number and lactation status into different traits was shown to increase genetic gain over the repeated days to calving trait. The heritability was 0.20, 0.18 and 0.25 for first joining days to calving, second joining days to calving and mature days to calving, respectively (Table 10). It should be noted that mature days to calving includes joining 3, 4 and 5+. This was determined by the genetic correlation calculated from a model that includes a separate variance component for each joining but estimation of the correlation between two joining at a time (Model_2), correlations with the first two joinings were the lowest and joining 3, 4 and 5+ generally had high correlations and combining them reduced the standard error (Table 8). The heritability estimated for traits that separated out first and second joining were higher than the overall days to calving (0.12), demonstrating in this Tropical Composite population that separating joining number will increase the amount of additive variance estimated and reducing bias.

Further analysis was done separating non-lactating cattle; the heritability estimate for these traits was 0.39 for second joining days to calving lactating and 0.17 mature days to calving lactating (Table 9). There is an increase in heritability estimates when days to calving is treated as multiple traits compared with an overall repeatable trait (DC, 0.12), this will allow for greater genetic gain. These increases of heritability indicate modelling across joining number and lactation status is improved and allows for more genetic improvement in a breeding program. These improvements are particularly important to a breeding program for a low heritable and hard to measure trait of high economic value.

4.2. Heifer Days to Calving

The heritability estimate of heifer days to calving (DC1) herein (0.20, Table 10) was similar to those reported by Johnston et al. [1], who estimated the heritability of first joining days to calving to be 0.13 in Tropical Composites and 0.22 in Brahmans. Slightly lower heritability was estimated in Angus cattle at 0.10 [13] and Brahman cattle at 0.09 [24]. In this study, cattle were first joined at 12–15 months; however, in other studies with tropically adapted cattle, heifers were not joined until 24–28 months. Further, Brahman cattle reach puberty at a later age than tropical composites [2,25], further explaining the difference in heritability estimates reported by Johnston et al. [1]. The genetic correlation could explain the higher heritability estimates of DC1 with the age of puberty. Johnston et al. [21] found a genetic correlation between age of puberty and DC1 to be 0.79 in Brahmans, with Johnston et al. [21] estimating a much lower correlation found in tropical composites (0.10); however, this is likely due to the age of puberty difference between the two breed types. The high correlation in Brahmans indicates that DC1 and the age of puberty influencing each other which results from similar biological mechanisms. Johnston et al. [1] also estimated the genetic correlation between DC1 and lifetime weaning rate, which was −0.54 in Brahmans and −0.57 in Tropical Composites. These genetic correlations indicate that DC1 are genetically positively associated with each other. Heifer days to calving is an important trait in the breeding program that is both heritable and positively associated with the age of puberty and lifetime weaning rate.

4.3. Second Joining Days to Calving

Second joining days to calving (DC2) heritability estimates (0.18, Table 10) were similar to those reported by Johnston et al. [1], at 0.17 in Tropical composites and 0.20 in Brahmans. Lower heritability estimates were reported in Angus to be 0.11 [13] and in Brahmans to be 0.15 [24]. The difference in the heritability estimates could be due to breed and age differences. A further model was run for DC2, separating lactating and non-lactating cattle. The heritability estimate of second joining days to calving lactating (DC2_Wet) much higher (0.39) than for DC2 when dry cows were included (Table 10, 0.18). A previous study separated DC2 lactating cows, and the heritability estimates were 0.49 in Brahman and 0.35 in Tropical composites which is similar to results herein for DC2_Wet. [1]. It should be noted that Johnston et al. [1] reported heifers that were not joined until 24–28 months, whereas in this study, heifers were joined at 12–15 months. This increase in heritability when only lactating cattle were included indicates that days to calving should be analysed separately for lactating and non-lactating cattle. Johnston et al. [1] also concluded that heritability estimates for all female fertility traits associated with the second joining were higher in lactating cows than in all cows. Indicating that more genetic progress can be made when only lactating animals are included in female fertility traits.

The difference between DC2_Wet and DC2_Dry heritability representing lactating and non-lactating estimates respectively could be due to post-partum anoestrus from either the threshold energy balance effect or the suckling effect that occurs in *Bos taurus indicus* cattle [26,27]. This post-partum effect prevents the cattle from cycling while lactating, with a previous study stating that in Droughtmaster cattle weaning was required to break anoestrus [28]. Lactating cattle combine growth and lactation in their second joining, which imposes greater energy requirements compared with non-lactating cattle in their second joining. These energy requirements are often not fulfilled when cows graze low-quality pastures, such as in Northern Australia and have a greater detrimental effect on post-partum reproduction in first lactation cows [29]. Johnston et al. [1] estimated the genetic correlation between days to calving second joining and lactation anoestrous interval to be 0.70 in Brahmans and 0.67 in Tropical composites. These high genetic correlations indicate that similar genes control both days to calving second joining and lactation anoestrus interval. Further, Johnston et al. [1] estimated genetic correlations between second joining days to calving and lifetime weaning rate to be −0.96 in Brahmans and −0.76 in Tropical composites. The high genetic correlations indicate that second joining days to calving is a good indicator of lifetime fertility.

4.4. Difference between Lactating and Non-Lactating Cows in Days to Calving

Lactating cows will often lose more body condition than non-lactation cows due to the energy cost of lactation and the low digestibility of tropical pasture [30,31]. Neville [31] calculated that lactating beef cattle require 38–41% more energy for maintenance compared to non-lactating. In addition to the extra energy requirements, in *Bos taurus indicus* cattle, an additional post-partum anoestrus effect is associated with suckling or the close presence of offspring [26,27]. Consequently, a model was fitted for days to calving for the animals second joining onwards that treated days to calving in lactating and non-lactating cows as separate traits. There was a low negative correlation (−0.10) between wet and non-lactating cows, demonstrating that lactating and non-lactating days to calving records are genetically different, and they were, therefore, analysed as separate traits. The low correlations could be driven by post-partum anoestrus. The post-partum anoestrus effect could be due to the calf's suckling effect and the threshold of energy balance.

Energy balance is important in lactating cattle and can affect their post-partum anoestrus interval, causing the genetic difference between lactating and non-lactating days to calving. Therefore, lactating cows that keep getting back into calf (lower days to calving records) also have a lower threshold energy balance meaning they start cycling earlier than the lactating cows that struggle to get pregnant (larger days to calving records). Wolcott et al. [32] found in tropical composite cows at their second joining, lactating cows had significantly lower weight and body condition than non-lactating cows. Therefore, lactating and non-lactating cows, particularly in their second joining, have different requirements affecting post-partum anoestrus, re-enforcing the importance of treating them as different traits. In addition to the energy threshold balance, a suckling effect also contributes to post-partum anoestrus in *Bos indicus* cattle [26,27]. As this composite population includes *Bos taurus indicus* cattle, post-partum anoestrus could be due both to energy balance and a suckling effect, causing lactating days to calving to be genetically different from non-lactating days to calving.

Cattle with longer post-partum anoestrus intervals spend longer periods not cycling, including when they are lactating, which results in the cattle not conceiving in the joining period. These cattle will then become non-lactating in the next joining season. However, the most profitable system is those with more cows lactating yearly. Post-partum anoestrus was highly heritable, with estimates ranging from 0.42–0.51 in Brahmans and 0.26–0.63 in tropical composites [1,33]. It has been reported to have a high genetic correlation with days to calving [33]. This demonstrates that the genetic difference between lactating and non-lactating cattle could be due to the post-partum anoestrus interval, providing further evidence that days to calving should be treated as separate traits for lactating (wet) and non-lactating (dry).

It is important to also look at the effect of the breeding values and the ranking of individual animals. A breeding value comparison between DC2+_Dry and DC2+Wet (Figure 1, r = 0.32), showed significant re-ranking of animals. These re-ranking of animals provide further evidence that lactating and non-lactating days to calving should be treated as separate traits, with lactating days to calving the main trait of interest.

Most breeding programs aim to improve the lifetime weaning rate, meaning more calves on the ground. Therefore, lactating days to calving is important in improving the breeding objective, representing lactating cows getting back into calf. Despite lactating days to calving being important, non-lactating days to calving does not provide information that would help improve the breeding objective. Treating lactating and non-lactating cows as two distinct traits would benefit the producer as the most profitable system will be those with more cows lactating yearly.

4.5. Genetic and Phenotypic Correlation for Days to Calving

4.5.1. DC1 and DC2 Correlations

The genetic and phenotypic correlation between DC1 and DC2 was −0.06 and −0.22, respectively (Table 14). Johnston et al. [1] estimated the genetic correlation between joining one and two for days to calving to be 0.55; however, these were mated at 24–28 months of age and will have different growth requirements compared to the current study. Here, further analysis was conducted by splitting DC2 based on lactation status (lactating/non-lactating). DC2_Dry had a positive low (0.13) genetic correlation and a negative phenotypic correlation with DC1 (Table 14). In contrast, DC2_Wet had a moderate negative genetic correlation of −0.42 and a phenotypic correlation of −0.04 (Table 14). There are two reasons for these genetic correlations, post-partum anoestrous in lactating cows and age of puberty in heifers. The genetic correlation between DC1 and DC2_Wet indicates that DC1 is both a measure of fertility and puberty whereas DC2_Wet is a measure of fertility and post-partum anoestrus. The genetic correlation estimates between DC1 and DC2 (including DC2_Wet and DC2_Dry) would suggest that treating them as distinct traits is essential.

4.5.2. DC1 and DC3+ Correlations

The genetic correlation between DC1 and DC3+_Wet was 0.65 (Table 14); thus, genetic gain in one trait would result in genetic gains in the other. As DC1 is an early in-life trait, selecting for DC1 would allow for genetic improvement in DC3+_Wet earlier, increasing the overall production of the breeding program. Indicating that DC1 is a good indicator of lifetime female fertility. The high genetic correlation could be caused by two factors, post-partum anoestrus in mature cows or the age of puberty in heifers [1,26,27]. Post-partum anoestrus has a greater effect on primiparous cows than in mature cows; hence, the genetic correlation is more favourable between DC1 and DC3+_Wet than with DC1 and DC2_Wet [29]. The genetic and phenotypic correlations would indicate that DC1 is a good indicator for DC3+_Wet and allows for selection earlier in life.

4.5.3. DC2 and DC3+ Correlations

The genetic correlation between DC2 and DC3+ was low and negative (Table 14); similar correlations were found between DC2_Dry with DC3+ and DC3+_Wet. These low negative correlations indicate that second joining non-lactating animals would not improve DC3+ and, therefore, are not a good measurement of lifetime fertility. Second joining non-lactating animals had lower days to calving values in the second joining compared with DC3+; this could be due to the effect of post-partum anoestrus. Therefore, DC2_Dry is not a good indicator of mature (joining three onwards) days to calving and thus cannot be used to increase genetic gain. Furthermore, DC2_Dry is not a trait of interest in the breeding program as the most profitable system is those with more lactating cows yearly.

A different trend was found in DC2_Wet, which was moderate to highly genetically correlated with DC3+ (0.41) and DC3+_Wet (0.69) (Table 14). The higher correlations between second joining lactating days to calving and mature joining traits indicate that DC2_Wet is a good representation of how an animal will perform over its lifetime. Wolcott et al. [32] found that cattle, particularly in their second joining, have different requirements due to post-partum anoestrus, which could explain the genetic correlation between DC2_Wet and DC3+_Wet. The genetic correlations estimated between DC2_Wet and DC3+_Wet indicate they are highly correlated; however, they should be treated as separate traits. This would enable early selection while improving lifetime fertility and maximising genetic gain. However, DC2_Dry and DC3+_Dry should be excluded from the analysis and values not be reported back to the breeder as these are not the traits of interest as it is more profitable to have more lactating cows every year.

4.6. Recommended Days to Calving Traits

Response to selection was calculated for each individual trait. The response to selection was calculated over one generation with a selection intensity of one. The trait that with the most negligible response to selection was the overall days to calving trait (Table 10), indicating that this is the least desirable days to calving trait. The trait with the smallest response to selection is the trait that will make the least amount of genetic gain. As the overall days to calving trait had the lowest response to selection, separating traits based on lactation status and joining number will increase genetic gain. Days to calving is a complex trait combining many different biological effects so consideration is needed for modelling this trait to maximise response to selection. Selection indexes are important to consider when developing traits as they will account for genetic correlations and how one trait will affect the other.

Three different indexes were calculated based on three different scenarios (Table 4). Index three (which included DC1, DC2_Wet and DC3+_Wet) had the greatest response of 6.08 days/generation, indicating that it is the best scenario or the scenario that will make the most genetic gain in this population. Index two (which included DC1, DC2+_Wet) had the second highest response to selection (3.32 days/generation); this demonstrates that separating the first joining will improve genetic gain. Unsurprisingly, the index with the lowest response to selection was index one or treating days to calving as one trait, further demonstrating that it is important to treat days to calving as separate traits based on lactation and joining to maximise genetic gain and productivity of the breeding program. These response calculations need to be taken into consideration when making recommendations on the breeding program.

Considering index theory and heritability estimates, treating days to calving as three separate traits (DC1, DC2_Wet and DC3+_Wet) compared to just one days to calving trait will result in the greatest genetic gain in this population (Table 13). A further multi-variate model was fitted with DC1, DC2_Wet and DC3+_Wet. The multi-variate model resulted in a different genetic correlation compared with the bivariate model between DC1 and DC2_Wet (Table 15). The heritability of these traits was 0.25, 0.40 and 0.30 from DC1, DC2_Wet and DC3+_Wet (Table 15). The genetic correlation between these traits also indicates that they should be treated as separate traits. Despite these results, treating days to calving as two traits still resulted in increased genetic gain compared with treating it as a single trait (Figure 2); therefore, if the dataset is small, it is recommended to use DC1 and DC2+_Wet to potentially reduce the amount of error. Implementing the separation of days to calving as three traits in this population will enable greater genetic improvement in female fertility, resulting in higher production.

4.7. Heterosis Effect on Days to Calving

Heterosis is the phenomenon that occurs when two genetically different breeds are crossed and produce offspring that outperforms the midpoint of their parents [34,35]. Heterosis tends to have a bigger impact on traits that are lowly heritable such as fertility [34,36].

Retained heterosis, is essential to the breeding program as composite breeding exploits heterosis without further crossing different breeds [34]. Herein heterosis as estimated from heterozygosity had a positive effect on fertility traits by decreasing days to calving in all models (−0.97 to −4.13 days/%, Table 10). It has been reported that heterozygosity fraction is positively and linearly related to cow fertility and lifetime productivity and can be used to optimise heterosis in a composite population [34,37]. The biggest impact of heterozygosity was on DC3+ where a 1% increase was associated with a 4.31 reduction in days to calving. Early in life days to calving measurements, DC1 and DC2 both had lower heterozygosity coefficient −1.90 and −1.07 days per percent increase in heterosis, respectively (Table 10). DC3+ having the highest heterosis coefficient could be due to the higher number of records (957) compared to DC1 and DC2 (648 and 636, respectively). Heterosis can increase the weight of calf weaned per cow exposed by 50% or more in *Bos taurus taurus* and *Bos taurus indicus* crosses, increasing the production of the breeding program [38]. Therefore, utilising heterozygosity coefficient estimates should improve the breeding program's overall fertility in composite populations.

4.8. Comparison of the Penalty Value

The penalty value commonly used for days to calving is to add 21 days to the largest value in the join group [10]. However, due to the breed type of these composite cattle and their environment, 32 days was used as the penalty value. An additional analysis was conducted to see the effects of changing the penalty value on the days to calving model. DC1 and DC2+ models were run using no penalty value, 21, 32, 43, 63 and 252 days. In both DC1 and DC2+, not using a penalty value affected the heritability estimate (Tables 5 and 6). DC1 with no penalty has a lower heritability compared to using any penalty value; this could be due to the smaller number of records. However, DC2+ with no penalty had a larger heritability compared to using any penalty value. This demonstrates the importance of including all joining data, as it will affect heritability estimates. Despite this, when a penalty value was applied to cows that were not pregnant (empty cows), there was little difference in the heritability estimate no matter the value for both DC1 and DC2+; however, there seemed to be a scaling effect due to the penalty value (Tables 5 and 6). Furthermore, the BLUPs between the models were all highly correlated (>0.90). These results demonstrate that the penalty value chosen does not affect the heritability and BLUPs; therefore, the value chosen for the penalty is a matter of convenience.

5. Conclusions

Days to calving is a trait used for genetic improvement of weaning rate and calving time. For the breeding objective of minimising days to calving in five joinings, in this population it is essential to estimate breeding values for three component traits: first joining (heifer) days to calving, second joining days to calving lactating and mature days to calving lactating. The results for heterozygosity fraction supported this. Selecting these three traits will allow for greater genetic gain in fertility, thus increasing production. As a commercial dataset was used, most estimates have large standard errors and further investigation using a larger dataset are required to support the results herein for further use in the industry. Despite the large standard errors, the results are comparable with published literature. It is important to use a penalty value for cows that are not pregnant as it allows for more complete records; however, the actual value of the penalty added does not affect heritability or BLUP estimates. As this is the first-time days to calving has been modelled this way and a commercial dataset was used, further study is required to confirm the genetic and phenotypic correlations between days to calving (separated into three traits, DC1, DC2_Wet and DC3+_Wet), particularly in other breeds. The relationship with production and days to calving traits should be investigated in composite breeds to allow for no detrimental effects.

Author Contributions: Conceptualisation, M.L.F. and W.S.P.; Methodology, M.L.F., R.A.M. and W.S.P.; Software, M.L.F., R.A.M. and H.O.; Validation, M.L.F.; Formal Analysis, M.L.F.; Investigation, M.L.F.; Resources, W.S.P.; Data Curation, M.L.F.; Writing—Original Draft Preparation, M.L.F.; Writing—Review and Editing, M.L.F., R.A.M., M.L.H., H.O. and W.S.P.; Visualization, M.L.F.; Supervision, H.O, M.L.H. and W.S.P.; Project Administration, W.S.P.; Funding Acquisition, W.S.P. All authors have read and agreed to the published version of the manuscript.

Funding: M.L.F. was supported by scholarships from the University of Adelaide Faculty of Sciences Divisional Scholarship and Popplewell Composites.

Institutional Review Board Statement: Ethical review and approval were not needed for this study as a commercial dataset followed standard animal farm practices.

Data Availability Statement: Third-Party Data. Restrictions apply to the availability of these data. Data were obtained from Popplewell Composites Pty Ltd. and are available with the permission of Greg Popplewell.

Acknowledgments: The Authors gratefully acknowledge the contributions of Greg Popplewell and all farm and technical staff involved in farm management practices and data collection.

Conflicts of Interest: The authors declare no conflict of interest.

References

1. Johnston, D.J.; Barwick, S.A.; Fordyce, G.; Holroyd, R.G.; Williams, P.J.; Corbet, N.J.; Grant, T. Genetics of early and lifetime annual reproductive performance in cows of two tropical beef genotypes in northern Australia. *Anim. Prod. Sci.* **2014**, *54*, 1–15. [CrossRef]
2. Chilcott, C.; Ash, A.; Lehnert, S.; Stokes, C.; Charmley, E.; Collins, K.; Pavey, C.; Macintosh, A.; Simpson, A.; Berglas, R. *Northern Australia Beef Situation Analysis: A Report to the Cooperative Research Centre for Developing Northern Australia*; CRC for Developing Northern Australia (CRCNA): Townsville, QLD, Australia, 2020.
3. Abeygunawardena, H.; Dematawewa, C. Pre-pubertal and postpartum anestrus in tropical Zebu cattle. *Anim. Reprod. Sci.* **2004**, *82*, 373–387. [CrossRef] [PubMed]
4. Frisch, J.; Munro, R.; O'Neill, C. Some factors related to calf crops of Brahman, Brahman crossbred and Hereford × Shorthorn cows in a stressful tropical environment. *Anim. Reprod. Sci.* **1987**, *15*, 1–26. [CrossRef]
5. Schatz, T.J.; Hearnden, M.N. Heifer fertility on commercial cattle properties in the Northern Territory. *Aust. J. Exp. Agric.* **2008**, *48*, 940. [CrossRef]
6. Pravia, M.I.; Ravagnolo, O.; Urioste, J.I.; Garrick, D.J. Identification of breeding objectives using a bioeconomic model for a beef cattle production system in Uruguay. *Livest. Sci.* **2014**, *160*, 21–28. [CrossRef]
7. Naya, H.; Peñagaricano, F.; Urioste, J.I. Modelling female fertility traits in beef cattle using linear and non-linear models. *J. Anim. Breed. Genet.* **2017**, *134*, 202–212. [CrossRef]
8. Thundathil, J.C.; Dance, A.L.; Kastelic, J.P. Fertility management of bulls to improve beef cattle productivity. *Theriogenology* **2016**, *86*, 397–405. [CrossRef]
9. Meyer, K.; Hammond, K.; Parnell, P.F.; Mackinnon, M.J.; Sivarajasingam, S. Estimates of heritability and repeatability for reproductive traits in Australian beef cattle. *Livest. Prod. Sci.* **1990**, *25*, 15–30. [CrossRef]
10. Graser, H.-U.; Tier, B.; Johnston, D.J.; Barwick, S.A. Genetic evaluation for the beef industry in Australia. *Aust. J. Exp. Agric.* **2005**, *45*, 913–921. [CrossRef]
11. Cammack, K.M.; Thomas, M.G.; Enns, R.M. Reproductive Traits and Their Heritabilities in Beef Cattle. *Prof. Anim. Sci.* **2009**, *25*, 517–528. [CrossRef]
12. Mucari, T.B.; Alencar, M.M.D.; Barbosa, P.F.; Barbosa, R.T. Genetic analyses of days to calving and their relationships with other traits in a Canchim cattle herd. *Genet. Mol. Biol.* **2007**, *30*, 1070–1076. [CrossRef]
13. Johnston, D.J.; Bunter, K.L. Days to calving in Angus cattle: Genetic and environmental effects, and covariances with other traits. *Livest. Prod. Sci.* **1996**, *45*, 13–22. [CrossRef]
14. Sargolzaei, M.; Chesnais, J.P.; Schenkel, F. FImpute-An efficient imputation algorithm for dairy cattle populations. *J. Dairy Sci.* **2011**, *94*, 421.
15. Vanraden, P.M. Efficient Methods to Compute Genomic Predictions. *J. Dairy Sci.* **2008**, *91*, 4414–4423. [CrossRef] [PubMed]
16. Facy, M.L.; Hebart, M.L.; Oakey, H.; McEwin, R.A.; Popplewell, G.I.; Pitchford, W.S. Evaluation of dominance in tropically adapted composite beef cattle. *Anim. Prod. Sci.* **2021**, *61*, 1811–1817. [CrossRef]
17. RStudio. *RStudio: Integrated Development Environment for R, 2022.7.1.554*; RStudio, PBC: Boston, MA, USA, 2022.
18. Butler, D.; Cullis, B.; Gilmour, A.; Gogel, B.; Thompson, R. *ASReml-R Reference Manual Version 4*; VSN International Ltd.: Hemel Hempstead, UK, 2017.
19. Hazel, L.N. The genetic basis for constructing selection indexes. *Genetics* **1943**, *28*, 476–490. [CrossRef] [PubMed]
20. Cameron, N. *Selection Indices and Prediction of Genetic Merit in Animal Breeding*; CAB International: Wallingford, UK, 1997.

21. Johnston, D.J.; Corbet, N.J.; Barwick, S.A.; Wolcott, M.L.; Holroyd, R.G. Genetic correlations of young bull reproductive traits and heifer puberty traits with female reproductive performance in two tropical beef genotypes in northern Australia. *Anim. Prod. Sci.* **2014**, *54*, 74–84. [CrossRef]
22. Wolcott, M.L.; Johnston, D.J.; Barwick, S.A.; Corbet, N.J.; Williams, P.J. The genetics of cow growth and body composition at first calving in two tropical beef genotypes. *Anim. Prod. Sci.* **2014**, *54*, 37. [CrossRef]
23. Johnston, D. Genetic Improvement of Reproduction in Beef Cattle. In Proceedings of the 10th World Congress of Genetics Applied to Livestock Production, Vancouver, BC, Canada, 17–22 August 2014; pp. 17–22.
24. Johnston, D.J.; Moore, K.L. Genetic Correlations between Days to Calving and Other Male and Female Reproductive Traits in Brahman Cattle. In Proceedings of the Association Advancement of Animal Breeding and Genetics, Armidale, NSW, Australia, 27 October–1 November 2019; pp. 354–357.
25. Johnston, D.J.; Barwick, S.A.; Corbet, N.J.; Fordyce, G.; Holroyd, R.G.; Williams, P.J.; Burrow, H.M. Genetics of heifer puberty in two tropical beef genotypes in northern Australia and associations with heifer- and steer-production traits. *Anim. Prod. Sci.* **2009**, *49*, 399. [CrossRef]
26. Entwhistle, K. *Factors Influencing Reproduction in Beef Cattle in Australia*; AMRC Review-Australian Meat Research Committee: Sydney, NSW, Australia, 1983.
27. Montiel, F.; Ahuja, C. Body condition and suckling as factors influencing the duration of postpartum anestrus in cattle: A review. *Anim. Reprod. Sci.* **2005**, *85*, 1–26. [CrossRef]
28. McSweeney, C.; Kennedy, P.; D'Occhio, M.; Fitzpatrick, L.; Reid, D.; Entwistle, K. Reducing post-partum anoestrous interval in first-calf *Bos indicus* crossbred beef heifers. 2. Responses to weaning and supplementation. *Aust. J. Agric. Res.* **1993**, *44*, 1079–1092. [CrossRef]
29. Ciccioli, N.H.; Wettemann, R.P.; Spicer, L.J.; Lents, C.A.; White, F.J.; Keisler, D.H. Influence of body condition at calving and postpartum nutrition on endocrine function and reproductive performance of primiparous beef cows. *J. Anim. Sci.* **2003**, *81*, 3107–3120. [CrossRef] [PubMed]
30. Rudder, T.; Seifert, G.; Burrow, H. Environmental and genotype effects on fertility in a commercial beef herd in central Queensland. *Aust. J. Exp. Agric.* **1985**, *25*, 489. [CrossRef]
31. Neville, W.E., Jr. Comparison of Energy Requirements of Non-Lactating and Lactating Hereford Cows and Estimates of Energetic Efficiency of Milk Production. *J. Anim. Sci.* **1974**, *38*, 681–686. [CrossRef] [PubMed]
32. Wolcott, M.L.; Johnston, D.J.; Barwick, S.A. Genetic relationships of female reproduction with growth, body composition, maternal weaning weight and tropical adaptation in two tropical beef genotypes. *Anim. Prod. Sci.* **2014**, *54*, 60. [CrossRef]
33. Olasege, B.S.; Tahir, M.S.; Gouveia, G.C.; Kour, J.; Porto-Neto, L.R.; Hayes, B.J.; Fortes, M.R.S. Genetic parameter estimates for male and female fertility traits using genomic data to improve fertility in Australian beef cattle. *Anim. Prod. Sci.* **2021**, *61*, 1863–1872. [CrossRef]
34. Bourdon, R. *Understanding Animal Breeding*, 2nd ed.; Prentice Hall: Upper Saddle River, NJ, USA, 2014.
35. Hallauer, A.R.; Miranda Filho, J.B. *Quantitative Genetics in Maize Breeding*, 1st ed.; Iowa State University Press: Ames, IA, USA, 1981.
36. Birchler, J.A.; Yao, H.; Chudalayandi, S. Unraveling the Genetic Basis of Hybrid Vigor. *Proc. Natl. Acad. Sci. USA* **2006**, *103*, 12957–12958. [CrossRef]
37. Basarab, J.A.; Crowley, J.J.; Abo-Ismail, M.K.; Manafiazar, G.M.; Akanno, E.C.; Baron, V.S.; Plastow, G. Genomic retained heterosis effects on fertility and lifetime productivity in beef heifers. *Can. J. Anim. Sci.* **2018**, *98*, 642–655. [CrossRef]
38. Gregory, K.E.; Cundiff, L.V. Crossbreeding in Beef Cattle: Evaluation of Systems1. *J. Anim. Sci.* **1980**, *51*, 1224–1242. [CrossRef]

 agriculture

MDPI

Article

Evolution of Genetics Organisations' Strategies through the Implementation of Genomic Selection: Learnings and Prospects

Robert Banks

Animal Genetics and Breeding Unit, University of New England, Armidale, NSW 2350, Australia; rbanks@une.edu.au

Abstract: Since its initial description in 2001, and with falling costs of genotyping, genomic selection has been implemented in a wide range of species. Theory predicts that the genomic selection approach to genetic improvement offers scope both for faster progress and the opportunity to make change in traits formerly less tractable to selection (hard-to-measure traits). This paper reports a survey of organisations involved in genetic improvement, across species, countries, and roles both public and private. While there are differences across organisations in what have been the most significant outcomes to date, both the increased accuracy of breeding values that underpins potentially faster progress, and the re-balancing of genetic change to include real progress in the hard-to-measure traits, have been widely observed. Across organisations, learnings have included the increasing importance of investment in phenotyping, and opportunities to evolve business models to engage more directly with a wider range of stakeholders. Genomic selection can be considered a more modular approach to genetic improvement, and its simplicity and effectiveness can transform both genetic improvement and the effectiveness of multi-disciplinary approaches to improving livestock and plant production, enabling potentially very significant increases in agricultural productivity, profitability and sustainability.

Keywords: genomic selection; implementation; strategy

Citation: Banks, R. Evolution of Genetics Organisations' Strategies through the Implementation of Genomic Selection: Learnings and Prospects. *Agriculture* **2022**, *12*, 1524. https://doi.org/10.3390/agriculture12101524

Academic Editor: Aimin Zhang

Received: 22 August 2022
Accepted: 20 September 2022
Published: 22 September 2022

1. Introduction

The core tools, or knowledge domains, deployed in all agriculture are genetics, nutrition, and management (including health considerations). To a large extent, Research and Development (R&D) and implementation have largely tended to proceed within these domains, albeit with often-stated goals of integration. The domains draw on separate types of knowledge, and this fact, coupled with limitations in scale and complexity of experimental design (reflecting the fact that R&D depends on investment of scarce funds) has tended to maintain the separation. Recent significant development in one area of measurement or data capture has transformed one of the three domains, and has begun to change how R&D can and does address the three domains and their integration. That development is the introduction of techniques for "reading" the genetic makeup of individuals–known as genotyping.

This paper explores how genotyping has been utilized from two perspectives: firstly, within the domain of genetics, and secondly, the way in which technology is changing how agricultural R&D is done, and what scope it can address. The introduction briefly describes the activity of genetics R&D and implementation, then outlines how potential applications of genotyping have been investigated, and the general principles of deployment. The central approach of this paper is then outlined: through structured interview with a range of practitioners of genetic improvement, common themes emerging through deployment in a range of species and for a range of goals are identified together with how strategies for genetic improvement are evolving. The paper then draws on the insights from those interviews and published results to consider how deployment of genotyping is changing agricultural R&D and its implementation.

The starting point for this approach is the genetics domain. In simple terms genetics in relation to agriculture is concerned with two effects: how choice of animals or plants from which to breed affects current performance, and how that choice affects performance in subsequent generations–this latter usually termed genetic improvement. Both choices (they can be applied in a single combined decision process) depend on identifying individuals with more favourable genetic make-up for some goal or goals. This simple description captures the "algorithm" of genetic improvement:

- define the goal–what aspects of performance or traits are important, and how we wish the change any or all of them: referred to as defining the breeding objective
- identify the individuals with superior genetic makeup for the breeding objective (i.e., for the traits in the objective): referred to as predicting breeding value
- selecting the individuals with superior genetic makeup and mating them in an optimal way to produce progeny

Harris and Newman [1] provide a comprehensive expansion of this algorithm.

While defining the breeding objective is the starting point of this approach, the second step has occupied the bulk of technical attention during the development of the theory and practice of genetic improvement over the last 7–8 decades. Individuals' genetic makeup, or merit, can be estimated from using two types of "clues" [for the description "clues", Brian Kinghorn, pers. comm.]–the relationships amongst them, based on known pedigree, and their individual performance, or phenotype, for traits of interest. Prior to 2001, the two significant developments in the theory of estimating genetic merit were selection index [2] and best linear unbiased prediction, or BLUP [3]. Both methods use phenotypic data, and where known, pedigree. BLUP offered advantages over selection index in accounting for genetic change through time, enabling use of heterogeneous data (both in phenotypes available per individual, and in pedigree relationships), and given appropriate relationship structure, unbiased estimation across multiple breeding units (such as flocks or herds, but extending to countries). BLUP methods were introduced into breeding programs in most livestock species in western countries from the 1980s, assisted by the increasing availability and price attractiveness of large-scale computing. Almost from the start of implementation, rates of genetic progress as estimated from the data increased in all species where BLUP methods were implemented (this is not to imply that progress had not been made prior to use of BLUP methods, but that it accelerated markedly).

The power of selection index and BLUP approaches were enhanced by the development of methods for utilisation of DNA data (genotypes), marked by the publication of the concept of genomic selection [4]. It is no exaggeration to say that this paper marks the start of a new, in some ways revolutionary phase in genetic improvement. Such description can be unfortunate where reality does not match "hype", but the method has indeed transformed approaches to genetic improvement wherever applied. Meuwissen et al. [4] outlined the basic logic of the approach; Goddard and Hayes [5] provide clear explanation of the application. The present paper does not go into technical detail, but a brief description of the basic elements, and summary of aspects of the research that has grown from the initial paper, will underpin the focus here.

Genomic selection depends on the ability to "read" the DNA of organisms–identifying the nucleotides present in an animal at known locations in the genome. The locations being read are referred to as Single Nucleotide Polymorphisms (or SNPs), and the ability to read SNP genotypes has grown exponentially in the last two decades, measured as the price per SNP. The "reads" at each SNP location together comprise the genotype of the individual, and genotype maps now routinely include tens of thousands of SNPs, up to full DNA base sequence.

The second essential requirement for genomic selection is a sample of individuals from the population of interest that have been genotyped, and on which some measure or assessment of performance has been collected. The measure may be a trait(s) measured on the individual itself (its phenotype) or trait(s) measured on known relatives of the individuals–for example, progeny groups. This group of measured and genotyped individuals is known as the genomic reference population.

With a genomic reference population established, it then becomes possible to genotype other members of the population, and look for similarity between their genotypes and those of the genomic reference population, and from that similarity infer the value of the genes of those individuals. To make the point very clear: the genomic reference is individuals with both known genotypes and measured phenotypes, and the remainder on the population (non-reference) can then be evaluated genetically from their genotype alone.

The size, and to a lesser extent the design, of the genomic reference population, is crucial in determining the accuracy with which genetic merit can be estimated for animals without phenotypes. This topic will be returned to later, but the key points in relation to that accuracy were covered by Goddard and Hayes [5]:

- accuracy of estimation of genetic merit (or breeding value) increases with increasing size of the reference population, but with diminishing returns to scale,
- accuracy of estimation of genetic merit (breeding value) is higher for traits with higher heritability,
- accuracy is affected by the level of relatedness of individuals within the reference population and between the reference population and the rest of the population–higher relatedness enabling higher accuracy, and
- the reference population needs to be maintained in terms of its relatedness to the current active population–for a population undergoing selection, the accuracy available from the reference population declines over time [6,7] (noting that drift and mutation will also contribute to reduction in relatedness over time).

Meuwissen et al. [8] set out the two opportunities presented by genomic selection:

- the accuracy of prediction of genetic merit that is possible, including for individuals with and without phenotypes, with tincreased accuracy providing scope for more rapid genetic improvement, and
- the decoupling of measurement and selection. This provides two sorts of opportunities:
    - to evaluate large numbers of individuals at the price of genotyping alone
    - to evaluate selection candidates for traits that cannot be measured on live animals or, or can only be measured some time after selection would ideally occur

Together these opportunities mean that genomic selection can be deployed to achieve faster genetic change in individual traits, genetic change in more traits simultaneously, or the two together, and enable screening of commercial populations to allocate individuals to their most appropriate management regimes and/or target markets.

At the time of the original paper [4], genotyping was not yet cheap enough for widespread implementation of genomic selection. Through the 2000s and beyond about 2008, this began to change rapidly, with the earliest wide-scale applications in dairy cattle. A major attraction there was in using genomic selection to evaluate young bulls, potentially replacing the longer and more expensive progeny testing systems that had been the basis of dairy cattle genetic improvement since the mid-1950s. Shaeffer [9] highlighted the potential cost-savings (92%) and increases in rate of progress (doubling) available, and almost immediately, dairy cattle breeding programs in the developed world switched to genomic selection. Since then, applications in other species has grown rapidly.

The concept of genomic selection has been termed a paradigm shift [8]. "Paradigm shift" refers primarily to the decoupling of trait recording and selection. The impact of the concept on the research community is evident in a Google count for the term "genomic selection"—163 million hits at 20 April 2022, compared to 2.4 million hits for BLUP at the same date. Similarly, the original paper now has nearly 5000 citations, compared to 1676 citations for the Henderson paper outlining BLUP [3]. These comparisons are not intended to imply relative merit–simply the level of engagement that genomic selection has generated.

This paper explores how a sample of practitioners have implemented genomic selection, what has been learnt to date, and how breeding strategies have evolved. A brief overview of areas of research and development around genomics and in particular genomic selection is provided as background or context for focus on implementation.

Firstly, there have been a number of useful collections of articles discussing genomic selection, with that in Animal Frontiers (6(1), 2016) providing an excellent starting point. Recent proceedings of the World Congress of Genetics Applied to Livestock Production [10,11] extend the coverage enormously–over 1600 papers across two conferences, with a high proportion in some way utilizing or addressing genomics. An excellent review of the status of genomic selection covering a range of theoretical and implementation issues is Misztal et al. [12], and Verbyla et al. [13] provides an excellent outline of the steps involved in implementation of genomic selection at commercial scale, including the R&D and application phases.

Secondly, areas of research and publication include:

- Methods and tools for analysis of genomic data, both for genomic selection and more broadly for exploration of the genomic architecture of traits
- Theoretical and more recently experimental investigation of changes at the gene or SNP level under genomic selection
- Theory and tools for use of genomic data jointly for selection and management of genetic variation
- Design of genomic reference populations on both within- and across-population basis (where population can be breed, region or country of origin)
- Pre- and post-implementation scoping and reporting of implementation in a wide range of species, including animals (terrestrial and aquatic), plants, insects, and microorganisms
- Traits, and methods of phenotyping, including production (both output and input), product quality, health- and disease-related, metabolic, and behavioural traits
- and Evaluation of breeding programs for effectiveness and economic efficiency, including for economically and/or environmentally challenging situations

Thirdly, and of relevance to the broader agriculture context within which genomic selection is practiced and genomic methods are used in research, studies of genotype-by-environment interaction, and more broadly of how genetic and non-genetic (e.g., nutritional, or management) choices work together in value chains, e.g., [14]. These latter reflect a paradigm shift additional to that noted by Meuwissen et al. [8]: as genotyping becomes more and more cost-effective, and as genomic reference populations are built using a mix of specific-purpose populations and field data, genotyping can describe the genetic makeup of recorded populations where there are identifiable environmental or treatment differences, which enables greatly increased scale in experiments in fields where traditionally experimental size has often been limited.

2. Materials and Methods

The foregoing is intended as a general introduction to genomic selection. Of potential interest both to practitioners across species, and to a wider, non-geneticist audience, are the questions:

- how has genomic selection been implemented in practice?
- with what results so far and what impacts on how genetic improvement is being practiced?

A survey was conducted to help address questions related to genomic selection that are of interest to plant and animal breeders as well as to a wider, non-geneticist audience.

The broad questions were examined via structured interviews with a diverse sample of practitioners. Twenty-three different organisations are represented, and all interviewees provided a mix of written and verbal response to the following questions. The organisations were chosen to represent four broad roles within genetic improvement, and from the author's experience of publication by the organisations. They reflect a range of species and countries, but to be clear, all have implemented genomic selection to some extent.

For the Results and Discussion, the interviewees have been grouped. The grouping outlined is somewhat arbitrary, but reflects diversity in two dimensions of enterprises or organisations involved in genetic improvement:

- Whether single enterprise or business, or including or servicing multiple enterprises or businesses. For example, an individual beef cattle stud is likely to be smaller in economic scale (e.g., turnover) than a breeding company, even though the number of animals (females) in the breeding population being managed may be similar. A breed association will provide some set of services to multiple such enterprises.
- Whether directly involved in collecting data and/or making selection decisions, or providing services supporting such decisions. Breeders and breeding companies (or programs) are directly involved in such decisions, whereas breed associations and national genetic evaluation systems provide supporting services for those decisions.

This grouping does not map precisely to economic scale or scale of breeding program. For example, an individual beef cattle stud is likely to be smaller in economic scale (e.g., turnover) than a breeding company, even though the number of animals (females) in the breeding population being managed may be similar. A breed association will provide some set of services to multiple such enterprises, but may be no bigger, or even smaller, than some of its member businesses (studs).

The full listing of interviewees within groups, with links, is included in Appendix B.

The interviewees are only sampled from two of the four possible cells within this 2-dimensional matrix. The breeding company/project group includes both private and public sector organisations, and as will become apparent, there is a trend for breed associations and national evaluation systems to make direct investments into at least data collection–within this simple classification, a role of the decision makers.

The questions included in the survey were:

1. What level of adoption or utilisation of genomic selection, i.e., genotyping for prediction of genetic merit, has been reached in your organisation –perhaps simplest answered as what proportion of selection candidates are genotyped, and in the case of a multi-stakeholder situation, what proportion of breeders or producers are genotyping some or all candidates for selection?
2. Has your organisation assisted with the uptake of genotyping–for instance via contribution to the cost of genotyping?
3. Has your organisation reviewed or changed breeding goals or indexes either as part of a planned process of implementation of genomic selection, or simultaneous with it?
    a. If so, has there been an increase in the number of traits being evaluated?
4. Has your organisation invested (or co-invested) in any specialised or designed phenotyping projects, either with members or separately?
5. Does your organisation provide any incentives for phenotyping, whether broadly or for specific traits?
6. Within the population you or your members work with, is there any evidence of changes of parameters of the response equation, such as:
    a. Reduction in average age of sires
    b. Increase in use of elite males and/or females via Artificial Insemination (AI) and Embryo Transfer (ET), partly or completely focussed on genotyped animals
    c. Increase in average accuracy of selection (in a general sense, this could reflect the average accuracy of genomic BVs in animals in the population)
    d. Increase(s) in rate of genetic progress, whether for some individual traits and/or overall merit index
    e. Changes in direction or rate of change in particular traits (an example I will be discussing is that of eating quality (EQ) in terminal sire sheep in Australia–where previously the genetic trend was unfavourable, driven by genetic correlations between EQ and traits contributing to efficiency of lean tissue growth

(Lean Meat Yield or LMY), but now the availability of genomic predictions for EQ as well as for LMY has resulted in reversal of the genetic trend for EQ

7. Can you comment on whether, and if possible to what extent, your organisation has drawn on government or industry funds in any aspect of the introduction of genomic selection? Additionally, if you have, do you anticipate that continuing into the future?
8. Has your organisation changed its strategy as a result of or in response to, the introduction of genomic selection? If possible, can you briefly describe the key changes underway?

Interviewees were contacted directly by the author. It was made clear that no confidential information was being sought, and that no evaluation or comparison of responses, in the sense of an organization performing in some way better or worse than others, was intended. Interviewees had opportunity to review the manuscript prior to submission.

Interviews were conducted by teleconference, but were not recorded electronically–rather, notes of the discussion were taken by the author. These notes have been distilled to generate the material reported here.

3. Results and Discussion

For this section, the interviewees have been grouped as outlined in Materials and Methods. Responses to the questions are summarized by the four groups as defined in Table 1. Direct comment or quotation from individual interviewees is minimal here, but key points from discussion are included.

1. What level of adoption or utilisation of genomic selection, i.e., genotyping for prediction of genetic merit, has been reached in your organisation–perhaps simplest answered as what proportion of selection candidates are genotyped, and in the case of a multi-stakeholder situation, what proportion of breeders or producers are genotyping some or all candidates for selection?

Table 1. Classification of Interviewees.

	Single Enterprise	Multiple Enterprise
Decision maker	Breeders Breeding Companies/projects	
Decision support		Breed associations National evaluation

Breeders: The beef breeders interviewed are all now genotyping 100% of young animals prior to selection, with the genotypes submitted for genetic evaluation via the BREEDPLAN system [15]. This level was not arrived at immediately genotyping became available: in all cases, some initial trialling was conducted allowing comparison of genomically enhanced estimates of genetic merit (EBVs in beef cattle in Australia) with standard BLUP EBVs [16].

The sheep breeders interviewed are genotyping lower proportions of their annual drop of candidates than beef breeders, likely reflecting the higher relative price of genotyping in sheep compared to cattle (prices for genotyping are similar per animal, but the sale value of rams is approximately one fifth that of beef bulls). However, the breeders interviewed indicated that the proportion of their flock being genotyped is rising and likely to continue to do so.

Breeding Companies: All breeding companies interviewed are genotyping 100% of candidates, with the exception of Tree Breeding Australia, where the reference population is being built through extensive genotyping across generations and evaluation of the potential utility of genomic information. As with the breeders, companies did not necessarily move to 100% genotyping of candidates immediately the technology became available–some research and development or learning process was involved. As will be discussed later, this in some cases included evaluation of different genotyping platforms and even development of custom genotyping assays or chips.

Breed associations: In the beef breeds interviewed, genotyping has reached high levels across their membership–broadly around 67–90% in young animals. While this information was not sought directly, there was an overall impression that adoption of genotyping was higher among larger breeders (i.e., in terms of numbers of animals and hence economic scale) within the breeds.

National evaluation systems: Observations varied between species and to an extent country. In dairy cattle, 100% of male candidates (i.e., young bulls) are now genotyped. Genotyping in females has grown more slowly, but an increasing proportion of "commercial" heifers–animals not automatically assumed to be likely to be dams of bulls–are being genotyped, reflecting increases in accuracy of genomic breeding values and the resulting increasing utility of such information in supporting management decisions. In beef and sheep, adoption varies widely between breeds in the countries represented, with a tendency for adoption to be higher and faster in breeds that have been more active users of science-based breeding methods and technologies.

Overall, adoption of genotyping for genomic selection has been quite rapid after initial trialing and research: not as dramatically as in dairy cattle, but rapid in terms of technology adoption in agriculture [17].

2. Has your organisation assisted with the uptake of genotyping–for instance via contribution to the cost of genotyping?

This question is really only relevant for breed associations and genetic evaluation systems. None of the breeds surveyed have offered any financial assistance with genotyping at the individual animal level, but all had negotiated pricing arrangements for their members with the genotyping providers.

In general, this was also the case with breeding companies and genetic evaluation systems, but a very important contribution has been made in almost all cases (breeds, breeding companies and genetic evaluation systems) via co-investment with breeders, and with industry and/or government, in R&D. Such R&D in broad terms has been aimed at establishing proof-of-concept while simultaneously building genomic reference data, and the assistance in most cases included subsidization of the cost of genotyping for breeders involved.

Extending this point, all the breeders (beef and sheep) interviewed had participated in such R&D, and their data has contributed to the relevant reference populations.

This R&D phase continues where new traits are under development: for example in Australian beef and sheep, large industry and government co-investment in phenotyping for methane output is underway, and similarly in the Australian dairy industry for fertility phenotypes via the GINFO project [18].

The challenge of maintaining appropriate levels of phenotyping, including for new traits, has become a significant strategic challenge for all the types of business consulted here, and approaches to meeting the challenge are under development [19].

3. Has your organisation reviewed or changed breeding goals or indexes either as part of a planned process of implementation of genomic selection, or simultaneous with it?
    a. If so, has there been an increase in the number of traits being evaluated?

This question was included to address the proposition that genomic selection would enable more effective selection for hard-to-measure traits, which in turn could lead to adjustment of breeding goals. Across the interviewees, no organization has specifically reviewed or changed breeding goals as part of a planned process of implementation of genomic selection, but in essentially all cases, that implementation has lead to new thinking regarding objectives and what traits could be important:

Breeders: Observations included an increased focus on some traits reflecting greater accuracy in EBVs for those traits, a desire to have more traits included in genomic evaluation, and that review of breeding goal(s) and indexes was enhanced by the increase in accuracy of EBVs for some traits, leading to greater focus on them.

Breed Associations: Similarly, among the breeds interviewed, two noted that review of breeding goals and indexes was occurring simultaneous with implementation of genomic selection, and that the implementation was increasing breed focus on collecting sufficient numbers of phenotypes to support useful genomic breeding values. The other noted that while the reviewing of breeding goals and indexes is a regular activity, implementation of genomic selection has encouraged consideration of formally including traits in the breeding goal that had previously been considered "important but too difficult to do anything about".

Breeding Companies: In general, implementation of genomic selection has allowed companies to shift emphasis in their breeding indexes more towards the balance implied by an already-defined breeding goal–in particular, to place more selection emphasis on hard-to-measure and/or low heritability traits. This is broadly similar to the observations for breeders and breeds, but noting that all the companies interviewed had formally defined breeding objectives that included traits not previously under strong selection pressure because of lack of data and/or low heritability. Although not a documented or objective reaction from those interviewed, there was a sense in the discussions that this re-balancing of selection pressure was something of a pleasant surprise–in essence, that genomic selection was in fact providing a benefit proposed theoretically, but not definitively proven anywhere prior to the implementation process.

National evaluation systems: The first observation for this group is that in almost call cases, some new traits were added to the routine evaluations as R&D generated sufficient phenotypes to enable genomic prediction. This in turn has led either to increased interest in or focus on hard-to-measure traits, informed simultaneous reviews of breeding goals or indexes, and/or stimulated revision of goals and indexes to explicitly incorporate new traits. An example of the latter is that Sheep Genetics now offers indexes that include Lean Meat Yield and Eating Quality breeding values, which have only become available via R&D and depend on genomic prediction for animals in ram-breeding flocks [Andrew Swan, pers. comm.] [20].

Extending this point, the changes–either rebalancing and/or introduction of new traits–have stimulated development of new extension tools and programs. These tools and programs have covered both new traits or indexes, as well as tools focused on genomic pedigree at the breed or across-breed level.

A more general observation relevant to this question is that in all four types of organization, implementation of genomic selection has stimulated more attention to the meaning and composition of breeding goals, and to the realization that expansion of breeding goals and introduction of new traits can be a conceptually simple and logical process. As this attention and realization grow, it seems likely to underpin quite considerable re-thinking of breeding goals, especially to include traits previously ignored as being too difficult to do anything about, and to incorporate traits relevant to more recent considerations, such as emissions and welfare-related traits [18,21]. To varying degrees, this is being actioned via more formal forward planning of R&D into potential traits and into phenotyping strategies, and into thinking much more deeply about the implications of, for example, more rapid change in fitness traits. This is somewhat analogous to new appreciation of the effectiveness of selection for disease resistance [22,23]–understanding that change in previously difficult traits can be much more effective than was thought should encourage broad and imaginative thinking about what a particular breeding program can realistically aim to achieve.

4. Has your organisation invested (or co-invested) in any specialised or designed phenotyping projects, either with members or separately?

Breeders: All breeders interviewed had participated in industry R&D programs involving collection of existing and/or new phenotypes, and genotyping. This invariably involved additional investment by the breeders, either via additional spending on recording equipment, additional labour, and/or investment in genotyping animals that would not otherwise have been genotyped. All such R&D programs involved some level of industry and/or government investment.

Breed Associations: As with the breeders, the breeds interviewed had all participated in industry R&D programs, including making significant financial investments from breed funds.

Breeding Companies: Similarly, the breeding companies surveyed have all invested in specific phenotyping projects, but the level of external co-investment (from industry and/or government) varied considerably. There is a trend to investment in phenotypes being informed by definition of the breeding goals, and therefore evolving as the breeding goals evolve (see point 3 above).

National evaluation systems: All the evaluation systems consulted have either initiated and/or participated in specific phenotyping projects, usually for a combination of existing and novel traits, moving to focus on novel traits as the value proposition for genomic selection for existing traits has become established.

There has been no single approach to the design of such projects, although some focus on ensuring involvement of higher genetic merit animals or breeding units is evident [24].

5. Does your organisation provide any incentives for phenotyping, whether broadly or for specific traits?

Breeders: This question is not relevant to breeders as individuals.

Breed Associations: All breeds interviewed have taken initial steps to encouraging phenotyping, either related to specific traits, or as some reduction in charges for enrolling animals into genetic evaluation. One breed has started offering incentives for submission of phenotypes for traits not currently well-recorded.

Discussions indicated that strategies in relation to this issue are likely to evolve [18].

Breeding Companies: This question is not directly relevant to breeding companies, except where the company has contractual agreements with suppliers of data, where the contractual arrangements will reflect the value of the data itself.

National evaluation systems: Incentives for phenotyping have been offered (provided) by national systems in a range of ways, including assistance with the cost of genotyping animals in particular priority herds or flocks, and similarly but for herds or flocks meeting recording level or quality standards, discounts on standard evaluation charges in return for defined levels of phenotyping.

The incentives provided through these mechanisms have in most cases overlapped with, or been a component of, R&D programs, but several organizations have or are considering moving to incorporating such incentives in their normal charging schedules [e.g., CDCB, Appendix A].

6. Within the population you or your members work with, is there any evidence of changes of parameters of the response equation (with possible examples listed):

There was considerable diversity in responses and discussions on this question, not so much in terms of whether changes in parameters contributing to rate of genetic progress had occurred or been observed, as in what those changes were. In part this likely reflects the way the question was posed, with sub-questions relating to different parameters, in addition to any variation reflecting the different natures of the organizations interviewed. Recognizing this diversity, some overview points are presented first, before reporting on the responses of the different organization types.

Firstly, all interviewees commented on increases in accuracy of breeding values–mostly in relation to young animals, but also for traits for which breeding values previously had only low accuracy. An interesting example of the latter is adult traits in sheep–weight and wool production of females. As interest grows in controlling maintenance costs of breeding females in extensively managed species, ability to restrict increases in adult weight while simultaneously increasing early life growth rate (for slaughter progeny) becomes increasingly important. To date, limited recording of adult weights has meant that these have simply increased as a correlated response to selection for early growth. This is a specific example of a more general issue of traits that for whatever reason have not been extensively recorded. Genetic evaluation of such traits (i.e., to generate estimates of genetic merit) has previously had low accuracy, limiting breeders' capacity to manage them genetically.

Secondly, and to varying extents, breeders are now making greater use of younger animals, reflecting increased accuracy of estimates of genetic merit for such animals. This has been most dramatic in dairy cattle breeding, where genomic selection has essentially replaced progeny testing schemes [9], but was commented on to some extent by all interviewees.

Thirdly, increases in either rate of genetic progress and/or rebalancing of genetic change across traits were most strongly noted in organisations with what seems reasonable to interpret as the strongest pre-existing focus on genetic improvement and breeding program design. For example, breeding companies with well-established and formally designed breeding programs (formally in the sense that detailed analysis of design options, including all steps in the breeding program design "algorithm" [1] most readily reported significant increases in either of these aspects. This was not so clear in the examples for beef and sheep, or for beef breeds, partly reflecting variation amongst breeders with multi-member organizations in their utilisation of genetic improvement methods and technologies.

Breeders: None of the beef or sheep breeders interviewed reported dramatic changes in rate of genetic progress, but did report what are likely early or leading indicators of significant acceleration: increased accuracy of breeding values in young animals including for hard-to-measure traits, and increased use of younger animals in reproductive programs (i.e., using Embryo Transfer on selected females). As confidence in genomic breeding values continues to build for these breeders, it seems likely that increased use of younger animals coupled with increases in accuracy across breeding goal traits will be reflected in significant acceleration of genetic progress.

Breed Associations: The general observation here was that to date there has been no "across the breed" increase in rate of genetic progress, but that early signs of change are apparent, and that there have been significant genetic changes in some specific traits introduced with the implementation of genomic selection. Retallick et al. [25] report examples of this for welfare-related traits in Angus cattle in the USA.

All breeds interviewed commented on increased extension effort through recent years, aimed at encouraging more effective use of genetic information amongst their members: this focus has coincided with the implementation of genomic selection.

Breeding Companies: Where the company interviewed had fully implemented genomic selection (as compared to being still in the process of trialing it), substantial increases in the rate of progress both overall and in hard-to-measure traits have been observed.

National evaluation systems: In broad terms, observations for the national evaluation systems reflect those for the other three interview groups: substantial acceleration reflecting reduced generation interval in dairy, considerable variation amongst breeders in the respective species–particularly beef, and increased response most noticeable in hard-to-measure traits. Examples of the latter which mirror the point noted above re welfare traits in US Angus, include:

- the dramatic change in genetic trend for traits related to meat eating quality in sheep in Australia [20], reversing previous unfavorable correlated responses to selection for growth and carcass leanness,
- the equally dramatic reversal of genetic trends for fertility in dairy cattle, which while not solely due to implementation of genomic selection, have been enhanced by it [Paul Van Raden, Andrew Cromie, Matt Shaffer–pers. comm.]

7. Can you comment on whether, and if possible to what extent, your organisation has drawn on government or industry funds in any aspect of the introduction of genomic selection? Additionally, if you have, do you anticipate that continuing into the future?

Responses were consistent across the four groups: to varying extents, all had participated in (or initiated) R&D projects focused on trialing aspects of genomic selection, and receiving industry and/or government R&D funding. Typically, such funding assisted with genotyping costs, development of analysis tools and data analysis. As evaluation of the technologies moved beyond the "proof of concept" phase, external assistance has increasingly focused on assisting with phenotyping of novel traits, such as methane emission levels [Sam Clark and Julius van der Werf, pers. comm.], and organizations fund more routine phenotyping and genotyping from internal sources.

All organizations indicated that where industry and/or government R&D funding is available, for instance to assist with development of novel traits, it would be applied for.

8. Has your organisation changed its strategy as a result of or in response to, the introduction of genomic selection? If possible, can you briefly describe the key changes underway?

Responses to this question were richly diverse, and to convey this rich diversity as fully as possible, individual comments and observations are presented below, without identifying the individual source.

Breeders:

- Greater use of young sires, more data capture from multipliers, and development of new traits
- Having greater scope to control direction is invigorating, and we have developed better understand of both the technology and our choices, leading to increased confidence in what we are aiming for and doing
- There is greater urgency to identify traits that need to be recorded and included in the evaluation
- We are reviewing our business structure aiming to get the best out of the technology
- We are developing new product lines–and can do so rapidly with genomics; and moving to having a genotype-only multiplier tier in our business
- We are focusing more on potential new traits, especially around easy care attributes and fertility

Breed Associations:

- Genomic selection has bought strategies related to genetic improvement more to the fore across our breed, and we are seeing increased engagement with our R&D programs; building strong Board understanding and support has been critical, expanding our range of collaborative R&D and strengthening our relationships with genotyping providers have been key strategic initiatives
- Minimizing costs of entry to genetic evaluation, in order to encourage participation– and phenotyping–has been an important initiative; and being open to data from a wider range of sources or partners have been important
- Managing the genomic pipeline has been critical, in parallel with faster development of new traits, increasing frequency of evaluations, being actively involved in chip design, and actively initiating R&D (rather than waiting for it to emerge from the R&D community) have all been significant strategic developments for our organization– aiming to lead rather than follow

Breeding Companies:

- We're now working with multipliers to increase accuracy of data collection for traits that can be recorded in that tier, and investing in new phenotyping, such as individual feed intake
- There has been a redirection of funds internally towards phenotyping, including phenotyping in end-users' operations; defining our phenotyping strategy–what data to collect, and how to collect it most efficiently, is now central
- Big changes are happening in our breeding program–making much more use of existing phenotypes, and this is stimulating re-defining our breeding direction; the innovation cycle is accelerating

- We're investing heavily in evaluating the technology, but we anticipate significant increase in our investments in phenotyping and genotyping
- Genotyping in the commercial level is being increased, this is changing our relationship with those stakeholders
- Our overall strategy has not changed, but we have identified new R&D

National Evaluation Systems:

- We are seeing increased use of sexed semen since the introduction of genomics, with producers making assigning females to different roles (breeding replacements, or breeding cross-bred progeny) based on their genetic merit
- Genotyping is helping increase engagement across the whole population, as well as bringing new opportunities such as breeding for reduced methane
- Investment in new high-value phenotypes is now clearly part of our R&D strategy, and the partnership with breeds becomes more important
- We are thinking more and more about "where will data come from in the future?", and are becoming more strategically engaged in managing data–what traits, ensuring quality of data
- Data from research projects is now potentially shared with industry, and we are developing new evaluations based on research data
- In the breeding programs where breeders have a role in design, there is heavier use of young males

A number of the observations made in response to this question repeat responses to earlier questions, and so it is possible to draw out some consistent messages, at two levels:

- firstly, relating to elements of the response equation

$$R = i.\ rIT.sigma(T)/L$$

where:

R is response to selection
i is standardized selection differentia
rIT is the correlation between the index on which selection is based on the breeding objective
T is the breeding objective
L is the generation interval

Changes in accuracy are typically the first effect of implementing genomic selection, and such changes increase the ratio rIT/L. rIT is accuracy itself, but L changes (reduces, at least potentially) because individuals can be evaluated earlier in life

Changes in accuracy also impact the direction of selection–which is captured by rIT.Sigma(T)–via increases in accuracy for previously hard-to-measure traits–observed as an opportunity to "re-balance" selection and genetic change.

Together, these changes underpin the opportunity to increase both the rate of genetic change and its value.

- Secondly, relating to overall strategy, a number of themes are clear:
    - phenotyping becomes a central concern: what traits to collect data on, how to collect the data (i.e., what individuals, what equipment, what data sources), and how to fund the phenotyping effort.
    - What opportunities are most significant for an individual organization to some extent depends on their pre-existing breeding program design. In broad terms, the first opportunity grasped seems to be to make more use of younger individuals, followed by the opportunity to re-balance selection.
    - All organisations had participated in, or initiated, an R&D phase aimed predominantly at validation–discovering how genomics works and what actual changes are seen. However, importantly, this phase seems to have been relatively short, at least as purely for validation: quite rapidly, new genotyping

and phenotyping effort becomes simply building gemomic reference data, and in most cases, expanding the traits being recorded. Extending this point, the organisations had essentially moved to having a continuing "R&D core" in their operations.

   - This last point in turn connects with further strategic changes: funding the phenotyping becomes a core business challenge, including to what extent organisations draw on or partner with industry or government funding; and, new data sources and hence potentially new business relationships are considered. This is most obvious in initiatives to capture more data from outside the breeding nucleus (for example, making use of data collected in the multiplier or even commercial tiers), and formalizing contractual relationships to support this.
   - A number of the organizations interviewed were beginning to evaluate opportunities to increase scale of leveraging from the nucleus–mainly via changed involvement with multipliers, exploiting the ease and in some cases lower cost of genomic evaluation.
   - A number of organisations reported use of approaches to managing inbreeding via some form of implementation of Optimal Contributions theory, and several stressed that such was an essential component of implementation of genomic selection. Reflecting the more precise knowledge of relationships available using genomic information, such information can inform design of the sampling for reference phenotyping and genotyping, and the mating design used in the nucleus [Mark Henryon, Anders Vernesen, Klara Verbyla, Brian Kinghorn—pers. comm.].

- Finally, while this was not a direct question, it was clear that in some instances, implementation of genomic selection benefited from and also stimulated synergies with other areas of R&D. A familiar example (to the author) is the interaction between meat science and genetics in the Australian sheepmeat industry R&D–meat science assisted with identification and definition of priority traits to collect phenotypes on, genetic analyses underpinned by genomic information shed new light on variation and co-variation in different aspects of meat quality and eating quality, and the joint focus helped raise industry awareness of opportunities for wide-scale change in product quality and offerings at the consumer level [26]. It seems likely that such synergies will grow–the need to capture the most useful phenotypes to enable appropriately balanced selection, will stimulate new interactions, with the growing interest in emissions reduction and in disease resistance being obvious areas for such interactions [22].

The last observation based on the interviews, responses, and insights, is that all these organisations seemed energized and stimulated by the involvement in implementation of genomic selection: they were getting new insights into what is possible, new ways of thinking about all aspects of their operations, new relationships and partnerships, and a very real sense of excitement at the prospect of faster and more valuable genetic progress.

What about impact?

No questions were asked in any interview about economic evaluation, and no attempt is made here to either review or estimate economic impacts. Ex-ante studies exist, e.g., [27], but a telling indicator of confidence in impact, or at least in success, is that all interviewees were increasing investment in phenotyping, and usually in genotyping, and in increasing scale (of the breeding program and/or commercial production) too. Faster genetic improvement with more comprehensively defined breeding goals should by definition deliver increased value from breeding programs. Analyses of economic impact, at least for programs involving any industry or public funding, can be expected in the coming years.

What about numerically smaller breeds, or less affluent industries or countries?

All the organisations interviewed here are operating in wealthy countries, and as has been reported, usually with some level of industry and/or government investment in at least R&D stages of the implementation of genomic selection. In addition, they either operate in industries based on large populations of animals (or plants), or have built such populations themselves through commercial growth (as an example, the US Angus breeding sector evaluates over 600,000 new animals per year, providing bulls to a commercial sector of several million animals [Kelli Retallick and Steve Miller, pers comm.].

Scale–both financial and numerical–is important because:

- Phenotyping and genotyping, to generate and maintain the genomic reference population, cost money
- One of the attractions of genomic selection is that such evaluation effort can be focused on a relatively small number of individuals per year (there is no precise formulation to define the number, but around 1–2000 new individuals recorded and genotyped per year will provide a strong basis for selection)

If either funds are extremely limited, and/or the total population is not much larger than the number needed to build a useful genomic reference, possible responses include:

- Organizing the breeding program to exploit the additional accuracy generated by knowledge of the genomic relationships, and aim to grow commercial scale based on faster and more valuable genetic progress. This requires a level of "whole population" coordination, to generate maximum value from the program
- Developing medium- or longer-term investment relationships with government and/or benevolent investors–drawing on justification from the scale and distribution of benefits that would not otherwise accrue to the wider farming population.

These challenges and possible responses are not in fact unique to the implementation of genomic selection–they apply equally to use of any breeding program where performance recording is required (i.e., all breeding programs). What is different is that genomic selection makes more obvious both the investment required and the significance of capturing returns at commercial scale to leverage that investment. Marshall et al. [28] provide some examples of implementation of genomic selection in African livestock, and comment on these challenges.

Wider implications–speculations

The focus of this paper has been on how organisations have implemented genomic selection and what they have learned and changed in doing so. At the same time, it is clear from the responses that the organisations are not interning to return to "pre-genomics"–if anything, as noted above, there is likely to be growth in scale of operation. This can reasonably be interpreted as reflecting increased effectiveness in genetic improvement, and indeed, a number of the interviewees reported quantified increases in rate and value of progress.

Even allowing that this sample of organisations are among lead users of genetic improvement technologies and methods, there seems no reason to doubt that genomic selection will be increasingly implemented to enhance existing genetic improvement schemes across species and countries, notwithstanding that it involves increased, or at least redirected, investment and the requirement for systems to handle genomic data. What might this mean for R&D generally, and for agriculture more broadly?

Regarding R&D, the point made above about synergies with other disciplines can be generalized: the core R&D activity of genomic selection is the measurement of large numbers of individuals for traits of interest, with experimental design embedded via the combination of knowledge of relationships (via pedigree, whether genomic or not) and of fixed effects (such as location, time, prevailing conditions etc). Genomic reference populations also need to be maintained or refreshed through time, with new phenotyping. Therefore, genomic reference populations provide an excellent platform for diverse research:

- They are by design reflective of the genetic variation in the population of interest, and so any finding is "validated" for that population, and for the medium-term future (in comparison with doing experiments on samples that reflect current commercial generations but previous genetic generations)

- Adding treatments–comparisons of diet or drug regime, for example–can be done in a balanced manner, at whatever scale is affordable, which effectively leverages the overhead costs of the program across more R&D outcomes
- Additionally, by automatically including both a "genetic" and a "non-genetic" dimension, they assist researchers and industry to better appreciate the joint influence of these two dimensions on what can be achieved in terms of performance, including obviously the extent of any genetics-by-environment and/or genetics-by-management interactions.

More generally, the acceleration of genetic progress under genomic selection will inevitably stimulate growth of thinking in two dimensions:

- Firstly, industry and the broader community will come to see genetic improvement as a more powerful, or more precisely rapid, means of change than has been generally appreciated. Despite the fact that changes under selection have been dramatic over periods of one or more decades, e.g., [29], and can mean continuous productivity or profitability improvement that more than matches the cost-price squeeze [30], genetic improvement is still widely spoken of as slow and steady, with the note that it is cumulative added as a generally misunderstood "add-on". However, when the facts of what can be achieved become more widely appreciated, there seems no reason that focus will not increasingly be paid to how much to invest in phenotyping, which is the fuel of genomic selection. The transformation of dairy breeding–completely reversing previous trends for fertility, the impact of selection for disease resistance in several species [22], and the dramatic increase in scale already apparent in some industries [Matt McDonagh, pers. comm.], signal a technology with rapid, immediate and valuable effect.
- A specific example is the appreciation that genetic improvement can contribute to reduced emissions in agriculture, and can do so without requiring a permanent treatment cost (of some mitigant) at the animal level [31].
- Secondly, the appreciation that essentially any problem can be tackled provided appropriate phenotyping is in place, coupled with the "naturalness" of change via genomic selection, could significantly improve confidence that challenges of sustainable food production with minimal undesirable side-effects can be overcome. This will stimulate more thinking about the true, or complete, breeding goals–expanding beyond simply producing more of the main product(s) to much more "holistic" goals incorporating product, product quality, reduction in disease, reduced environmental footprint, etc. [32], consistent with the more comprehensive approach to valuing changes that has been labelled "Doughnut Economics" [33].

Rational optimism about what can be done to produce more food (and other animal and plant products), of higher quality and health-supporting standard, with less environmental impact can make a real contribution to humanity actually tackling these challenges with a sense of hope.

4. Conclusions

Genomic selection has been implemented in an increasing number of situations since it was first described, building on continuing research into genotyping tools, traits, and methods of utilizing genomic information in breeding. The growing implementation provides practical support for or validation of the benefits proposed for the technology, which centre around opportunity for faster genetic progress, particularly in traits previously difficult to improve (the "hard-to-measure" traits).

The organisations interviewed here reflect a diversity of species, country, scale and position and role within a market economy. Differences are reported in the path to implementation–mainly in rapidity of adoption (in dairy for example, transition to genomic selection was almost instant in some countries) and in the process of evaluation, whether private, public or a mixture.

Despite this diversity, the interviewees report successful implementation in a range of situations, including increased accuracy of estimates of genetic merit, increased rates of genetic progress, and/or growing adoption. At the same time, a number made clear that that such changes were not solely due to implementation of genomic selection, the implementation being simultaneous with other R&D and/or extension activities, which contributed to the changes. Where pre-existing breeding program design was advanced, implementation of genomic selection did accelerate progress, essentially making better use of and adjusting the existing breeding program design.

Other themes that emerge, and which reflect opportunities available essentially anywhere, include:

- Increased accuracy of selection (noted above) and the ability to re-balance selection, both contribute to the opportunity to make genetic progress more valuable–both in terms of speed but also in economic value of each unit of change. (In terms of the response equation, this simply means that either i/L and/or riT.sigmaT can be increased, the first ratio expressing speed of change, and the second more the direction and value of change).
- New challenges emerge, or become more obvious:
  - definition of breeding goals becomes even more important both because genomic selection offers the "opportunity" to move faster in the wrong direction, and at the same time offers scope to improve traits previously intractable to selection
  - investment in phenotyping becomes more obviously the underlying rate-limiting parameter, in turn highlighting any challenges in return on investment for breeders or other stakeholders, and potentially requiring new investment relationships, particularly in extensive, multi-stakeholder industries
- Synergies with other disciplines: genomic selection leverages data, and so the more data that is relevant to income and cost in the species production system the better–meaning that drawing on knowledge of a range of disciplines will almost invariably increase value creation.
- The significance of the observation that genomic selection decouples recording from selection takes time to be appreciated, but is profound. As that appreciation grows, the opportunities in terms of both what can be changed through genetic improvement, and the rate at which those traits can be changed, are very clearly re-energising and engender a sense of excitement about opportunities. Genetic improvement can be seen more and more clearly as a "design" enterprise [34], design in the sense of being able to think expansively about what changes are possible.

Overall, the success of implementation, coupled with the appreciation of the scope for genetic improvement, and the recognition that genomic selection is technically benign–it simply increases the effectiveness of the "breed the best to the best" approach which has underpinned agriculture for millennia, together generate a more positive outlook on agriculture–much can be done to make the best and most valuable use of scarce resources, and can be done with much more effective consideration of direct and indirect consequences (direct and correlated responses), as long as we think carefully about what changes we want to make.

Funding: This research received no external funding.

Acknowledgments: The willingness of the individuals to be interviewed and to share insights into their organisations is gratefully acknowledged. The individuals, and links to their organisation websites, are listed in Appendix B.

Conflicts of Interest: The author declares no conflict of interest.

Appendix A

The fee schedule for the Council of Dairy Cattle Breeding can be accessed at https://redmine.uscdcb.com/attachments/download/13496/CDCB-Fee-Schedule-Update-06-22-2021.pdf (accessed on 20 April 2022).

In the context of this paper, the important point to note is that fees are adjusted for persons or organisations supplying defined phenotypic data on animals.

Appendix B

The individuals interviewed for this paper, and the relevant web links are here grouped as in the body of the paper.

Breeders	
Ian Locke, Wirruna Poll Herefords	https://wirruna.com/
Tom Gubbins, Te Mania Angus	https://www.temaniaangus.com/
Lucinda Corrigan, Rennylea Angus	https://www.rennylea.com.au/
Alf Collins, CBV Brahmans	https://cbv.com.au/
Mark Mortimer, CentrePlus Merinos	https://www.centreplus.com.au/
Tom Bull, Lambpro	https://www.lambpro.com.au/
John Keiller, Cashmore-Oaklea	https://www.cashmoreoaklea.com.au/
Breed Associations:	
Andrew Byrne and Peter Parnell, Angus Australia	https://www.angusaustralia.com.au/
Kelli Retallick and Steve Miller (now AGBU), Angus Genetics Inc.	https://www.angus.org/agi
Matt McDonagh, Australian Wagyu Association	https://www.wagyu.org.au/
Breeding Companies or Projects:	
Anders Vernesen and Mark Henryon, Danish Agriculture nd Food Council–re Danbred	https://danbred.com/about-us/
Pieter Knap, PIC	https://www.pic.com/
Matthew Cleveland, ABS Global	https://www.absglobal.com/
Lewis Rands, Peter Kube and Klara Verbyla, re SALTAS	https://tassalgroup.com.au/our-planet/our-operations/
Rachel Hawken, Cobb-Vantress	https://www.cobb-vantress.com/
Tony McRae, Tree Breeding Australia	http://www.treebreeding.com/
Wallace Cowling, Faba bean breeding	https://research.aciar.gov.au/rapidcookingbeans/news/brio-news-release
National evaluation centres and organisations:	
Andrew Cromie, Irish Cattle Breeding Federation	https://www.icbf.com/
Laurent Journaux, INRAE and IDELE–now GenEval	https://www.geneval.fr/english
Hamish Chandler, Meat and Livestock Australia	www.mla.com.au
Joao Durr, Council on Dairy Cattle Breeding	https://www.uscdcb.com/
Mark Thallman, USDA Meat Animal Research Center	U.S. Meat Animal Research Center: USDA ARS
Matt Shaffer, DataGene	https://datagene.com.au/

References

1. Harris, D.L.; Newman, S. Breeding for profit: Synergism between genetic improvement and livestock production (a review). *J. Anim. Sci.* **1994**, *72*, 2178–2200. [CrossRef] [PubMed]
2. Hazel, L.N. The genetic basis for constructing selection indexes. *Genetics* **1943**, *28*, 476. [CrossRef] [PubMed]
3. Henderson, C.R. Sire evaluation and genetic trends. *J. Anim. Sci.* **1973**, *1973*, 10–41. [CrossRef]
4. Meuwissen, T.H.; Hayes, B.J.; Goddard, M.E. Prediction of total genetic value using genome-wide dense marker maps. *Genetics* **2001**, *157*, 1819–1829. [CrossRef] [PubMed]
5. Goddard, M.E.; Hayes, B.J. Mapping genes for complex traits in domestic animals and their use in breeding programmes. *Nat. Rev. Genet.* **2009**, *10*, 381–391. [CrossRef] [PubMed]

6. Habier, D.; Fernando, R.L.; Dekkers, J.C.M. The impact of genetic relationship information on genome-assisted breeding values. *Genetics* **2007**, *177*, 2389–2397. [CrossRef] [PubMed]
7. Dekkers, J.C.M.; Su, H.; Cheng, J. Predicting the accuracy of genomic predictions. *Genet. Sel. Evol.* **2021**, *53*, 55. [CrossRef] [PubMed]
8. Meuwissen, T.H.; Hayes, B.J.; Goddard, M.E. Genomic selection: A paradigm shift in animal breeding. *Anim. Front.* **2016**, *6*, 6–14. [CrossRef]
9. Schaeffer, L.R. Strategy for Applying Genomic Selection in Dairy Cattle. *J. Anim. Breed. Genet.* **2006**, *123*, 218–223. [CrossRef] [PubMed]
10. World Congress of Genetics Applied to Livestock Production. 2014. Available online: http://www.wcgalp.org/proceedings/2014 (accessed on 20 April 2022).
11. World Congress of Genetics Applied to Livestock Production. 2018. Available online: http://www.wcgalp.org/proceedings/2018 (accessed on 20 April 2022).
12. Misztal, I.; Lourenco, D.; Legarra, A. Current Status of Genomic Evaluation. *J. Dairy Sci.* **2020**, *98*, skaa101. [CrossRef] [PubMed]
13. Verbyla, K.L.; Kube, P.D.; Evans, B.S. Commercial implementation of genomic selection in Tasmanian Atlantic salmon: Scheme evolution and validation. *Evol. Appl.* **2022**, *15*, 631–644. [CrossRef] [PubMed]
14. Burrow, H.M.; Bindon, B. Genetics research in the cooperative research centre for cattle and beef quality. *J. Exp. Agric.* **2005**, *45*, 941–957. [CrossRef]
15. Graser, H.-U.; Tier, B.; Nicol, D.C.; Scarth, R.D.; Hammond, K. Group BREEDPLAN: An across herd evaluation procedure for NBRS. In Proceedings of the Sixth Australian Association of Animal Breeding and Genetics Conference, Perth, WA, Australia, 9–11 February 1987; Volume 6, pp. 123–130.
16. Dominik, S.; Porto-Neto, L.R.; Reverter, A.; Corrigan, L. Benefits of genomic information in the Angus industry. In Proceedings of the Association for the Advancement of Animal Breeding and Genetics, Armidale, NSW, Australia, 27 October–1 November 2019; Volume 23, pp. 504–507.
17. Alston, J.M.; Wyatt, T.J.; Pardey, P.G.; Marra, C.M.; Chan-Kang, C. *A Meta-Analysis of Rates of Return to Agricultural R&D—Ex Pede Herculem?* Research Report; International Food Policy Research Institute: Washington, DC, USA, 2000; Volume 113.
18. Pryce, J.E.; Shaffer, M. Cost benefit analysis of a dairy genomic reference population. In Proceedings of the World Congress on Genetics Applied to Livestock Production Massey University of New Zealand, Auckland, New Zealand, 11–16 February 2018; Volume Genetic Gain—Genotyping & Phenotyping Strategies, p. 200. Available online: http://www.wcgalp.org/proceedings/2018/cost-benefit-analysis-dairy-genomic-reference-population (accessed on 20 April 2022).
19. Banks, R.G. An approach to coordinating and encouraging investment in phenotypes. In Proceedings of the 44th ICAR Conference, Leeuwarden, NL, The Netherlands, 26–30 April 2021; pp. 215–220. Available online: https://www.icar.org/wp-content/uploads/2021/10/ICAR-Technical-Series-25-virtual-meeting-2021-Proceedings.pdf (accessed on 20 April 2022).
20. Meat and Livestock Australia (2021) L.GEN.1704 Final Report. Available online: https://www.mla.com.au/research-and-development/reports/2021/advanced-genetic-evaluation-tools-and-systems-enabling-faster-and-more-valuable-genetic-gain-in-the-red-meat-industries-final-report/ (accessed on 20 April 2022).
21. Rowe, S.J.; Hickey, S.M.; Johnson, P.L.; Bilton, T.P.; Jonker, A.; Bain, W.; Veenvliet, B.; Pilel, G.; Bryson, B.; Knowler, K.; et al. The contribution animal breeding can make to industry carbon neutrality goals. In Proceedings of the Association for the Advancement of Animal Breeding and Genetics, Adelaide, SA, Australia, 2–4 November 2021; Volume24; pp. 15–18.
22. Bijma, P.; Hulst, A.D.; De Jong, M.C.M. A quantitative genetic theory for infectious diseases. In Proceedings of the World Congress on Genetics Applied to Livestock Production, Rotterdam, The Netherlands, 3–8 July 2022.
23. Bishop, S.C.; Stear, M.J. Modelling responses to selection for resistance to gastro-intestinal parasites in sheep. *Anim. Sci.* **1997**, *64*, 469–478. [CrossRef]
24. Van der Werf, J.H.J.; Kinghorn, B.P.; Banks, R.G. Design and role of an information nucleus in sheep breeding programs. *Anim Prod Sci* **2010**, *50*, 998–1003. [CrossRef]
25. Retallick, K.J.; Lu, D.; Garcia, A.; Miller, S.P. Genomic selection in the US: Where it has been and where it is going? In Proceedings of the World Congress on Genetics Applied to Livestock Production, Rotterdam, The Netherlands, 3–8 July 2022.
26. Pethick, D.W.; Warner, R.D.; Banks, R.G. Genetic improvement of lamb—Industry issues and the need for integrated research. *Aust. J. Agric. Res.* **2006**, *57*, 591–592. [CrossRef]
27. Banks, R.G.; Van der Werf, J.H.J. Economic evaluation of whole genome selection, using meat sheep as a case study. In Proceedings of the Association for the Advancement of Animal Breeding and Genetics, Barossa Valley, SA, Australia, 28 September–1 October 2009; Volume 18, pp. 430–433.
28. Marshall, K.; Gibson, J.P.; Mwai, O.; Mwacharo, J.M.; Haile, A.; Getachew, T.; Mrode, R.; Kemp, S.J. Livestock Genomics for Developing Countries—African examples in practice. *Front. Genet.* **2019**, *10*, 297. [CrossRef] [PubMed]
29. Hill, W.G. Estimation, effectiveness and opportunities of long term genetic improvement in animals and maize. *Lohmann Inf.* **2008**, *43*, 3–20.
30. Banks, R.G. Challenges in investing in genetics and biotechnology for the Australian extensive livestock industries. In Proceedings of the Association for the Advancement of Animal Breeding and Genetics, Noosa, QLD, Australia, 25–28 September 2005; Volume 16, pp. 372–377.

31. Hegarty, R. World Congress on Genetics Applied to Livestock Production, Rotterdam, The Netherlands, 3–8 July 2022. Available online: https://www.eaap.org/event/wcgalp-2022-world-congress-on-genetics-applied-to-livestock-production/ (accessed on 20 April 2022).
32. Knap, P.W. Pig Breeding Goals in Competitive Markets. In Proceedings of the 10th World Congress on Genetics Applied to Livestock Production, Vancouver, BC, Canada, 17–22 August 2014; Available online: https://wcgalp.org/system/files/proceedings/2014/pig-breeding-goals-competitive-markets.pdf (accessed on 20 April 2022).
33. Raworth, K. *Doughnut Economics: Seven Ways to Think Like a 21st-Century Economist*; Chelsea Green Publishing: Hartford, VT, USA, 2017.
34. Banks, R.G. Who Benefits from Genetic Improvement? In Proceedings of the World Congress on Genetics Applied to Livestock Production, Vancouver, BC, Canada, 17–22 August 2014; Volume Genetic Improvement Programs: Breeding Objectives, Economics of Selection Schemes, and Advances in Selection Theory, p. 009. Available online: http://www.wcgalp.org/proceedings/2014/who-benefits-genetic-improvement (accessed on 20 April 2022).

 *agriculture* 

Article

Optimization of Dairy Cattle Breeding Programs with Genotype by Environment Interaction in Kenya

Peter K. Wahinya [1,2,*], Gilbert M. Jeyaruban [1], Andrew A. Swan [1] and Julius H. J. van der Werf [3]

[1] Animal Genetics & Breeding Unit, University of New England, Armidale, NSW 2351, Australia
[2] Department of Agricultural Sciences, Karatina University, P.O. Box 1957-10101, Karatina 10101, Kenya
[3] School of Environmental and Rural Science, University of New England, Armidale, NSW 2351, Australia
* Correspondence: pwahiny2@une.edu.au

Abstract: Genotype by environment interaction influences the effectiveness of dairy cattle breeding programs in developing countries. This study aimed to investigate the optimization of dairy cattle breeding programs for three different environments within Kenya. Multi-trait selection index theory was applied using deterministic simulation in SelAction software to determine the optimum strategy that would maximize genetic response for dairy cattle under low, medium, and high production systems. Four different breeding strategies were simulated: a single production system breeding program with progeny testing bulls in the high production system environment (HIGH); one joint breeding program with progeny testing bulls in three environments (JOINT); three environment-specific breeding programs each with testing of bulls within each environment (IND); and three environment-specific breeding programs each with testing of bulls within each environment using both phenotypic and genomic information (IND-GS). Breeding strategies were evaluated for the whole industry based on the predicted genetic response weighted by the relative size of each environment. The effect of increasing the size of the nucleus was also evaluated for all four strategies using 500, 1500, 2500, and 3000 cows in the nucleus. Correlated responses in the low and medium production systems when using a HIGH strategy were 18% and 3% lower, respectively, compared to direct responses achieved by progeny testing within each production system. The JOINT strategy with one joint breeding program with bull testing within the three production systems produced the highest response among the strategies using phenotypes only. The IND-GS strategy using phenotypic and genomic information produced extra responses compared to a similar strategy (IND) using phenotypes only, mainly due to a lower generation interval. Going forward, the dairy industry in Kenya would benefit from a breeding strategy involving progeny testing bulls within each production system.

Keywords: genotype by environment; breeding strategies; selection index; response

Citation: Wahinya, P.K.; Jeyaruban, G.M.; Swan, A.A.; van der Werf, J.H.J. Optimization of Dairy Cattle Breeding Programs with Genotype by Environment Interaction in Kenya. *Agriculture* **2022**, *12*, 1274. https://doi.org/10.3390/agriculture12081274

Academic Editors: Heather Burrow and Michael Goddard

Received: 7 June 2022
Accepted: 18 August 2022
Published: 21 August 2022

Publisher's Note: MDPI stays neutral with regard to jurisdictional claims in published maps and institutional affiliations.

1. Introduction

Animal breeders are often challenged to carry out selection in the presence of genotype by environmental interaction (GxE). GxE affects sire and dam rankings among environments, consequently impacting on selection across environments and the optimal design of breeding programs [1]. GxE is also important among the dairy industries in developing countries where, to a large extent, genetic improvement relies on imported semen and herds vary in terms of input and output [2]. Often, the breeding goals of local dairy farmers and the breeding organizations that control semen supply are not always well aligned, ultimately affecting the rate of genetic progress in semen importing countries [3–6]. In this situation, local breeding programs involving genetic evaluation and progeny testing of sires within the country are advisable.

An effective genetic improvement program is lacking in Kenya due to various constraints, including small herd size, inadequate animal performance and pedigree recording, organizational challenges, and a lack of standardized methods of genetic evaluation [5,7]. A functional local breeding scheme would provide motivation to achieve higher participation of dairy farmers in pedigree and performance recording [8]. This would also facilitate farmers to select their breeding stock and produce replacement stock through a genetic evaluation within production systems. A breeding program with large-scale farms as the nucleus has been recommended as a solution to the small herd sizes, recording, and organizational challenges in Kenya [8,9]. However, this strategy could result in biased selection as is suggested by Wahinya [2], and Ombura [10], due to the fact that the large-scale farms are intensive with high input and output production systems. Under intensive systems, the scaling effect due to the spread in breeding values influences sire and index rankings [11]. Wahinya [2], recommended selection among animals evaluated within the target production systems as an alternative to the current selection based on intensive production systems. To maximize the overall gains, three strategies including: selection in one environment, selection within each environment, and selection on an index combining information in each environment were evaluated to determine the optimum strategy. These strategies have been applied in different studies to optimize dairy cattle breeding programs for different environments while accounting for GxE [1,12–15]. The local dairy cattle breeding program in Kenya has not been optimized for the different environments with genotype by environment interaction. Genomic information is not considered in the current national selection scheme and the potential of a multiple-trait genomic index to optimize genetic improvement for multiple environments with the presence of GxE in Kenya is not known.

Using selection index theory, different strategies based on sire proving can be evaluated to identify an optimum strategy to maximize the overall genetic gain in the three production systems. A deterministic simulation was therefore used in this study to evaluate and recommend an optimum dairy cattle breeding strategy to maximize the overall genetic gain for low, medium and high dairy production systems in Kenya.

2. Materials and Methods

2.1. Breeding Objective

A single dairy cattle breeding program with three production systems represented in the overall breeding objective was simulated to optimize genetic gain. The production systems were defined as low, medium, and high production systems, categorized based on milk yield occurring within a standard lactation [2]. The low, medium, and high production systems differ in terms of inputs and outputs as detailed in Wahinya [16]. Genetic improvement was defined by the selection of six traits including milk yield (MY, kg) which was the total milk yield in a lactation, butterfat yield (FY, kg), the total butterfat yield in a lactation, age at first calving (AFC, days), the age in days at the time of first calving, calving interval (CI, days), the time interval between subsequent calving events, mature weight (MWT, kg), the live weight at maturity, and survival rate (SR) which is the average probability of an animal to survive between lactations. The economic importance of these traits has been shown by Wahinya [16]. Revenue from dairy cattle is mainly derived from milk and the sale of animals. Fat yield influences the energy requirements, thus the amount of feed required. Fertility traits, including age at first calving and calving intervals have an influence on the days in milk and the number of calves for replacement or sale in the productive lifetime of a cow. Cull for age cows and cull heifers are also marketed based on their live weight. Cow survival between lactations is of economic importance in the tropics where disease and significant mortality rates are a constraint [17]. These traits were chosen to account for the current situation in Kenya, characterized by minimal recording within the dairy industry. To account for G x E, each trait was considered as a different trait in the three production systems.

2.2. Population Structure

The population consisted of a nucleus where elite dams and sires are selected and used as parents for the next generation of selection candidates. All dams in the nucleus were assumed to have phenotypes to provide information for genetic evaluation of the selection candidates. The nucleus consisted of three populations including dams in low, medium, and high production systems. To evaluate the effect of different nucleus sizes, we simulated nucleus populations with 500, 1500, 2500 and 5000 dams, of which the performances were recorded annually in each of the three production systems. Two-hundred and nineteen test bulls were assumed across the three production systems.

The population consisted of overlapping generations. Dams and test bulls were spread across eight age classes. Annually, 10 bulls and 300 cows (100 in each production system) were selected to produce the next generation. Each of the 10 selected bulls was progeny tested with 5, 10, 15, and 30 daughters per year. The daughters were considered to attain sexual maturity in their second year and therefore their first offspring were born in the third year (36 months) with a lifetime period of eight years (up to the sixth lactation). Therefore, progeny information was available when the bulls were five years and above. A 50:50 sex ratio was assumed for calves at birth while the calving rates were assumed to be 0.67, 0.74, and 0.77 under the low, medium, and high production systems, respectively. The survival rates under low (0.90), medium (0.93), and high (0.94) production systems were used to calculate the number of dams available for selection at different age classes up to eight years. The commercial population was assumed to have non-recorded dams and it relied on the sires selected in the nucleus for genetic improvement.

2.3. Breeding Strategies

Sires and dams were selected annually by truncation selection using multi-trait index selection. Progeny and existing sires and dams were used as selection candidates to produce the next selection of candidates. Candidates were considered for selection after all the information needed for selection decisions was available. In the simulation, we assumed an animal model for genetic evaluation considering all the genetic relationships. Male selection candidates were evaluated based on their half-sib sisters, daughters and dams information while females were evaluated on their own performance records, half-sib sisters and parent's information. To reduce bull maintenance cost and loss of selection candidates due to involuntary culling, we assumed a situation where semen was collected and stored. Bulls were therefore culled after two years. Genetic evaluation and selection of male and female candidates was varied to represent different selection strategies. We considered several strategies to maximize genetic gain in the overall objective with three production systems.

The breeding program aimed to maximize genetic gain in the overall objective (ΔH) with genetic gains in each of the three production systems:

$$\Delta \mathrm{H} = \Delta \mathrm{H}_{\mathrm{Low}} + \Delta \mathrm{H}_{\mathrm{Medium}} + \Delta \mathrm{H}_{\mathrm{High}}$$

where $\Delta \mathrm{H}_{\mathrm{Low}}$, $\Delta \mathrm{H}_{\mathrm{Medium}}$ and, $\Delta \mathrm{H}_{\mathrm{High}}$ are the genetic gains in the low, medium, and high production systems, respectively. The proportions of cows in the low (0.30), medium (0.33), and high (0.37) production systems in Wahinya [2], were used to weight the gains in the respective production systems for the population size. The breeding goal ($\mathrm{H_i}$) within each (Low, Medium, and High) breeding program was defined as:

$$\mathrm{H_i} = \mathbf{v}'\mathbf{a}$$

where $\mathbf{v}'$ and $\mathbf{a}$ are vectors with economic weights and true breeding values for the six traits in the breeding objective under the ith production system: low, medium, and high production systems, respectively. Four different breeding strategies were simulated in this study: (1) one breeding program with progeny testing all bulls in the high production system only (HIGH), (2) one joint breeding program with progeny testing all bulls in each of three environments (JOINT), (3) and three environment-specific breeding programs (sub-programs) each with testing of bulls only within each environment (IND). A fourth strategy similar to IND was simulated to evaluate the effect of genomic information on genetic improvement (IND-GS).

HIGH strategy consisted of one breeding program with progeny testing of bulls in the high production system. The aim was to improve the breeding objective with the six traits in the high production system. Selection of candidate sires was therefore based on the selection index under the high production system. The economic weights for the traits under the low and medium production system were therefore set to zero. The low and medium production systems obtained a correlated response from selection in the high production system.

JOINT strategy consisted of one breeding program with progeny testing of all test bulls in the three production systems. The aim was to improve the breeding objective with eighteen traits representing the six traits in all three production systems simultaneously. Economic weights specific for each production system were obtained from Wahinya [16].

IND strategy consisted three separate breeding sub-programs, one for each production system. Test bulls were progeny tested and selected within their sub-program of origin. The aim was to improve the breeding objective with six traits within each breeding program separately. The number of test bulls used in each production system was equated to the relative proportion of the population of cows under each production system multiplied by the total number of test bulls. Proportions of 0.30, 0.33, and 0.37 were assumed for the low, medium, and high production systems, respectively [2].

IND-GS strategy was similar to IND. The only difference was that phenotypic and genomic information were used to select males and females. The breeding objective therefore had twelve traits, one extra trait for each of the six traits in the IND strategy to represent the genomic information. All dams within the three production systems were assumed to be genotyped and phenotyped to form the reference population.

2.4. Prediction of Genetic Gain

Response to selection was predicted by deterministic simulation based on selection index theory using the SelAction software. SelAction predicts genetic gains at equilibrium accounting for overlapping generations, a build-up of pedigree information [18], and reduction of genetic variance due to selection [19]. Further details about the features and the theoretical background of the software are described in Rutten [20]. Selection was simulated by truncation with overlapping generations, while the annual genetic gain due to selection was estimated as in Ducrocq and Quaas [21]. Genomic selection was simulated by adding an extra trait to represent the marker information [22,23]. Marker information was modelled using a trait with a heritability of 0.999, correlated to each trait. The genetic correlation between the marker and each trait was the accuracy of genomic EBV ($r_{g\hat{g}}$). The accuracy of genomic information depends on the size of the reference population (n_p), the effective number of loci for which the effects have to be estimated (n_G), and the correlation between the true breeding value of a genotyped individual with its phenotypic record (r). This was calculated as [22,24]:

$$r_{g\hat{g}} = \sqrt{\frac{\lambda r^2}{\lambda r^2 + 1}}$$

where $\lambda = n_p/n_G$, n_p is the number of individuals in the reference population with both phenotypic records and genotypic information and n_G depends on the historical effective population size (N_E) and was estimated as $n_G = 2N_E L$, where L is the size of the genome in Morgan. Since individuals in the reference population are genotyped and phenotyped, r is equal to the square root of heritability of the trait and therefore, $r^2 = h^2$. The environmental correlation between the marker information and the original trait was set to zero based on the assumption that genotypes can be observed without error, the marker information is fully heritable and has no residual variance [23]. Genetic and phenotypic correlations $\left(r_{\hat{Q}_1\hat{Q}_2}\right)$ between the genomic EBVs were calculated as in Dekkers [22].

Table 1 shows the assumed genetic and phenotypic standard deviations, economic weights, heritabilities, genetic, and phenotypic correlations for traits under the low, medium and high production systems. The estimated accuracy of the genomic information for the breeding objective traits with different reference populations under the low, medium, and high production systems is shown in Table 2.

Table 1. Genetic (σ_a) and phenotypic standard deviations (σ_P), economic weights (EW) and genetic parameters; heritabilities—diagonal, genetic—below diagonal and phenotypic—above diagonal correlations for traits under the low, medium and high production systems.

	Trait [1]	σ_a	σ_P	EW [2]	Low						Medium						High					
					MY	FY	AFC	CI	MWT	SR	MY	FY	AFC	CI	MWT	SR	MY	FY	AFC	CI	MWT	SR
Low	MY	285.94	626.1	20.43	0.21	0.83	0.02	−0.01	0.31	0.16												
	FY	9.94	29.70	51.44	0.65	0.11	−0.09	0.08	0.00	0.16												
	AFC	77.73	156.48	−4.62	−0.38	−0.22	0.25	0.00	−0.11	−0.04												
	CI	33.3	130.85	−114.69	−0.11	0.02	0.03	0.06	0.03	0.00												
	MWT	14.53	31.97	−5.95	0.23	0.11	−0.09	−0.41	0.21	0.00												
	SR	0.36	1.11	399.26	0.24	0.20	−0.01	−0.22	0.01	0.02												
Medium	MY	467.32	923.12	18.35	0.42	0.56	−0.22	−0.53	0.16	0.26	0.26	0.84	0.02	0.02	0.31	0.16						
	FY	26.97	60.47	56.84	0.56	0.33	−0.09	0.03	0.10	0.20	0.54	0.20	0.15	0.08	0.00	0.16						
	AFC	66.58	129.67	−5.73	−0.14	−0.12	−0.06	0.24	−0.15	−0.01	−0.05	0.34	0.26	0.03	−0.11	−0.04						
	CI	15.81	97.56	−180.42	−0.46	−0.01	0.29	0.05	−0.40	−0.03	0.34	−0.04	−0.12	0.03	0.03	0.00						
	MWT	29.65	54.14	−6.48	0.23	0.10	−0.09	−0.40	0.06	−0.26	0.16	0.10	−0.15	−0.40	0.30	0.00						
	SR	0.34	1.06	486.4	0.27	0.19	−0.01	−0.25	0.01	0.50	0.26	0.19	0.01	−0.23	0.01	0.02						
High	MY	613.03	1226.38	20.31	0.64	0.66	−0.25	−0.07	0.12	0.22	0.75	0.65	−0.21	0.14	0.12	0.26	0.25	0.81	−0.01	0.04	0.31	0.16
	FY	28.66	56.84	61.41	0.62	0.84	−0.12	0.06	0.11	0.21	0.61	0.58	−0.10	0.03	0.11	0.21	0.73	0.25	0.01	0.08	0.00	0.16
	AFC	13.61	60.75	−7.88	−0.05	−0.12	−0.64	0.51	−0.13	−0.04	−0.47	−0.15	0.32	−0.44	−0.13	0.00	−0.28	−0.14	0.05	0.03	−0.11	−0.04
	CI	13.72	68.01	−298.59	0.00	0.15	−0.50	0.08	−0.35	0.03	0.51	0.15	0.06	0.62	−0.34	−0.01	0.43	0.14	0.09	0.04	0.03	0.00
	MWT	29.65	54.14	−7.8	0.21	0.11	−0.11	−0.45	0.51	0.01	0.17	0.11	−0.16	−0.42	0.51	0.00	0.16	0.12	−0.13	−0.39	0.30	0.00
	SR	0.34	1.06	605.56	0.26	0.21	0.02	−0.24	0.03	0.50	0.22	0.20	0.00	−0.25	0.03	0.50	0.22	0.22	−0.01	−0.19	0.01	0.02

[1] MY—lactation milk yield (kg); FY—butterfat yield (kg); AFC—age-at-first calving (days); CI—calving interval (days); MWT—mature weight (kg); SR—cow survival (%). Source of genetic parameters: [2,16,25–28]. The standard errors for the genetic parameters ranged between 0—0.5. Source of economic weights: [16].
[2] EV—economic weights (Kenya Shillings). The genetic correlation matrices were bended to make them positive definite using the Higham [29], algorithm in R software [30].

Table 2. Accuracies of genomic information for the breeding objective traits depending on the size of the reference populations under the low, medium and high production systems.

System	Reference Population	Trait [1]					
		MY	FY	AFC	CI	MWT	SR
Low	500	0.10	0.08	0.12	0.06	0.11	0.03
	1500	0.18	0.13	0.20	0.10	0.18	0.06
	2500	0.23	0.17	0.25	0.13	0.23	0.07
	5000	0.31	0.24	0.34	0.18	0.32	0.10
Medium	500	0.12	0.10	0.12	0.04	0.10	0.03
	1500	0.20	0.18	0.20	0.07	0.18	0.06
	2500	0.26	0.23	0.26	0.09	0.23	0.07
	5000	0.35	0.31	0.35	0.13	0.32	0.10
High	500	0.12	0.12	0.05	0.05	0.10	0.03
	1500	0.20	0.20	0.09	0.08	0.18	0.06
	2500	0.25	0.25	0.12	0.10	0.23	0.07
	5000	0.34	0.34	0.16	0.14	0.32	0.10

[1] MY—lactation milk yield (kg); FY—butterfat yield (kg); AFC—age-at-first calving (days); CI—calving interval (days); MWT—mature weight (kg); SR—cow survival (%).

3. Results

3.1. Response to Selection

The responses to selection per year under the low, medium, and high production systems for each of the different breeding strategies for six traits assuming a nucleus with 500 dams are shown in Table 3. A positive response was predicted for lactation milk yield (5.37 to 19.49 kg), lactation fat yield (0.12 to 0.78 kg), mature weight (0.02 to 0.05 kg), and survival rate (0.002 to 0.004%) within all breeding strategies and production systems. Age at first calving (−0.03 to −1.53 days) under all production systems and calving interval (−0.18 to −0.41 days) under the low production system had negative responses, which is desirable according to their economic weight. Responses in lactation milk yield, fat yield, mature weight, and survival rate increased across production systems with the level of production. There was no clear trend for the fertility traits (age at first calving and calving interval). Relying on one breeding program based on the high production system (HIGH) generated less responses under the low and medium production systems compared to the strategies based on evaluating bulls and cows within each of the production systems. The JOINT strategy with one joint breeding program with bull testing within the three production systems had the highest responses observed for most of the traits under all production systems. The IND-GS strategy involving the use of genomic information to test bulls within each of the production systems had slightly higher responses compared to a similar strategy that did not use genomic information (IND).

Table 3. Response to selection per year for six traits under the low, medium, and high production systems: total economic gain within each system and overall gain with four selection strategies and a nucleus with 500 dams.

Traits [1]	HIGH[2]			JOINT [2]			IND [2]			IND-GS [2]		
	Low	Medium	High	Low	Medium	High	Low	Medium	High	Low	Medium	High
MY	5.37	9.44	14.26	8.48	14.53	19.49	6.48	10.60	13.39	6.59	11.45	13.69
FY	0.20	0.48	0.61	0.25	0.67	0.78	0.12	0.36	0.58	0.12	0.39	0.59
AFC	−0.33	−0.36	−0.07	−1.15	−0.46	−0.03	−1.51	−0.28	−0.07	−1.53	−0.25	−0.07
CI	−0.17	0.05	0.08	−0.41	0.02	0.15	−0.18	0.09	0.08	−0.18	0.11	0.08
MWT	0.01	0.02	0.03	0.03	0.04	0.05	0.02	0.02	0.03	0.02	0.02	0.03
SR	0.002	0.002	0.002	0.004	0.004	0.004	0.002	0.002	0.002	0.002	0.002	0.002

Table 3. *Cont.*

Traits [1]	HIGH[2]			JOINT [2]			IND [2]			IND-GS [2]		
	Low	Medium	High	Low	Medium	High	Low	Medium	High	Low	Medium	High
TEG [1] (KES)	141.57	194.41	304.28	240.35	305.38	402.49	166.92	200.04	285.93	169.67	214.92	292.13
OG [1] (KES)		640.25			948.22			652.89			676.72	

[1] MY—lactation milk yield (kg); FY—lactation fat yield (kg); AFC—age at first calving (days); CI—calving interval (days); MWT—mature weight (kg); SR—survival rate (%). [2] HIGH—one production system breeding program with bull testing in High environment only; JOINT—one joint breeding program with bull testing in three environments; IND—three environment-specific breeding programs each with testing of bulls within each environment; IND-GS—three environment-specific breeding programs each with testing of bulls within each environment using genomic information; TEG—total economic gain; OG—overall objective gain.

3.2. Effect of Nucleus Size and Number of Progeny per Sire on Response

The effects of increasing the size of the nucleus from 500 to 5000 in response to selection are shown in Figure 1. Response for all the traits under the three production systems increased (−1.74 to 2.65 phenotypic standard deviations) for all strategies with an increase in the size of nucleus. However, the rate of increase in response is not linear.

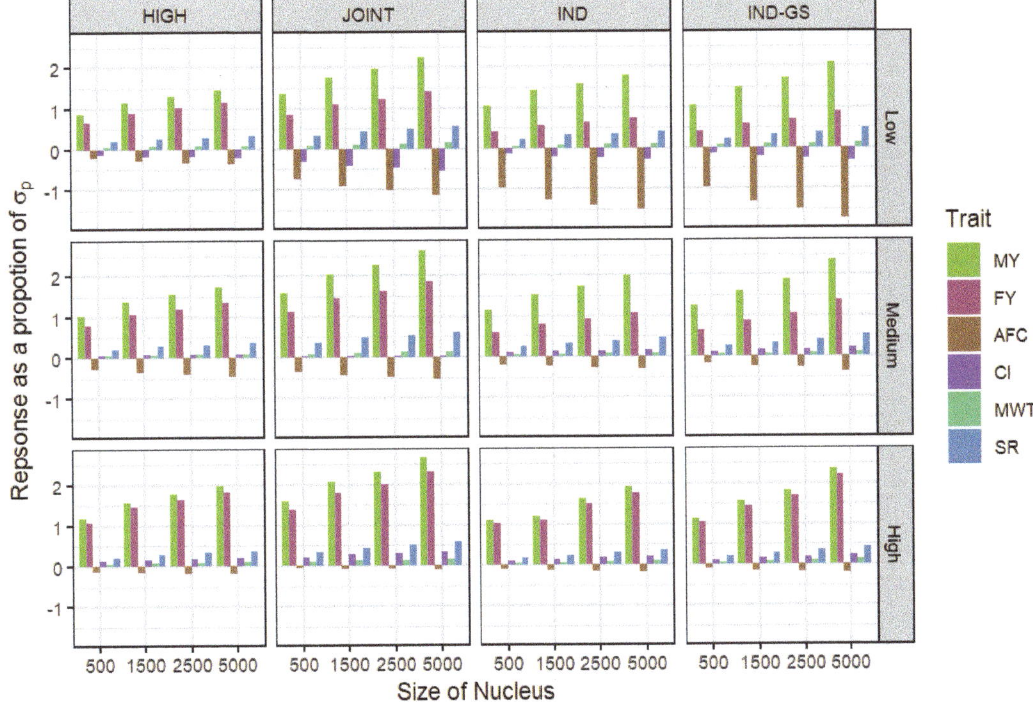

Figure 1. Comparison of response to selection per year as a proportion of the phenotypic standard deviations expressed as a percentage under the low, medium and high production systems with different strategies. MY—lactation milk yield (kg); FY—lactation fat yield (kg); CI—calving interval (days); MWT—mature weight (kg); SR—survival rate (%). HIGH—one production system breeding program with bull testing in one environment; JOINT—one joint breeding program with bull testing in three environments; IND—three environment-specific breeding programs each with testing of bulls within each environment; IND-GS—three environment-specific breeding programs each with testing of bulls within each environment using genomic information.

4. Discussion

A joint breeding program with bull testing within each of the three production systems (JOINT) produced the highest response among all the three strategies using progeny testing due to higher accuracy of the index and higher variance of the overall breeding objective. The response predicted under the low and medium production systems from selection in the high production system (HIGH) is lower compared to other strategies where bull testing is carried out within each of the three production systems (Table 3). The JOINT strategy is a favorable strategy compared to having separate breeding programs. The extent to which the three production systems would select the same sire(s) is dependent on the genetic correlations between the breeding objectives of the three production systems. The correlations between the breeding objectives under the low and medium, low and high, and medium and high production systems are 0.79, 0.66, and 0.77, respectively [16]. A strategy where bulls are tested within each of the production systems would help selection of more robust animals to maintain diversity without necessarily developing specialized lines. This would also lead to an increase in the effective population size [31].

Genomic selection has greatly transformed animal breeding and significantly impacted dairy cattle genetic improvement, especially in developed countries. This has widened the gap between countries implementing genomic selection and semen importing countries [32]. Several studies have recommended the potential of a genomic selection scheme to provide a higher rate of genetic improvement for small-sized nucleus breeding programs in developing countries [33–35]. Combining phenotypic and genomic information had a minimal effect on the response compared with the use of progeny phenotype only (Table 3). This shows that genomic selection cannot compete with traditional selection when the number of phenotypic records is limited, unless in a situation where the generation interval is significantly reduced by using genomic selection only [23,33]. The reduction of the generation interval, however, comes at a cost of reduced accuracies. The accuracies of genomic breeding values predicted in this study could be low due to the low to moderate heritabilities (Table 2) for the traits used in this study [36], and the small reference population. Regardless of this, genomic selection schemes are still attractive and could be beneficial for multi-trait selection with limited phenotypic records considering that traditional breeding schemes still need many phenotypes and long generation intervals for progeny testing. This is shown in Wahinya [37], where using correlated genomic information lead to a higher overall economic response compared to progeny testing for a nucleus with 5000 dams. Correlated genomic information could not be implemented in this study due to a limitation of number of traits in SelAction software. Genomic information could also be used for parentage assignment and breed composition determination, which is particularly beneficial to enhance the pedigree for genetic evaluation [35,38].

The dairy industry in Kenya would benefit from a higher response achieved by increasing the size of the nucleus. A large nucleus allows a higher selection intensity, young bulls can also be evaluated with more daughters, and it also minimizes inbreeding. A large nucleus would also help to address the structural weakness of the current breeding program due to a few herds contributing breeding males [39]. To create a larger nucleus it would require a considerable effort to persuade many herds to participate by providing pedigree and performance records to the recording organization. This has been a constant challenge in the developing dairy industries where pedigree and performance recording is already minimal and erratic. One of the main reasons linked to this is the failure of the recording scheme to meet the farmer's expectations and to offer noticeable returns [7]. Nevertheless, in practice, farmers still need records within their herds to make management decisions. As shown in this study, the current performance recording herds can be used to drive genetic gain for the commercial herds and the national dairy herd. This however, requires an efficient way to evaluate animal performance including as much information provided by the farmers [2]. A platform that is conspicuously missing for the current performance recording system also needs to be developed to provide feedback and quality information. Good examples can be learnt from other developed dairy industries that have applied digital strategies and education to provide quality information and tools for better herd improvement decisions.

5. Conclusions

This study shows that a strategy based on bull testing within production systems would be more beneficial compared to bull testing solely in high production environments. A higher rate of genetic improvement would also be achieved by increasing the size of the nucleus and the number of progeny per sire. A selection strategy using genomic information is promising with a large reference population. Application of these recommendations will be difficult but possible with the right level of investment, backed by innovative solutions, digital strategies, and education to encourage pedigree, and performance recording in developing countries.

Author Contributions: Conceptualization, P.K.W.; methodology, P.K.W.; validation, P.K.W.; formal analysis, P.K.W.; writing—original draft preparation, P.K.W.; writing—review and editing, P.K.W., G.M.J., A.A.S. and J.H.J.v.d.W.; supervision, G.M.J., A.A.S. and J.H.J.v.d.W. All authors have read and agreed to the published version of the manuscript.

Funding: P.K.W. was financially supported by the University of New England (Armidale, Australia) International Postgraduate Research Awards (IPRA) to pursue PhD studies at the Animal Genetics and Breeding Unit.

Institutional Review Board Statement: Not applicable.

Informed Consent Statement: Not applicable.

Acknowledgments: The authors thank the Dairy Recording Services of Kenya (DRSK, Nakuru) in Kenya for providing data and Kim Bunter for reviewing the paper. This work was conducted as part of a PhD thesis titled: "Wahinya, P.K. (2020). Strategies for genetic improvement of dairy cattle under low, medium and high production systems in Kenya. (PhD thesis), University of New England, Armidale, Australia.

Conflicts of Interest: The authors declare no conflict of interest.

References

1. Mulder, H.A.; Bijma, P. Benefits of Cooperation Between Breeding Programs in the Presence of Genotype by Environment Interaction. *J. Dairy Sci.* **2006**, *89*, 1727–1739. [CrossRef]
2. Wahinya, P.K. Estimation of genetic parameters for milk and fertility traits within and between low, medium and high dairy production systems in Kenya to account for genotype-by-environment interaction. *J. Anim. Breed. Genet.* **2020**, *137*, 423–519. [CrossRef] [PubMed]
3. Ojango, J.; Pollott, G. The relationship between Holstein bull breeding values for milk yield derived in both the UK and Kenya. *Livest. Prod. Sci.* **2002**, *74*, 1–12. [CrossRef]
4. Muasy, T.; Peters, K.; Kahi, A. Effect of diverse sire origins and environmental sensitivity in Holstein-Friesian cattle for milk yield and fertility traits between selection and production environments in Kenya. *Livest. Sci.* **2014**, *162*, 23–30. [CrossRef]
5. Musani, S.; Mayer, M. Genetic and environmental trends in a large commercial Jersey herd in the Central Rift Valley, Kenya. *Trop. Anim. Health Prod.* **1997**, *29*, 108–116. [CrossRef]
6. Bichard, M. Genetic improvement in dairy cattle—An outsider's perspective. *Livest. Prod. Sci.* **2002**, *75*, 1–10. [CrossRef]
7. Wasike, C.B. Factors that influence the efficiency of beef and dairy cattle recording system in Kenya: A SWOT-AHP analysis. *Trop. Anim. Health Prod.* **2011**, *43*, 141–152. [CrossRef]
8. Kariuki, C.M. Multiple criteria decision-making process to derive consensus desired genetic gains for a dairy cattle breeding objective for diverse production systems. *J. Dairy Sci.* **2017**, *100*, 4671–4682. [CrossRef]
9. Kahi, A.K.; Nitter, G. Developing breeding schemes for pasture based dairy production systems in Kenya I. Derivation of economic values using profit functions. *Livest. Prod. Sci.* **2004**, *88*, 161–177. [CrossRef]
10. Ombura, J. An assessment of the efficiency of the dairy bull dam selection methodology in Kenya. *Livest. Res. Rural. Dev.* **2007**, *19*, 10.
11. Santos, B. Factors Affecting Rankings of Dairy Bulls across New Zealand Dairy Farm Systems. In Proceedings of the World Congress on Genetics Applied to Livestock Production, Vancouver, BC, Canada, 17–22 August 2014; p. 4.
12. James, J.W. Selection in two environments. *Heredity* **1961**, *16*, 145–152. [CrossRef]
13. Mulder, H.A.; Bijma, P. Effects of genotype × environment interaction on genetic gain in breeding programs1. *J. Anim. Sci.* **2005**, *83*, 49–61. [CrossRef] [PubMed]

14. Slagboom, M. Genomic selection improves the possibility of applying multiple breeding programs in different environments. *J. Dairy Sci.* **2019**, *102*, 8197–8209. [CrossRef] [PubMed]
15. Smith, C.; Banos, G. Selection within and across populations in livestock improvement. *J. Anim. Sci.* **1991**, *69*, 2387–2394. [CrossRef] [PubMed]
16. Wahinya, P.K. Breeding objectives for dairy cattle under low, medium and high production systems in the tropics. *Animal* **2022**, *16*, 100513. [CrossRef] [PubMed]
17. Bebe, B.O. Smallholder dairy systems in the Kenya highlands: Cattle population dynamics under increasing intensification. *Livest. Prod. Sci.* **2003**, *82*, 211–221. [CrossRef]
18. Dekkers, J.C.M. Structure of Breeding Programs to Capitalize on Reproductive Technology for Genetic Improvement. *J. Dairy Sci.* **1992**, *75*, 2880–2891. [CrossRef]
19. Bulmer, M. The effect of selection on genetic variability. *Am. Nat.* **1971**, *105*, 201–211. [CrossRef]
20. Rutten, M.J.M. SelAction: Software to Predict Selection Response and Rate of Inbreeding in Livestock Breeding Programs. *J. Hered.* **2002**, *93*, 456–458. [CrossRef]
21. Ducroco, V.; Quaas, R. Prediction of Genetic Response to Truncation Selection across Generations. *J. Dairy Sci.* **1988**, *71*, 2543–2553. [CrossRef]
22. Dekkers, J.C.M. Prediction of response to marker-assisted and genomic selection using selection index theory. *J. Anim. Breed. Genet.* **2007**, *124*, 331–341. [CrossRef] [PubMed]
23. Van Grevenhof, E.M.; van Arendonk, J.A.; Bijma, P. Response to genomic selection: The Bulmer effect and the potential of genomic selection when the number of phenotypic records is limiting. *Genet. Sel. Evol.* **2012**, *44*, 26. [CrossRef] [PubMed]
24. Daetwyler, H.D.; Villanueva, B.; Woolliams, J.A. Accuracy of Predicting the Genetic Risk of Disease Using a Genome-Wide Approach. *PLoS ONE* **2008**, *3*, e3395. [CrossRef] [PubMed]
25. Kahi, A.K.; Nitter, G.; Gall, C.F. Developing breeding schemes for pasture based dairy production systems in Kenya II. Evaluation of alternative objectives and schemes using a two-tier open nucleus and young bull system. *Livest. Prod. Sci.* **2004**, *88*, 179–192. [CrossRef]
26. Musingi, B.M.; Muasya, T.K.; Kahi, A.K. Genetic Parameters for Measures of Longevity in Kenyan Sahiwal Cattle. *Livest. Res. Rural. Dev.* **2018**, *30*, 96. Available online: http://www.lrrd.org/lrrd30/6/muasy30096.html (accessed on 24 February 2020).
27. Ilatsia, E.D. Sahiwal cattle in semi-arid Kenya: Genetic aspects of growth and survival traits and their relationship to milk production and fertility. *Trop. Anim. Health Prod.* **2011**, *43*, 1575. [CrossRef] [PubMed]
28. Haile-Mariam, M.; Bowman, P.J.; Goddard, E.M. Genetic and environmental relationship among calving interval, survival, persistency of milk yield and somatic cell count in dairy cattle. *Livest. Prod. Sci.* **2003**, *80*, 189–200. [CrossRef]
29. Higham, N.J. Computing a nearest symmetric positive semidefinite matrix. *Linear Algebra Its Appl.* **1988**, *103*, 103–118. [CrossRef]
30. R Core Team. *R: A Language and Environment for Statistical Computing*; R Foundation for Statistical Computing: Vienna, Austria, 2021.
31. Goddard, M. Optimal Effective Population Size for the Global Population of Black and White Dairy Cattle. *J. Dairy Sci.* **1992**, *75*, 2902–2911. [CrossRef]
32. Ducrocq, V. Development of genomic selection in dairy cattle in two emerging countries: South Africa and India. In Proceedings of the Annual Meeting of the European Association for Animal Production (EAAP), Tallinn, Estonia, 28 August–1 September 2017.
33. Kariuki, C.M. Optimizing the design of small-sized nucleus breeding programs for dairy cattle with minimal performance recording. *J. Dairy Sci.* **2014**, *97*, 7963–7974. [CrossRef]
34. Marshall, K. Marker-based selection within smallholder production systems in developing countries. *Livest. Sci.* **2011**, *136*, 45–54. [CrossRef]
35. Marshall, K. Livestock Genomics for Developing Countries—African Examples in Practice. *Front. Genet.* **2019**, *10*, 297. [CrossRef] [PubMed]
36. Hayes, B. Accuracy of genomic selection: Comparing theory and results. *Proc. Assoc. Adv. Anim. Breed. Genet.* **2009**, *18*, 34–37.
37. Wahinya, P.K.; Swan, A.A.; Jeyaruban, M.G. Proposed genetic improvement strategies for dairy cattle in Kenya. In Proceedings of the Association for the Advancement of Animal Breeding and Genetics, Adelaide, Australia, 2–4 November 2021; pp. 414–418.
38. Ojango, J.M.K. Genetic evaluation of test-day milk yields from smallholder dairy production systems in Kenya using genomic relationships. *J. Dairy Sci.* **2019**, *102*, 5266–5278. [CrossRef]
39. Muasya, T.K.; Peters, K.J.; Kahi, A.K. Breeding structure and genetic variability of the Holstein Friesian dairy cattle population in Kenya. *Anim. Genet. Resour. Ressour. Génétiques Anim. Recur. Genéticos Anim.* **2013**, *52*, 127–137. [CrossRef]

 agriculture

Review

Promoting Sustainable Utilization and Genetic Improvement of Indonesian Local Beef Cattle Breeds: A Review

Nuzul Widyas [1,*], Tri Satya Mastuti Widi [2], Sigit Prastowo [1], Ika Sumantri [3], Ben J. Hayes [4] and Heather M. Burrow [5]

1 Department of Animal Science, Universitas Sebelas Maret, Surakarta 57127, Indonesia
2 Faculty of Animal Science, Universitas Gadjah Mada , Yogyakarta 55281, Indonesia
3 Department of Animal Science, University of Lambung Mangkurat, Banjarbaru 70714, Indonesia
4 Queensland Alliance for Agriculture and Food Innovation, University of Queensland, 306 Carmody Road, St Lucia 4067, Australia
5 Faculty of Science, Agriculture, Business and Law, University of New England, Armidale 2351, Australia
* Correspondence: nuzul.widyas@staff.uns.ac.id

Abstract: This paper reviews the literature relevant to the breeding of cattle grazed in tropical environments and particularly Indonesia. The aim is to identify new breeding opportunities for cattle owned by Indonesia's smallholder farmers, whilst also conserving unique local cattle beef breeds. Crossbreeding has been practiced extensively in Indonesia, but to date there have been no well-designed programs, resulting in many mixed-breed animals and no ability to determine their genetic composition, productive capabilities or adaptation to environmental stressors. An example of within-breed selection of Bali cattle based on measured live weight has similarly disregarded other productive and adaptive traits. It is unlikely that smallholder farmers could manage effective crossbreeding programs due to the complexities of management required. However, a tropically adapted composite breed(s) could perhaps be developed and improved using within-breed selection. Establishing reference population(s) of local breeds or composites and using within-breed selection to genetically improve those herds may be feasible, particularly if international collaborations can be established to allow data-pooling across countries. The use of genomic information and a strong focus on all economically important traits in practical breeding objectives is critical to enable genetic improvement and conservation of unique Indonesian cattle breeds.

Keywords: beef cattle; tropical environments; crossbreeding; within-breed selection; genomic selection; productive traits; resistance to environmental stressors; reference populations; breed conservation

Citation: Widyas, N.; Widi, T.S.M.; Prastowo, S.; Sumantri, I.; Hayes, B.J.; Burrow, H.M. Promoting Sustainable Utilization and Genetic Improvement of Indonesian Local Beef Cattle Breeds: A Review. *Agriculture* **2022**, *12*, 1566. https://doi.org/10.3390/agriculture12101566

Academic Editor: Ligang Wang

Received: 14 July 2022
Accepted: 22 September 2022
Published: 28 September 2022

1. Introduction

About 120 million Indonesians, or ~11% of the country's total population, live on less than USD2 per day, with another ~40% of Indonesia's population vulnerable to falling into poverty as their income hovers marginally above the national poverty line. The agricultural sector employs two thirds of Indonesia's poor and hence, it represents a vitally important component of Indonesia's economy.

Demand for beef in Indonesia has been increasing due to growth in population and household income. However, demand has been outstripping supply, and the self-sufficiency ratio at a national level has hovered around 65% over the past 10 years, requiring 30–40% of beef to be met by imports, mainly live cattle and frozen beef from overseas [1].

About 6.5 million smallholder farmers living in rural areas across Indonesia produce ~90% of the beef produced in Indonesia, while the remaining ~10% of beef production is delivered by a small number of commercial farmers (<1% of all beef farmers) and large beef cattle companies concentrated primarily in Java [2]. A very strong opportunity, therefore, exists to strengthen Indonesia's beef sector, to improve the productivity and profitability of smallholder beef farmers and to also improve the livelihoods of Indonesia's rural poor.

Beef cattle production in Indonesia is increasing through ongoing improvements of cattle health and growth and reproduction rates and reducing cattle death rates through improved cattle management (for example, the use of forage tree legumes in intensive and extensive cattle production systems) [3–9]. Genetic improvement of the existing cattle population would also be a very feasible opportunity if effective options to identify genetically superior breeding cattle could be developed and implemented [2].

This paper, therefore, undertakes a review of the scientific literature relevant to genetic improvement programs in tropical environments with a specific aim of identifying the best opportunity or opportunities to genetically improve beef cattle in Indonesia. It examines the value of existing crossbreeding and within-breed selection programs in Indonesia as well as the potential role of new genomic (DNA-based) tools to improve on-farm productivity and business profitability. An additional aim in designing new breeding programs for Indonesia must be the conservation of indigenous cattle genetic resources where their ongoing viability may be endangered. Hence, the review also considers their conservation, with the aim of making recommendations on the best option(s) available for smallholder cattle farmers in Indonesia in the foreseeable future.

2. Indonesian Beef Production and Marketing Systems

Knowledge of specific beef production and marketing systems is required for the design of relevant breeding objectives underpinning effective genetic improvement programs. In Indonesia, smallholder farmers use a wide range of crop residues and by-products to feed and manage cattle through intensive or extensive production systems. Intensive systems, typical of areas where availability of grazing land is scarce, use stalls to house the cattle and cut-and-carry feeding systems, primarily to fatten sale cattle. Under extensive systems, cattle are free-grazing. Extensive systems apply only where greater land areas exist (e.g., eastern Indonesia) and are generally used for breeding and growing young cattle prior to the sale for fattening.

Extensive research in eastern Indonesia shows that cattle numbers, beef production, reproduction and farm profitability can all be significantly increased for cow-calf systems and cattle fattening operations that are closely integrated with dryland farming systems [3,5–9]. In those regions, Bali cattle (*Bos javanicus*) are most often used by smallholder farmers, although some crossbred animals of unknown breed composition and bred using artificial insemination (AI) are used for fattening. In Central Java and other areas of Indonesia, where cattle grazing under palm oil plantations is strongly encouraged by the Government of Indonesia [10], crossbred cattle are preferred by the farmers although the economic yield is similar to that of local cattle. This is because crossbred cattle are taller and cattle traders buy cattle primarily on the basis of their visual appearance, with larger body-framed animals (regardless of weight) attracting higher prices [11,12].

Cattle purchased by traders across Indonesia are subsequently slaughtered at local butcher shops and abattoirs, for purchase by consumers through local wet markets. However, ongoing research in Indonesia shows that very strong potential exists to establish new beef markets for Indonesian cattle slaughtered in modern commercial abattoirs that will reward smallholder farmers for the value of the product they deliver to consumers through large supermarkets, hospitality venues and tourist hotels across Indonesia [Dahlanuddin et al., University of Mataram, unpublished].

Although the sale of cattle through local traders is currently the main outlet for cattle from smallholder farmers, it is suggested that in the future, well-designed genetic improvement programs should target the specifications provided by these potential new markets as part of their breeding objective.

3. The Need for Cattle That Are Well Adapted to Tropical Beef Production Systems

Indonesia is a tropical country with an innately harsh environment. This means that Indonesian cattle are routinely exposed to numerous environmental stressors such as ecto-parasites (cattle ticks; horn-, buffalo- and screw-worm flies; other biting insects), endo-parasites (gastro-intestinal helminths or worms), seasonally poor nutrition, high heat and humidity and diseases that are often transmitted by the parasites, with such exposure likely to increase significantly due to climate change over coming years. Each of these stressors has the potential to seriously impact the survival, growth and reproduction of the cattle. The impact of each stressor on production and animal welfare is often multiplicative rather than additive, particularly when animals are already undergoing physiological stress such as lactation, e.g., [13–16]. Under Indonesia's cattle production systems, it is generally not possible to control these stressors through cattle management strategies due to lack of access to treatments (either due to their high cost and/or non-availability in Indonesia) or the unwillingness of the Indonesian farmers to use them, often for cultural reasons. Even if intervention strategies were feasible, the treatments per sé often cause their own problems. For example, chemical treatments to control parasites generate concern about residues in beef products. As well, the parasites acquire resistance to the chemical treatments, creating additional parasite-control problems [17]. In intensive feedlot systems, high heat and humidity, even in the absence of other stressors, can become critically important for both production and animal welfare reasons. In such cases, management interventions may be possible, but they are difficult and/or expensive to implement, particularly in poorly adapted cattle. Hence, the best method of reducing the impacts of these stressors to improve productivity and animal welfare is to breed cattle that are productive in their presence, without the need for managerial interventions. In Indonesia, this means using cattle that are very well adapted to the harsh tropical environments, usually because they or their ancestors have evolved in those climates.

Differences in the performance of cattle breeds reared in tropical climates are often masked by the effects of environmental stressors on productive attributes [18,19]. In particular, British and European breeds perform differently in both temperate and tropical environments, with the European breeds growing faster to a larger mature size than British breeds when the feed supply is unlimited, but because of their larger breed size, they also take longer to conceive and calve than the British breeds. In tropical environments when feed is limited or of poor quality, the European breeds perform more poorly than the British breeds.

Hence, as summarized by [18] and adapted by [20], for most purposes in tropical environments, cattle breeds can be categorized into general breed types or groupings including:

- *Bos taurus* (British/European breeds that evolved in cooler British and European climates, e.g., Angus, Hereford/Limousin, Simmental);
- *Bos indicus* (breeds that evolved on the Indian sub-continent, e.g., Ongole, Brahman, Nelore);
- Tropically adapted taurine breeds (southern African Sanga, e.g., Afrikaner, Mashona; West African humpless, e.g., N'dama; and Criollo breeds of Latin America and the Caribbean, e.g., Romosinuano);
- Tropically adapted *Bos indicus* × *Bos taurus* composite breeds (e.g., Santa Gertrudis, Braford, Charbray);
- Tropically adapted taurine × British/Continental composite breeds (e.g., Bonsmara, Belmont Red, Senepol);
- East African zebu breeds (e.g., Boran); and
- The first cross (F_1) between *Bos indicus* and *Bos taurus*, which has attributes that are different from other breed types, particularly in harsher environments.

Additionally, and of direct relevance to Indonesia, *Bos javanicus* (Bali cattle) and *Bos banteng* (Banteng cattle) are believed to have each evolved from *Bos bibos* ancestors, but independently of these other breed types [21]. Bali cattle are hypothesized to have differences in their Y chromosomes relative to other species of cattle [22]. They can be crossed with *Bos taurus* and *Bos indicus*, though the male offspring are hypothesized to be infertile [23].

Comparative rankings of some of the different breed types for different characteristics in temperate and tropical environments based on [18] and adapted by [20] are shown in Table 1. Because of the paucity of direct breed-type comparisons from most tropical and sub-tropical areas, the rankings in these regions are largely based on results from Belmont Research Station and from associated research programs in northern Australian beef industry herds [18,20]. Comparisons in temperate areas are largely derived from the Meat Animal Research Center in Nebraska, USA and related studies in warmer environments such as Florida, USA [18,20]. There are no known direct breed comparisons between *Bos javanicus* (Bali cattle) and the other breed types, so the rankings in Table 1 are inferred from breed performance summarized by [21].

Table 1. Comparative rankings of different breed types for productive traits in temperate and tropical environments and for adaptation to some stressors of tropical environments [18,20] (the higher the number, the better the performance for the trait, and the greater the resistance to environmental stressors).

Breed Type	*Bos taurus*		Adapted *Bos taurus*	*Bos indicus*		F_1 Brahman × British	*Bos javanicus* (Bali) [f]
	British [g]	European [e,g]	Sanga [g]	Indian [g]	African [g]		Indonesian
Temperate [a]							
Growth	4	5	3	3	2	4	3
Fertility	5	4	4	3	4	5	4
Tropical [a]							
Growth	2	2	3	4	2	4	3
Fertility	2	2	5	3	4	5	5
Mature size	4	5	3	4	3	4	3
Meat quality [b]	5	4	5	3	4	4	unknown
Resistance to environmental stressors							
Cattle ticks [c]	1	1	4	5	5	4	4
Worms [d]	3	3	3	5	4	4	unknown
Eye disease	2	3	3	5	4	4	unknown
Heat	2	2	5	5	5	5	5
Drought	2	1	5	5	5	4	5

[a] A temperate environment is assumed to be one free of environmental stressors, while tropical environment rankings apply where all environmental stressors are operating. Hence, while a score of, for example, 5 for fertility in a tropical environment indicates that breed type would have the highest fertility in that environment, the actual level of fertility may be less than the actual level of fertility for breeds reared in a temperate area, due to the effect of environmental stressors that reduce reproductive performance. [b] Principally meat tenderness. [c] *Rhipicephalus (boophilus) microplus*. [d] Specifically *Oesophagostomum*, *Haemonchus*, *Trichostrongylus* and *Cooperia* spp. [e] Data from purebred European breeds are not available in tropical environments and responses are predicted from the CSIRO Rockhampton crossbreeding data. [f] No direct comparisons available, so rankings should be regarded as indicative only. [g] British breeds include for example Hereford, Angus and Shorthorn; European breeds include for example Charolais, Simmental and Limousin; Sanga breeds include for example Afrikaner, Mashona and Ndama; *Bos indicus*—Indian breeds include for example Ongole, Brahman and Nelore; *Bos indicus*—African breeds include for example Boran.

The comparative rankings shown in Table 1 suggest that potentially the best breeds for use in Indonesian cattle production systems are likely to include the tropically adapted taurine breeds and *Bos javanicus* (as either pure breeds or composites that retain sufficient levels of resistance to environmental stressors) or crossbreeding programs that focus on combining both productive and adaptive traits.

4. Indonesian Cattle Breeds

As the second largest biodiversity, Indonesia possesses a collection of native and local cattle breeds. These types of cattle have in the past been considered to be less efficient [24] and less productive [25] relative to the imported genetically improved *Bos taurus* (taurine) breeds. However, the perceptions of those earlier studies are not based on scientifically valid breed comparisons under tropical production systems. Hence, recommendations about the use of imported genetically improved *Bos taurus* (other than in well-designed crossbreeding programs based on tropically adapted cow breeds) are unlikely to be valid under Indonesian beef production environments.

The most common local breeds farmed by beef producers in Indonesia are Bali (*Bos javanicus*), Peranakan Ongole Grade (PO, *Bos indicus*) and Madura cattle, an ancient stabilized cross possibly based on combinations of *Bos bibos*, *Bos indicus* and *Bos taurus*, but with the breeds of origin yet to be accurately determined.

To date, no well-designed studies have been undertaken to actually determine whether the perceived low productivity in local cattle is due to poor management and inadequate nutrition (that will impact to an even greater degree on the poorly adapted taurine breeds being introduced to Indonesia), or because they are yet to undergo any well-designed genetic improvement programs, or both.

4.1. Bali Cattle (Bos javanicus)

Bali cattle (*Bos javanicus*) are native Indonesian cattle, believed to have possibly originated from wild Banteng and domesticated on Bali island [26,27]. Based on archaeological data from Indonesia there is no evidence that cattle were introduced in Java or Bali before the advent of Hinduism, suggesting that Bali cattle have been domesticated for less than 2000 years [28]. This helps explain why the physical characteristics of domestic Bali cattle are so similar to those of the wild Banteng (Figure 1). Bali cattle are visibly smaller relative to *Bos indicus* or *Bos taurus* breeds, but they are inferred to be more robust than the *Bos taurus* breeds in coping with stressors of harsh tropical environments and poor-quality feed (Table 1). Several reports document their good reproductive performance [21,27,29], though it should also be noted these reports are not based on scientifically valid direct comparisons with other breeds of cattle. Bali cattle are also reportedly highly resistant to infections by ticks and tick-borne diseases, but highly susceptible to Jembrana disease [30,31]. Jembrana is a disease with symptoms similar to Malignant Catarrhal Fever (MCF), capable of infecting cattle, buffalo, swine and sheep. In Bali cattle, the most visible clinical symptoms include high fever and severe diarrhea, which can lead to death [32]. Unfortunately, none of these reports on the performance of Bali cattle is based on well-designed experimental comparisons of their performance relative to other breeds, so the need for genuine breed comparisons remains for future research.

Figure 1. Images of Bali (*Bos javanicus*) breed animals: (**A**) Bali bull at final weight prior to slaughter; and (**B**) Bali cow and calf.

Nowadays, Bali cattle populations are spread across the Indonesian archipelago, mainly in the islands of Bali, Sumatra, Kalimantan, Java and East and West Nusa Tenggara. In Bali, they are managed in intensive and semi-intensive systems, whilst in Sumatra and Kalimantan they mostly integrated with palm oil plantations. In East and West Nusa Tenggara they are reared on a mix of intensive (cut-and-carry, stall-based) and extensive pasture systems [29].

Although Bali cattle are smaller than other *Bos indicus* and *Bos taurus* breeds, and even though there are no results from well-designed breed comparisons yet available, it can be interpreted that *Bos javanicus* cattle are well adapted and productive in Indonesia's tropical beef production systems. Additionally, because they are a unique species of cattle they warrant ongoing conservation. Ideally in the near future, the genome of these cattle will be sequenced for the first time, to allow a better understanding of their unique attributes relative to other *Bos* spp. (e.g., *Bos indicus*, *Bos taurus*), as well as to assist in the design of within-breed selection programs to enhance their conservation. We are currently exploring opportunities through Indonesian and international agencies to secure funding to sequence the unique Indonesian cattle breeds and enable this comparison.

4.2. Ongole Grade Cattle (PO)

Peranakan Ongole cattle (PO) are derived from Ongole or Sumba Ongole [33]. Towards the end of the 19th century, several Indian *Bos indicus* breeds were imported into Indonesia for use as dual-purpose, draught/beef animals, with the Ongole (known elsewhere as Nelore) deemed the most suitable [28]. Hence, around 1914 all purebred Ongole cattle in Indonesia were sent to the Island of Sumba, where they were managed as a purebred herd, with the herd gradually expanding and purebred Ongole bulls subsequently exported from Sumba to other Indonesian islands to be used for crossbreeding purposes [28]. Ongole cattle are a specific breed from within the *Bos indicus* species, though a recent study [34] found that *Bos javanicus* contributes about 6–7% of the average breed composition of PO cattle. PO cattle are now mainly found in the island of Java (90%), with the remainder located in Lampung, South Sumatra, North Sumatra, Central Sulawesi and North Sulawesi provinces [35]. PO cattle were developed for draught purposes. Hence, they are large with strong bodies (Figure 2) as well as being docile and tolerant to heat and other environmental stressors [27], though again there are no scientifically valid comparisons relative to other cattle breeds.

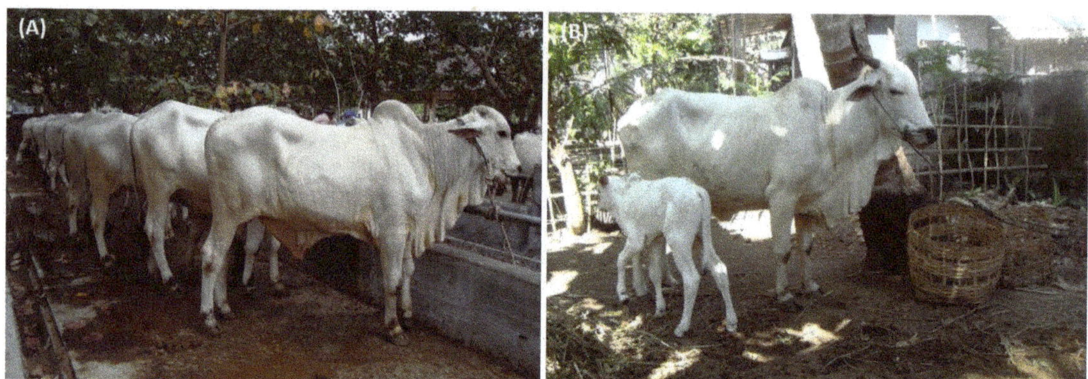

Figure 2. Images of Ongole grade (*Bos indicus*) breed animals: (**A**) Ongole bulls in a traditional Indonesian cattle market; and (**B**) Ongole cow and calf.

The characteristics of productive and reproductive traits of PO cattle are summarized in Table 2, but as with the performance of Bali cattle, the results are not based on scientifically valid comparisons of performance relative to other breeds managed under the same production conditions.

Table 2. Productive and reproductive characteristics of Ongole grade (PO) cattle (Note these figures are not based on scientifically valid breed comparisons, so such comparisons can only be inferred from the different studies).

Trait	Number of Animals	Mean ± sd	Production System [Reference]
Birth weight (kg)	59	30.91 ± 1.97	Govt. facility, Kendal, Central Java [36]
	309	31.09 ± 5.31	Farmers, Kebumen, Central Java [37]
Weaning weight (kg)	59	110.10 ± 11.20	Govt. facility, Kendal, Central Java [36]
	165	102.10 ± 12.19	Govt. facility, Grati, East Java [38]
	1819	98.87 ± 1.97	Farm, Tanjungsari, Lampung [39]
Yearling weight (kg)	165	127.30 ± 0.27	Govt facility, Grati, East Java [38]
First mating (month)	1407	23.06 ± 0.93	Farmers, Kebumen, Central Java [40]
	35	22–28	Govt facility, Ciamis, West Java [41]
First calving (month)	1407	32.46 ± 0.90	Farmers, Kebumen, Central Java [40]
Post-partum mating (month)	1407	4.37 ± 0.64	Farmers, Kebumen, Central Java [40]
	103	2.38 ± 0.86	Farmers, Tuban, East Java [42]
Service per conception	1407	1.97 ± 0.20	Farmers, Kebumen, Central Java [40]
	55	2.11	Govt facility, Ciamis, West Java [43]
	60	1.18 ± 0.39	Farmers, Lamongan, East Java [44]
Calving interval	103	16.94 ± 2.19	Farmers, Tuban, East Java [42]
	55	15.39 ± 2.43	Govt facility, Ciamis, West Java [43]

PO cattle are mostly kept by smallholder farmers in Java under low-input/low-output production systems, with low average daily gains of around 0.25 kg/day [45]. However, under a controlled research environment with fermented sorghum based feed, the average daily gain of PO cattle ranged from 0.8 to 0.9 kg/day [46], while Ongole cattle directly imported from Sumba and reared under grain-based feed in the same research centre based on a diet of 12.5% crude protein yielded an average daily gain of 0.94 kg/day [47].

Over recent decades the main genetic improvement strategy for PO cattle has focused on upgrading the breed to Brahman or Brahman cross [27], using Brahman semen to inseminate the existing PO females. As both PO and Brahman are *Bos indicus* breeds, the practice in Indonesia has been to name the resulting crossbred progeny as Brahman crosses. If the female crossbreds are subsequently inseminated again with Brahman semen, the progeny closely resemble the Brahman breed and hence, locally, they are known as Brahman.

Over the last two decades there has also been a marked change in the purpose of this breed, with an important use of the Ongole breed now being as dams in crossbreeding programs with exotic breeds such as Simmental and Limousin, using imported semen and AI. The F_1 crossbreds derived from this type of cross have proven to be highly productive in both tropical and temperate environments, because the very great genetic diversity between *Bos indicus* and *Bos taurus* ensures heterosis or hybrid vigour is maximized (Table 1). However, careful consideration needs to be given to how the F_1 crossbreds are subsequently used in breeding programs because of the strong possibility they will be joined with an inappropriate third breed, resulting in progeny that are poorly adapted to the stressors of tropical environments.

Based on the performance of PO cattle inferred from Table 2 (albeit in the absence of scientifically valid breed comparisons), their productive and reproductive characteristics suggest their performance does not differ markedly from other *Bos indicus* breeds. This in turn suggests that conservation of the PO as a pure breed may not be justified, particularly when it is considered the breed has been genetically improved through well-designed and commercially focused breeding programs in Brazil (where it is known as Nelore) over many generations. Rather, their best use may be as a dam breed in well-designed crossbreeding programs or in formation of composite breed(s) that specifically match the requirements of cattle production systems in Indonesia, as described in more detail in subsequent sections of this paper.

4.3. *Madura Cattle*

Madura cattle were initially believed to be derived from ancient crossbreeding of Bali and/or the wild ox *Bos bibos* and zebu cattle either in Madura or in Java, with the crossbreeding believed to have occurred ~1500 years ago when Indian culture was introduced to Indonesia [48]. In 1977, those authors [48] suggested that phenotypic evidence indicated Madura cattle could have been derived from three-way crosses between *Bos bibos* spp., *Bos indicus* and *Bos taurus* types. This theory has received substantial support from [49] based on the geographical distribution of bovine haemoglobin beta (Hbb) alleles in Southeast Asian cattle. That study on the genetic components of Indonesian cattle confirmed the mitochondrial DNA of Madura cattle was a mix of zebu (*Bos indicus*) and Banteng, while the Y chromosome contained traces of zebu and *Bos taurus* breeds [26]. A three-way cross that excluded Bali cattle may also explain why Madura bulls do not appear to suffer the same fertility problems as are believed to occur for crosses between Bali cattle and *Bos taurus* or *Bos indicus* [23].

Hence, Madura cattle are a unique, stabilized composite of different *Bos* spp., with the precise composition of *Bos* spp. still to be determined. Determination of the specific breed composition will best be achieved through use of genomic sequence information, as suggested in a later section of this paper.

Madura cattle are small- to medium-sized animals, with very homogenous but unique characteristics [28]. Full details of their characteristics are provided by [48,49]. They are reported to be one of the best draught cattle breeds and are very well adapted to the local conditions and traditional management systems of Indonesia [50], though again, the reports are not based on scientifically valid breed comparisons. Madura cattle are now distributed in East Java, Kalimantan, Sulawesi and East and West Nusa Tenggara Islands in Indonesia [51] and are popular for their perceived good beef quality and their ability to grow under harsh tropical environments [52], although another report suggests Madura cows only produce sufficient milk for their calves to grow slowly [28]. They are also reputed to be more resistant to Jembrana disease than Bali cattle [53]. Hence, they are very well accepted in Indonesia's dryland farming systems, particularly with regard to what are perceived to be higher growth and reproduction rates relative to Bali cattle [50].

Other than the common Madura cattle that have no cultural function (and where no sustained genetic improvement has occurred), there are two types of Madura cattle, namely Sonok and Karapan. Karapan is a bull racing event where Madura bulls are used. Sonok is also a cultural event, but where female Madura cattle are used in conformation contests (Figure 3).

Past selection within both the Sonok and Karapan lines has therefore been based on attributes that improve the ability of the animals to compete in these cultural events. Karapan cattle are characterized for their strength, agility and aggressiveness. By contrast, female Sonok cattle are judged by conformation traits, such as height at their withers, colour, body conformation, body condition, health and harmonious walking in a pair [50]. Bulls in the Sonok population must be descendants from cows that participate in Sonok contests. They are selected on phenotype, especially body conformation and some "beauty" standards (coat colour, horn shape, eye shape, etc.). Cows that perform well in Sonok contests and their male descendants are very popular for breeding purposes.

Hence, these two traditional festivals became the means for traditional selection and conservation of Madura cattle, and that may continue into the future unless the farmers are encouraged to change their breeding objectives to focus on profitability traits to service commercial beef markets. Caution should be taken though regarding these traditional breeding practices. As animal records are non-existent and the demand for superior males for breeding purposes is very high, there are reasons to be suspicious that the inbreeding level may be high. An unpublished study by Prastowo and Widyas in 2018 on the SRY genes of 18 Sonok bulls showed those genes can be narrowed down to just three haplotype groups. The productive and reproductive characteristics of Madura cattle are shown in Table 3.

Table 3. Productive and reproductive characteristics of Madura cattle (note these studies are not based on scientifically valid breed comparisons).

Trait	Mean ± s.d.	Number of Animals	Breed (Type)	Reference
Birth weight (kg)	16.89 ± 2.86	60	Madura	[53]
	16.25 ± 2.37	77	Madura male	[54]
	17.08 ± 2.40	60	Madura female	[54]
Weaning weight (kg)	90.32 ± 3.42	186	Madura	[55]
	97 ± 13.77	84	Madura	[56]
	151.58 ± 36.98	24	Madura (Sonok) female	[57]
Yearling weight (kg)	122.17 ± 6.15	186	Madura	[55]
	120 ± 10.86	84	Madura	[56]
	247.55 ± 40.48	11	Madura (Sonok) female	[57]
Mature weight (kg)	294.3 ± 43.0	30	Madura (Karapan) female	[54]
	279.1 ± 89.0	30	Madura female	[54]
	392.3 ± 60.4	37	Madura (Sonok) female	[54]
First mating (month)	198.5 ± 0.81	59	Madura	[52]
	22.63	39	Madura (Karapan)	[58]
	23.40 ± 4.17	291	Madura (Sonok)	[59]
First calving (month)	29.96 ± 0.81	59	Madura	[52]
	33.92 ± 3.88	291	Madura (Sonok)	[59]
Post partum mating (month)	3.44 ± 0.17	59	Madura	[52]
	2.57	39	Madura (Karapan)	[58]
	3.40	25	Madura (Sonok)	[58]
	1.48 ± 0.09	59	Madura	[52]
	1.47	39	Madura (Karapan)	[58]
	1.59 ± 0.3	291	Madura (Sonok)	[59]
Open Days (day)	5.53 ± 0.65	291	Madura (Sonok)	[59]

Currently there is insufficient information to determine whether conservation of the Madura breed as a pure breed is warranted because, depending on the genomic composition of the breed, it may be possible to regenerate the breed through crossbreeding. However, if the breed can be demonstrated scientifically to be more resistant to Jembrana disease [51], and have better growth and reproductive performance than Bali cattle [48], then conservation and selection of the breed based on existing populations would be much more useful than attempting to regenerate the breed.

Ongoing use of the breed also depends on whether future uses of the Madura breed are primarily commercially market-driven (where objectively measured traits in the breeding objectives are important and hence, their performance relative to other cattle breeds must be a strong consideration) or whether future breeding programs remain focused on traditional or cultural uses of the breed. The breeding objectives for these different uses of the cattle differ substantially, so if future breeding programs remain focused on cultural uses, then conservation of the breed would be justified due to the lengthy history of genetic improvement within the breed based on selection for cultural attributes, with perhaps a 'desired gains' approach being used to derive economic weightings for these cultural attributes in future (see Section 6.1 for further discussion of different types of breeding objectives).

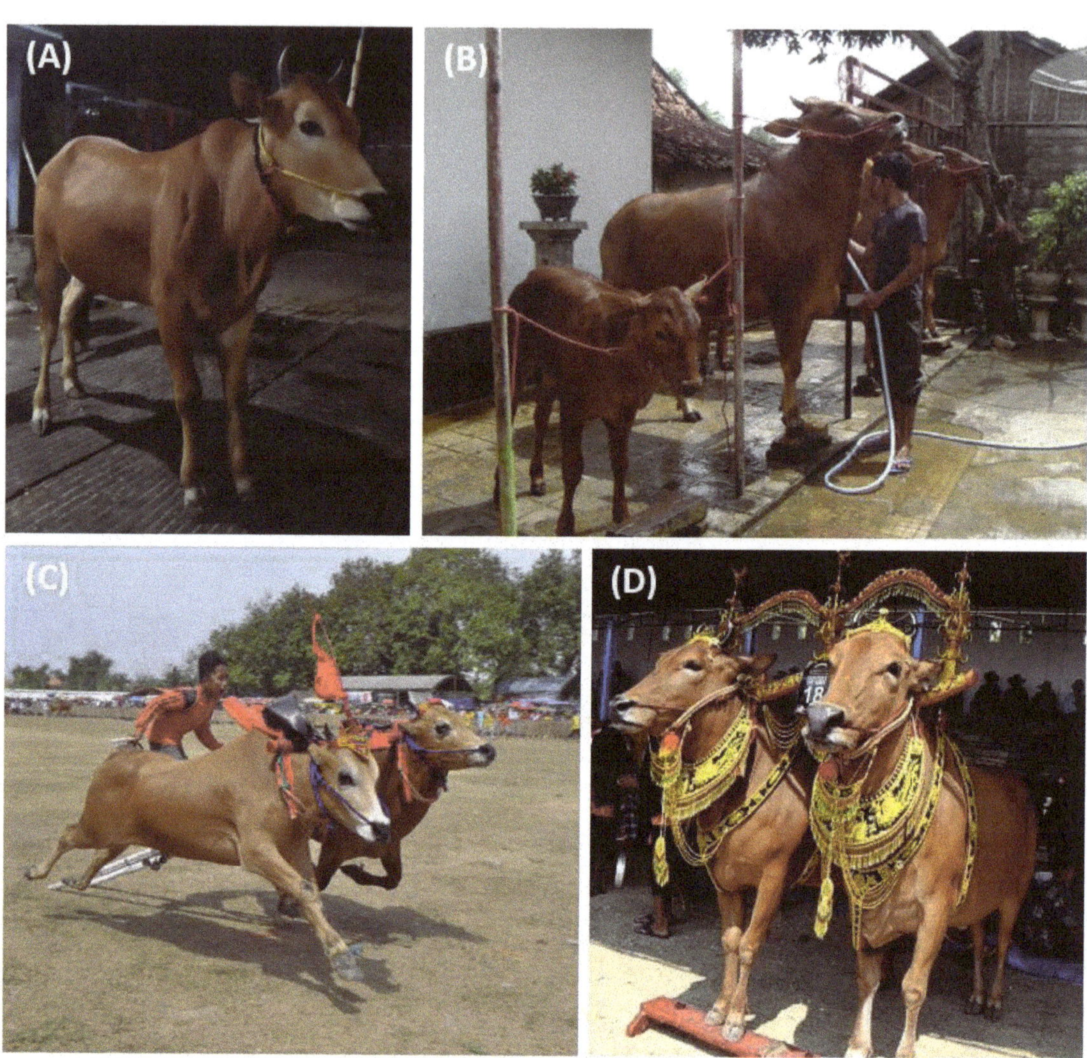

Figure 3. Images of Madura breed animals: (**A**) Madura bull–these bulls are typically selected for their strength, agility and aggressiveness to participate in Karapan cultural events; (**B**) Madura cows and calf; (**C**) Madura bulls participating in a Karapan contest; and (**D**) Madura cows used in Sonok cultural events, where they are traditionally selected on their conformation traits as well as their ability to walk harmoniously as a pair in these Sonok events.

Additional information about the Madura's breed composition through genomic analyses and ideally, additional phenotypic information of the Madura relative to the performance of other cattle breeds in well designed, controlled experiments would assist in decisions about the need to conserve the breed.

4.4. Challenges to Improving These Indonesian Cattle Breeds

Indonesia has established initiatives aimed at conservation and genetic improvement of Indonesian local cattle but those initiatives are unable to operate as needed to ensure genetic improvement is focused on both cattle productivity and adaptation. By way of example, the Bali cattle breeding centre located on the island of Bali conducts progeny tests to identify the "best" sires within the population [60]. Proven bulls from the breeding centres are then sold to AI centres, with semen from the bulls sold to farms across Indonesia. However, such progeny tests only rank the bull candidates based on the growth performance of their offspring, completely ignoring other economically important attributes in the offspring such as reproductive performance, beef quality and resistance or tolerance to environmental stressors. Together with an Indonesian law that forbids the importation of any cattle to Bali to ensure "pure" Bali cattle are conserved on Bali Island, the export of the best genetic resources from Bali each year (via semen sales from the AI centre to all areas of Indonesia) may have actually decreased the genetic merit of Bali cattle. By contrast, in the extensive production systems of Indonesia's palm oil plantations, Bali cattle are allowed to roam and mate naturally, with no formal breeding program applied and mating amongst relatives being common, potentially resulting in high levels of inbreeding.

The island of Madura is also subjected to the same law that forbids the importation of cattle to the island. However, unlike Bali, there is no facility responsible for breeding Madura cattle. Rather, cultural events have been primarily responsible for stimulating smallholder farmers to select female cattle based on body conformation, appearance and behavior and bulls on their strength and aggressiveness. Such selection was undertaken traditionally with no formal knowledge of animal breeding, though farmers are aware of the drawbacks of inbreeding even though that cannot be avoided completely due to limited population size. The use of visual appraisal to select breeding cattle has improved the cultural value of cattle, particularly in the Sonok population [61], with their productive performance generally regarded as comparable to their respective crossbred populations [50].

Hence, to improve the productivity and adaptation of cattle herds across Indonesia, well designed genetic improvement programs (crossbreeding and/or within-breed selection) need to be implemented, with a strong focus on economically important productive, adaptive and fitness traits and consideration also given to conservation of unique cattle breeds through development of practical breeding objectives, as suggested by [62] Nielsen et al. (2014).

5. Achieving Genetic Improvement by Crossbreeding

Crossbreeding has been widely used to improve livestock productivity in many countries to capture the advantages of heterosis or hybrid vigour resulting from crossing genetically diverse breeds [2,25]. Key reasons for crossbreeding include [63]:

- Generation of direct heterosis or hybrid vigor, measured as the extra performance of the crossbreds relative to their parental breeds. The percentage increase in performance ranges from 0–10% for growth traits and 5–25% for fertility traits [64], but the effect of heterosis on the total production system can be greater than this, as effects accumulate over traits e.g., [65]:
- The averaging of breed effects, to breed an animal of intermediate size to fit a particular management cycle or market demand and could involve either regular systems of crossing or the creation of composite breeds e.g., [65];
- Maternal heterosis where crossbred cows exhibit considerable heterosis in their ability to rear fast-growing, viable offspring;
- Sire-dam complementation, where the crossbreeding system aims to use breeding cows of small to intermediate mature size (though not so small for dystocia to be a problem) as well as being fertile. When a large terminal sire breed is used over a smaller dam, the proportion of available feed directed to growing animals is increased for the benefit of the entire production system, particularly where breeding occurs through AI; and
- Breed complementarity whereby breeds used for the crossbreeding program are specifically selected so the strengths of one breed are used to complement or mask the weaknesses of another breed.

The advantages of crossbreeding are greatest in tropical beef production systems because of the need to use breed types with greater diversity (e.g., *B. indicus* × *B. taurus*) to achieve adaptation in the offspring. This in turn maximizes the amount of heterosis in the cross [15,16,51,66,67]. However well-designed crossbreeding programs that reliably deliver improvements in productivity (other than the first crosses between unrelated breeds) are also very difficult to implement by most farmers globally due to the need to avoid reductions of heterosis (also known as recombination loss) in second and subsequent generation crossbreeds. This is particularly true in Indonesia, where sustainability of crossbreeding programs is frequently challenged by constraints such as poor adaptation of the crossbred progeny to the local environment or lack of logistical support [25]. Even in advanced countries with very large herds and sophisticated infrastructure, cattle breeders find it difficult to ensure segregation of bulls of one breed from cows of another to ensure appropriate crossbreeding. Hence, in those cases they have often preferred to develop stabilized composite breeds comprising appropriate admixtures of the different breed types.

In the following discussions on options for crossbreeding using Indonesia's cattle breeds, we acknowledge a considerable wealth of crossbreeding knowledge exists from other tropical areas of the world such as South and Central America, southern USA and northern Australia. However, we have elected not to present data from those studies because of the vastly different production systems, the significantly greater expertise and skills of the farmers in those different regions and their increased access to technologies for measurement and data capture relative to those of smallholder farmers in Indonesia.

To examine the feasibility of various crossbreeding programs in Indonesia's smallholder farming systems, Table 4 provides a brief examination of some selected crossbreeding designs and makes recommendations around which types of programs might be successful in Indonesia. In crossbreeding theory, there are many additional combinations of different cattle breeds, but at a practical level, none of the more complex systems is feasible for smallholder farmers in Indonesia and hence, they are deliberately excluded from Table 4.

Table 4. Examples of simple crossbreeding designs and the feasibility (as assessed by the authors) of managing them by Indonesia's smallholder beef farmers [25].

Crossbreeding System and Key Features	Feasibility of Implementation in Indonesia
F_1 cross: bulls of Breed A (e.g., *Bos indicus*) are joined to cows of unrelated Breed B (e.g., *Bos taurus*) to generate F_1 progeny (AB). F_1 AB progeny are then either sold or joined to a third unrelated breed as described in the next cross	Generating F_1 crosses between unrelated breeds using AI is feasible (see discussion below) but selling all F_1 females will cause problems for smallholder farmers unless they can purchase new purebred Breed B heifers to retain sufficient, productive females in their herds. If the F_1 females are used for breeding, then it is essential they be joined to a well-adapted, unrelated third breed of bull to ensure their progeny are sufficiently well adapted to environmental stressors
Backcross: F_1 AB females are backcrossed to a bull of either Breed A or breed B, generating progeny that are 25% Breed A, 75% Breed B or vice versa. In subsequent joinings, 75% Breed B females are joined to purebred Breed A bulls, generating progeny that are 62.5% Breed A, 37.5% Breed B or vice versa	The major problem with a backcross program in Indonesia is that, in addition to some loss of heterosis in the F_2 *et seq.* progeny, almost inevitably some progeny in F_2 *et seq.* generations (i.e. those comprising 62.5% *Bos taurus*) will lack the resistance to environmental stressors required to ensure their productivity in smallholder farming systems
Terminal three-breed cross: bulls of an unrelated Breed C (e.g., *Bos javanicus*) are joined with F_1 female AB progeny to generate progeny comprising 25% Breed A, 25% Breed B and 50% Breed C	If issues associated with retaining a productive breeding herd described above can be overcome, then a 3-breed cross is expected to be as productive as the F_1 AB cross, However all 3-breed cross progeny would need to be sold or used to form a composite breed. This means that, as with the F_1 cross above, smallholder farmers would need to purchase replacement females to maintain the size of their breeding herds. Such a system is unlikely to be feasible amongst smallholder farmers in Indonesia
Formation of composites or stabilized crossbreeds: an alternative to the 3-breed cross described above could be to inter-se mate bulls and females having the same (or additional) breeds in their genetic composition e.g., join F_1 Breed A × B bulls with F_1 Breed C × Breed D females to produce F_2 progeny with 25% of each of Breeds A, B, C and D and then continue to inter-sé join the 25% A × B × C × D progeny over subsequent generations	Providing the progeny of these crosses were sufficiently resistant to the stressors of tropical climate, development and maintenance of a composite breed in this type of design would capture the benefits of heterosis and the system could be managed in a relatively straightforward way by smallholder farmers. Ideally, a progeny test program could be introduced in Indonesia to identify genetically superior animals in such a stabilized composite breed (see later section in this paper). And if the third unrelated breed type included a representative breed(s) from the tropically adapted taurine breeds (Table 1), then expected recombination losses due to epistasis would be significantly reduced relative to F_2 *et seq.* crosses between *Bos indicus* × *Bos taurus*

The Indonesian government initiated crossbreeding programs based on AI and imported semen in the early 1980s [51]. Although those programs have continued to be supported by a number of government incentives since then, significant improvements in the productivity of beef herds in the country are yet to be demonstrated [2], probably because F_2 *et seq*. generation crossbred progeny experienced recombination losses due to the difficulties in maintaining appropriate crosses of the different parental breeds in smallholder farmer herds. Some successes in improving individual animal productivity were achieved, but there has been no widespread impact on increasing cattle growth rates or the national cattle population through the improved reproductive performance of crossbred females [51], with no economic benefits at the farm level and no observable improvement in the adaptation of the crossbred offspring to the environmental stressors [51].

The lack of control of these Indonesian crossbreeding programs at the farm level has led to the emergence of unidentified mixed breeds of cattle [65]. Without good knowledge of the composition of these cattle in terms of 'adapted' versus 'productive' genetics, it has not been possible to achieve the primary aim of crossbreeding in Indonesia i.e., to maximize animal productivity without compromising cattle adaptation to the tropical environment [19]. Furthermore, these poorly designed crossbreeding programs may have jeopardized the ongoing conservation of unique local cattle genetic resources, which need to be retained to maintain biodiversity.

Some examples of simple crossbreeding systems based on Bali and Madura cattle in Indonesia are summarized below to allow recommendations to be made on the future directions of genetic improvement of cattle herds in Indonesia (either through crossbreeding and/or within-breed selection–see later sections of this paper.)

Crossbreeding has occurred between *Bos javanicus* (Bali) and *Bos taurus* or *Bos indicus* breeds outside Bali Island. The crossbred progeny had improved growth performance relative to pure Bali cattle [68–71], but field observations reported the male progeny had reproductive problems. Another problem was related to high mortality rate of calves under extensive production systems [29]. Causes of these high mortality rates have not been identified but field reports from oil palm plantations suggest that female Bali cattle have poor mothering abilities as they often left their new-born calves under those extensive production systems, thereby contributing to the high calf mortality rates. Previous studies also showed a decline of population size and genetic quality of Bali cattle represented by a decrease of growth-related traits (such as body size and weight) over generations [27,72].

Table 5 reports results of crossbreeding experiments based on progeny of Bali females and F_1 crosses with Simmental, Limousin, Brahman and PO bulls. The crossbreeding results are derived from independent experiments in AI centres in Papua and Nusa Tenggara provinces, while the exotic breed purebred data are derived from other AI centres. Hence the results are not directly comparable across experiments and should therefore be interpreted with care. The F_1 crossbred progeny were heavier at weaning (standardized to 205 days) and yearling ages, had higher average daily gains from the age of 7 to 12 months and they had larger body measurements compared to purebred Bali progeny. Table 5 was derived from [71] and compares economically important traits in beef cattle from different studies. It was aimed to give an initial awareness that crossbreeding using Bali cattle as dams to genetically improve their productivity requires greater consideration. It was concluded that the benefits of hybrid vigor or heterosis just for these growth traits were maximized due to the genetic diversity of the parental breeds [71].

Table 5. Growth performance of the progeny of purebred Bali cows and F_1 crosses with unrelated sire breeds.

Breed	Weaning Weight (Kg)	Yearling Weight (Kg)	Average Daily Gain [1] (Kg/day)	Body Measurement (cm) [2]		
				Wither Height	Body Length	Chest Girth
Bali (in Bali)	85.61 ± 14.30 [1]	135.09 ± 24.22 [1]	0.36 ± 0.24	107 ± 7.58	112.6 ± 11.67	149.2 ± 8.64
Simmental	235.50	430.00	-	-	-	-
Limousin	225.00	394.50	-	-	-	-
Brahman	223.50	372.00	-	-	-	-
PO	102.13	134.30	-	-	-	-
F_1 Simmental × Bali	131.61	179.21	0.26 ± 0.12	119.8 ± 4.76	123 ± 2.55	155.2 ± 14.15
	105.65 ± 5.27 [3]	200.43 ± 21.52 [3]				
	121.67 ± 16.18 [4]	230.98 ± 30.23 [4]				
F_1 Limousin × Bali	128.75	176.80	0.38 ± 0.06	119.2 ± 2.95	120.4 ± 10.04	153.6 ± 3.85
F_1 Brahman × Bali	115.90	157.60	0.45 ± 0.21	120.71 ± 4.31	127.43 ± 13.21	163 ± 12.9
F_1 PO × Bali	111.66	148.25	0.41 ± 0.25	120.4 ± 3.36	124 ± 7.04	339.2 ± 43.61

[1] [71]; [2] [70]; [3] [68] in lowland; [4] [68] in highland.

In contrast to growth traits for male progeny in Table 5, the female F_1 Bali crossbreds had poorer reproductive performance than their purebred Bali contemporaries. A similar occurrence of infertility was reported in other groups of male crossbred offspring [73]. The reproductive performance of F1 Bali × Simmental and F1 Bali × Limousin females is worse compared to similar groups of purebred Bali female cattle that were reared by smallholder farmers with similar production systems but not in the same contemporary group as the crossbred females (Table 6). The crossbreds had longer days open and increased calving intervals, lower pregnancy and calving rates, and higher pre-weaning mortality rates, resulting in the crossbreds having lower overall reproductive efficiency compared to purebred Bali females reared by smallholder farmers with similar production systems. These results are atypical of most crossbreeding studies elsewhere, where generally the greatest heterosis is achieved in traits with the lowest heritabilities, such as female reproductive performance, meaning that F_1 crossbreds usually out-perform either of the parental breeds. Reasons for these results in Indonesia could be due to the need for significantly greater feed inputs in the much larger crossbred females. In beef cattle production systems, female reproductive performance is a key trait and considered to be the most important factor economically [74], especially for smallholder farmer cow-calf production systems. Poor reproductive performance of breeding cows results in major economic losses, due to the additional expenses needed for feed, labor, breeding and animal health costs as well as the costs of calf losses [74].

It is possible that chromosome number imbalance from crossing of Bali cattle (*Bos javanicus*) with *Bos taurus* and *Bos indicus* species may have resulted in infertility of the female crosses, not only in males as suggested by [23]. However, it should also be noted that the reproductive performance of all breeds in this table are higher than the reproductive performance reported by most studies undertaken in extensive tropical pastoral production systems elsewhere in the world, suggesting that genetic aberrations are unlikely to be the reason for this lower performance of the crossbreds. It is possible the high reproduction rates simply reflect the low numbers of animals owned by smallholder farmers and hence, improved management of individual breeding cows is not only feasible but also very likely. Although the studies are not directly comparable, based on these results and the high economic weighting of reproductive performance in most cow-calf breeding objectives, smallholder farmers would likely improve the productivity and profitability of their herds by continuing to breed purebred cattle rather than trying to manage complex crossbreeding programs with exotic large European bull breeds.

Table 6. Reproductive data performance of Bali and F_1 crossbred females, where the purebred cows were reared by smallholder farmers in similar production environments [(a)].

Reproductive Performance	Bali	F_1 Bali × Simmental	F_1 Bali × Limousin
Days open (days)	81 ± 10	125 ± 23	-
Calving interval (days)	363 ± 20	412 ± 36	387
Pregnancy rate (%)	94.2 ± 3.8	78.9 ± 5.3	-
Calving rate (%)	88.5 ± 4.1	76.3 ± 6.3	75
Pre-weaning mortality (%)	8.7 ± 0.4	17.2 ± 1.2	5.6
Observation year	2015	2015	2019
Reference	[69]	[68]	[75]

[(a)] It should be noted that the purebred cows were not managed in the same contemporary groups as the crossbred cows, which in turn were recorded in independent studies. The numbers of progeny per sire are not available due to the lack of calving records across all three studies. Females ranged in age from 4 to 8 years but the study did not differentiate performance across the different ages and nor was age at first calving recorded (an important variable that may favour purebred Bali cattle because the F_1 crossbreds would have greater nutritional requirements than Bali cattle over their lifetime and those needs are unlikely to be achieved under smallholder farmer systems. A lack of sufficient nutrition during the dry season is also believed to be responsible for the high pre-weaning calf mortality rates of F_1 Bali × Simmental cows.

In a different study, several genes such as GH, FSHR, BMP15 were reportedly involved with reproductive function in Bali cattle, but the correlations between those genetic variations and reproductive performance were low [76] suggesting they are unlikely to be having a major impact on female reproductive performance. It is not known whether those variations may also be associated with reproductive performance in Bali crossbred cattle. Further studies based on genomic information from *Bos javanicus* and other cattle species is warranted to determine the role of genetics in reproductive performance of these types of cattle.

With regards to Madura cattle, the Government of Indonesia also recommended genetic improvement by crossbreeding with *Bos taurus* breeds such as Limousin [11]. The aim of such crossbreeding was to produce offspring with larger and heavier body sizes with higher selling prices, but which retained the preferred Maduranese traits such as dark red coat color. Table 7 shows the physical characteristics of F_1 Madura × Limousin females. However, the crossbreeding study was poorly designed and uncontrolled in practice [27] and hence, there is concern the uncontrolled crossbreeding will threaten the conservation of Madura cattle genetic resources due to the lack of adaptation of the crossbred animals [11].

Table 7. Physical characteristics of Madura × Limousin cows based on different observational studies from Madura Island comparing sub-populations of Madura cattle (Sonok, Karapan, non-selected Madura and Limousin × Madura). Results from the different studies reported in this table are not directly comparable and hence, caution should be taken when interpreting the results.

Physical Characteristics	Mean ± sd	Reference
Age (year)	4.5 ± 0.2	[50]
	>4.0	[52]
Body weight (kg)	400.1 ± 92.6	[50]
	406.6 ± 28.8	[52]
Height at the withers (cm)	125.7 ± 6.2	[50]
	126.0 ± 1.9	[52]
Body condition score	3.8 ± 0.7	[50]

The Value of Crossbreeding to Genetically Improve Indonesia's Cattle Herds

Based on this review of crossbreeding studies using Indonesia's cattle breeds, it is clear there are no comprehensive and well-designed crossbreeding studies undertaken to enable valid comparisons of different cattle breeds and crossbreeds and to measure the extent of heterosis and recombination loss under Indonesian beef production systems. Well-designed crossbreeding studies are still required to enable scientifically valid conclusions to be drawn about the role of crossbreeding in Indonesia, although it is unlikely that Indonesia's smallholder cattle farmers would be able to effectively manage the complexities of those types of studies. Additionally, in crossbreeding studies undertaken in Indonesia to date, there has been no distinction between different generations of crossbreeds, other than where progeny are known to be F_1 generation because they are bred using AI over cows of known pure breeds. The distinction between crossbred generations is critically important because of differences in the amount of heterosis and recombination loss in the different generations and crosses between different species, where the amount of both heterosis and recombination loss varies significantly depending on the generation and parental breeds used to generate the crosses.

Since the introduction of crossbreeding programs based on imported semen, it appears that most semen used in Indonesia has been derived from large European breeds such as Simmental and Limousin. Based on what is known of species performance (Table 1), these European sire breeds are not a logical choice for use in Indonesia's tropical production systems, firstly because of their known calving difficulties even when joined to the same cow breeds (meaning that dystocia can be anticipated as a problem if they are joined to the smaller Indonesian cow breeds) and secondly, because of their need for larger quantities of feed and poorer adaptation to tropical environments than other potential sire breeds.

It is therefore recommended that future crossbreeding programs be specifically designed around breeding objectives focused on high productivity and high adaptation to environmental stressors, to meet emerging commercial market requirements for high quality beef, unless cultural factors are expected to continue to be important as is likely in the case of Madura cattle. Under that scenario, the best option for the Madura breed would be to concentrate on within-breed selection, specifically focusing on the important cultural attributes.

6. Achieving Genetic Improvement by Within-Breed Selection

As described previously, the main beef cattle genetic improvement strategy implemented by smallholder cattle farmers in Indonesia to date has been crossbreeding. However, that strategy has not delivered the expected results, partly due to lack of consideration of appropriate breeding objectives that indicate cattle in Indonesia need to be both highly productive and very well adapted to the stressors of tropical environments. In the absence of well-defined breeding objectives, poor choices were made about the use of sire breeds in those programs, resulting in the generally poor performance of the crossbred progeny. Even with the development of appropriate breeding objectives and improved breed choices, it is not clear from the previous section that smallholder cattle farmers in Indonesia would be able to manage the many complexities of designed crossbreeding programs to enable them to achieve effective genetic improvement of their herds.

Due to the complex management requirements of crossbreeding systems, the opportunity to achieve genetic improvement using within-breed selection needs to be evaluated because genetic improvement has very significant potential to improve the productivity, profitability and adaptability of Indonesia's cattle herds.

To date, two substantially different approaches have been taken to genetically improve Indonesian cattle herds using within-breed selection. The most successful approach has been the improvement of the Madura breed for cultural purposes, where use of visual selection successfully improved cows' body conformation, appearance and behaviour and bulls' strength and aggressiveness. However, it is not clear whether there is an ongoing need to continue visually improving these cattle for cultural reasons or whether more commercially-oriented breeding objectives are likely to become more relevant in future. The second within-breed selection approach used in Indonesia is the ongoing use of progeny tests to select Bali bulls, but based only on the growth performance of their offspring while completely ignoring other economically important traits such as reproductive performance, beef quality and adaptation to environmental stressors [60], some of which are known to be negatively genetically correlated in other breeds of cattle.

It therefore appears that commercially-relevant, market-driven breeding objectives and effective within-breed selection services to enable selection of cattle based on those breeding objectives will need to be evaluated, and their feasibility determined, prior to establishing new systems from the beginning, if within-breed genetic improvement is to effectively achieve ongoing genetic improvement of cattle herds in Indonesia.

As described by [77], traditional genetic improvement programs based on measuring large numbers of pedigree-recorded animals in well-defined cohort groups for the full range of economically important productive and adaptive traits is generally impossible for smallholder farmers and particularly those in tropical environments such as in Indonesia, where environmental stressors encountered by the cattle significantly compound the difficulties of implementation. Even the Bali breed progeny test program described earlier in this paper can only obtain accurate pedigrees for, at most, 3 generations because the program is not undertaken in a sustainable way. The system used by that program evaluates 3–5 bull candidates in one year, with the candidates identified based on mass selection of growth traits. The bulls are then mated to multiple females to produce multiple offspring, with the progeny test then being used to rank the bulls on the growth performance of the offspring. For subsequent years, the same system is applied but without any genetic linkages created across the different years.

However, the opportunity now exists to use genomic data in conjunction with the use of digital information and communication technologies to develop new opportunities to improve the rates of cattle genetic gains by characterizing indigenous and crossbred animals for use in conservation, crossbreeding and within-breed selection programs, to improve economically important traits. Use of genomic information is costly, but keeping large numbers of animals over many generations to obtain accurate pedigrees and genetic parameters is also very costly with regards to both financial and time investments and is also generally not feasible for smallholder cattle farmers in Indonesia.

6.1. Developing Breeding Objectives for Indonesia's Cattle Smallholder Farmers

Breeding objectives define the "ideal" animals that a farmer wants to breed, with breeding objectives applying to both crossbreeding and within-breed selection programs. Generally, breeding objectives are defined by identifying the traits affecting the profit of the cattle business, as well as the importance of each trait to that profit. A breeding objective is specific to a particular market or group of similar markets, meaning it is important to understand the market requirements. Depending on the target market, some traits have greater importance than others (for example, live weights early in life as an indication of live weight at sale). However, the breeding objective also needs to consider factors that might promote or detract from achieving those important goals (for example, if cattle in Indonesia are not well adapted to the environmental stressors that are endemic in Indonesia, they will not grow to market weights as expected, a case in point being the overall failure of crossbreeding programs in Indonesia to achieve their potential highlighted in an earlier section of this paper.

In terms of the genetics of the traits, some traits are highly heritable or readily passed on from one generation to another and so greater progress towards breeding objectives can be achieved by targeting traits that are highly heritable (although there is also good evidence that strong genetic progress can be made by focusing on traits such as cow annual weaning rates that are lowly heritable). Greatest progress towards achieving the breeding objective will be achieved by focusing on traits of economic importance rather than traits associated with the personal preference of the breeder.

In the case of Madura cattle used for Sonok and Karapan festivals, where cultural or traditional attributes are important, an economic weighting could be derived based on the value of those cultural or traditional attributes to the sale price of the cattle, as well as potentially to the value of the cattle to the communities based on income from the festivals. Regardless of the traits included in the breeding objective, it will be important for the farmer to record the desired animal traits impacting on enterprise profitability and to estimate the relevant importance of each of those traits. From there, the economic impact of changing each important trait can be calculated from both financial and production data.

A sequential procedure to enable development of beef cattle breeding objectives is presented by [78] and includes four phases: (1) specify the breeding, production and marketing system; (2) identify the sources of business income and expenses; (3) determine the biological traits influencing the income and expenses; and (4) derive the economic value of each trait, which those authors recommended be based on discounted gene flow method. However, there are a number of alternative approaches such as simple profit equations, bio-economic models simulating the whole production system or use of desired gains approaches (for example with cultural or traditional attributes in Madura cattle) in combination with profit equations or bio-economic models (to adjust for undesirable genetic changes) that could be used [62]. Additional studies provide examples of the different applications of breeding objectives in both beef cattle cross- and straight-breeding programs [62,78–80].

Amongst smallholder cattle farmers in Indonesia, it will therefore be important that they not only focus on optimizing cattle productivity, but their production systems should also account for all traits related to productivity (growth, reproduction, product quality), adaptability (resistance or tolerance to environmental stressors), sustainability (animal health and welfare) [19,25,81–83] and even cultural and traditional attributes where they are economically important.

6.2. Requirements for Within-Breed Selection Programs for Indonesia's Smallholder Beef Farmers

Conducting a within-breed genetic improvement program for smallholder farmers in Indonesia is not straightforward. However, there is clear evidence from many countries that within-breed selection for a range of economically important productive and adaptive traits in the breeding objective has resulted in permanent genetic improvement of those traits, directly benefiting not only the pure breeds under selection, but also animals in crossbreeding and composite development programs. This is particularly true with the use of genomic (DNA based) information [84,85] in the breeding programs.

Although many successes have been reported in the past, conventional genetic improvement programs that rely on measurement of large numbers of pedigree-recorded animals in well-defined and controlled cohort groups is time consuming, laborious, financially costly and in Indonesia, probably too complex for smallholder cattle farmers to manage. However, recent major advances in genomic technology (summarized by [77]), in conjunction with the use of new digital information and communication technologies that allow automated or semi-automated data collection, are now enabling very significant new opportunities to improve the productivity of livestock industries in countries like Indonesia through the use of genomic selection, which uses genome-wide genetic markers to estimate the genetic merit of individual animals [86–88].

Within Indonesia, there is a consortium of researchers from several universities and Indonesian government agencies with interests in conserving and genetically improving Indonesia's unique livestock breeds. Strong support and willingness to engage have also been expressed by potential international collaborators and funding agencies. Hence, a primary purpose of this paper is to demonstrate that possible solutions to implementing genetic improvement programs for smallholder cattle farmers do exist. The major requirements for setting up new within-breed selection program(s) in Indonesia are therefore briefly summarized below.

- *Accurately recorded phenotypes*: The main limitation to genomic and traditional within-breed selection in extensively managed livestock such as beef cattle is the difficulty and expense of measuring animals in appropriately sized contemporary groups for the full range of economically important productive and adaptive traits. Unless the genetic improvement programs are adequate in terms of contemporary group size and structure, the measurements will not enable useful predictions of genetic merit. A related paper [77] provides a detailed description of the phenotypes that should ideally be included in cattle breeding objectives and the feasibility of recording them in smallholder farmer herds in low-middle income countries. As indicated by [77], measurement of most phenotypes required for genetic improvement programs in smallholder herds is generally not feasible. Hence, those authors recommended establishing reference populations that are genetically linked, but managed separately, to smallholder cattle herds. The feasibility of setting up reference populations for this purpose in Indonesia is explored further below.

- *Genetic parameters for all traits in breeding objectives:* As indicated in an earlier section of this review, the only known within-breed selection program to focus on improvement of an objective trait in Indonesia is that undertaken for Bali cattle and based only on cattle growth. Heritabilities for live weights at different ages and genetic correlations between them are reported by [89–91]. However, there are no known genetic parameters for other traits for beef cattle reared under Indonesian production systems. There are, though, many reports from the scientific literature providing heritabilities and genetic correlations between most economically important traits, though the estimates based on adaptive traits and new traits (e.g., carcass and beef quality, efficiency of feed utilization, methane emissions) are more limited than those derived for growth and reproductive performance. Hence, the best approach to establish within-breed selection programs for use in Indonesia would be to initially use published estimates from other tropical and sub-tropical environments, with the aim of ultimately estimating relevant parameters from Indonesia's cattle herds in future. This approach would enable within-breed selection programs for the full range of economically important traits to commence, once the other challenges to establishing such programs have been overcome. A positive aspect of such an approach is that benefits would accrue not only to within-breed genetic improvement, but also to crossbred populations, as reported by [84].
- *Data analyses and estimation of genomic breeding values*: The previously-mentioned related paper [77] also provides a detailed description of optimal models for predicting genomic estimated breeding values (GEBVs) and the value of genomic data in replacing recorded pedigree information. It additionally compares the positive and negative aspects of different approaches (BayesR, GBLUP) to estimating GEBVs, as applied specifically to genetic improvement programs for low-middle income countries, as well as describing the benefits of imputing full sequence information from lower density Single Nucleotide Polymorphism (SNP) panels.
- *Establishing infrastructure and human capacity building*: As outlined by [77], two problems of major significance to genetic evaluation systems for smallholder farmers in low-middle income countries are: (i) the lack of infrastructure required to undertake on-farm management and phenotyping of animals, laboratory testing of animal samples, data capture and storage and lack of computing facilities and appropriate software programs; and (ii) lack of human capacity, particularly in areas of technological capability and data analysis and interpretation. As those authors describe, new developments in the use of information and communication technologies are starting to address the issues of on-farm management and phenotyping of animals, whilst the development of livestock reference populations and formation of international collaborations currently hold greatest promise of addressing the remainder of these challenges.
- *Possibility of establishing cattle reference populations in Indonesia*: The related paper [77] also provides detailed descriptions of the role and value of livestock reference populations that have been established in multiple developed and developing countries globally, specifically to overcome the difficulty and expense of obtaining accurate phenotypes in large and well-designed cohort groups. The value of forming such reference populations in Indonesia would be increased considerably if they could

also be combined with international collaborations to enable sharing of computing platforms, data storage, analytical software, joint data analyses and human capacity development in all aspects of genetic improvement programs as also described by [77]. To date though, there are currently no known reference populations for smallholder beef cattle in any low-middle income country, primarily due to lack of the significant funding required for their establishment and the significant length of time required to achieve genetic improvement in those herds, which have an average generation interval of 4–6 years [77]. For Indonesia to establish such populations, this would mean not only identifying new sources of funding to support establishment of such populations, but potentially an even greater challenge in securing sufficiently large areas of suitable cattle grazing land in such a highly urbanized country to enable adequate numbers of beef cattle to be managed and recorded within well-designed cohort groups. The land area challenge and potential solutions will be examined in greater detail as part of a design phase, assuming the new sources of funding can be achieved.

- *Could international collaborations help overcome these challenges?*: As described in detail by [77] and summarized previously in this paper, international collaborations with genetic evaluation providers servicing smallholder farmers in other countries in tropical areas would help Indonesia overcome most of the challenges currently facing Indonesia. Additional benefits from international collaborations would include the need for fewer animals with recorded phenotypes and for less common cattle breeds (e.g., Bali and Madura cattle) where data are very limited, as evidenced by recent cross-country studies where pooled data were used to accurately estimate GEBVs for tick resistance in African and Australian breeds of cattle with limited data. The exceptional challenges that would remain for Indonesia are perhaps the most challenging though, as they are based on the need for new funding to help establish and maintain the resource populations over perhaps 10–20 years and access to the areas of land on which those populations would need to be managed. However, if the resource populations were able to be established, it is anticipated that, after the initial 10–20 years of operation, new business models would be implemented to allow farmers and other beef value chain participants who benefit directly from the genetic improvement to assume control and ongoing funding of the populations, as is now starting to occur in other countries.

7. The Critical Role of Genomic Information in Designing and Implementing New Beef Cattle Genetic Improvement Programs in Indonesia

As summarized by [77], potentially the greatest opportunity to significantly improve the productivity of all livestock industries in low-middle income countries in tropical environments is through the use of genomic information, with recent significant advances in genomic technologies greatly improving that potential. However, we also note that the value of genomic information for livestock breeding depends on the availability and quality of essential phenotypes for the full range of economically important productive and adaptive traits important in tropical environments.

Given the relatively poor success of earlier cattle breeding programs in Indonesia as summarized in this review, it is suggested the best way to design new cattle breeding programs (within-breed selection, formation of composites, crossbreeding programs and/or conservation of unique cattle breeds) would be for Indonesian researchers to establish international collaborations and work with those collaborators to develop the opportunities provided by genomic technologies and to simultaneously develop the capacity of Indonesian researchers in the use of those technologies.

In our view, this would best begin with sequencing the genomes of Bali (*Bos javanicus*), Madura and PO breeds, and comparing their sequences directly with the genomic sequence of *Bos taurus* and *Bos indicus* breeds. This would ideally be undertaken in collaboration with researchers involved in the 1000 Bull Genomes project [92]. Thereafter, lower density

SNP panels would be used for ongoing genotyping of animals recorded in those breeding programs. Results from such comparisons would:

- Determine whether the Indonesian breeds are unique and hence, require conservation as pure breeds;
- Definitively confirm the breed composition of Madura and PO cattle, thereby assisting in decisions about how best to use those cattle in future genetic improvement programs targeting productivity of Indonesian cattle herds;
- Confirm whether chromosome imbalance will potentially cause problems with crossbreeding of *Bos javanicus* with *Bos indicus* or *Bos taurus*;
- Assist in the design of within-breed selection, crossbreeding and conservation programs for these breeds; and
- Provide the basic information needed for ongoing studies to identify genes or genomic regions in those cattle breeds that account for large proportions of genetic variation in economically important traits.

Thereafter, the international collaborations established by Indonesian researchers would potentially also assist in the design and implementation of new breeding programs, particularly if it was possible to establish the proposed reference populations to measure and manage the cattle herds required to achieve the targeted genetic improvement (with smallholder beef farmers using AI and semen from proven genetically superior sire to achieve improvement in their own herds).

Once established, those reference populations would routinely use low-density, low-cost SNP panels and imputation to full sequence to provide genomic data (including pedigree information) for all animals in the reference populations. The genomic information, combined with measured phenotypes, would then be used to estimate GEBVs for every animal in the population and allow the effective design of within-breed selection programs that also prevent the rapid and large increases in inbreeding that are occurring in some cattle populations in Indonesia due to small population sizes.

Genomic information would also allow the design of crossbreeding programs based on precise knowledge of the breed composition of all animals to be used in the programs, thereby ensuring breeding females are joined to appropriate bulls in order to generate well adapted and productive crossbred progeny. This approach has been successfully used in controlled crossbreeding programs by smallholder dairy farmers in Africa and India [93–96].

In the relatively short-term future, the use of genomic (and other "-omics") information also has the potential to reduce and possibly replace the requirements for some difficult- or expensive-to-record phenotypes that currently represent a major constraint to effective implementation of genetic improvement programs in low-middle income countries. The greatest limitation to achieving this is the current lack of accurately-recorded phenotypes for those difficult- or expensive-to-record traits, against which the genomic information can be tested.

8. Conclusions

To respond to Indonesia's increasing need for greater supplies of beef, as well as to improve the productivity and profitability of Indonesia's smallholder cattle farmers, sustainable utilization and genetic improvement of Indonesia's local cattle are vitally important.

Based on this review of the literature relevant to Indonesia's smallholder beef breeding programs, it appears that to date, all crossbreeding and within-breed selection programs implemented in Indonesia (except for the visual selection for cultural attributes amongst Madura cattle) have not achieved the levels of genetic improvement that were initially targeted. Previous crossbreeding programs appear to have failed primarily through poor design of the programs as well as the complexity of management required for such programs, beyond the skill levels of smallholder farmers. The sole within-breed selection program based on objective measurements was restricted just to selection of bulls based on live weights, whilst ignoring other productive and adaptive traits known from the published literature to be genetically correlated (sometimes negatively).

Hence, to achieve the permanent genetic improvement required to improve the productivity and profitability of Indonesia's smallholder farmer cattle herds, and to achieve conservation of Indonesia's unique cattle breeds, the opportunity now exists to develop and implement entirely new breeding programs.

Regardless of how the local cattle breeds are utilized, genetic improvement programs with well-formulated breeding goals should be designed and implemented to improve the ongoing productivity and profitability of smallholder farmer herds, as well as to conserve unique Indonesian genetic resources.

Author Contributions: Conceptualization, methodology, and original draft: N.W. and H.M.B.; writing and editing: N.W., H.M.B. and S.P.; criticized the manuscript: T.S.M.W., I.S. and B.J.H. All authors have read and agreed to the published version of the manuscript.

Funding: The Australian Centre for International Agricultural Research (ACIAR) is acknowledged for provision of funding for open access publication of this manuscript through grant number GMCP/2020/149.

Conflicts of Interest: The authors declare they have no conflict of interest.

References

1. Rohaeni, E.S.; Sumantri, I.; Yanti, N.D.; Hadi, S.N.; Hamdan, A.; Chang, C. Understanding the farming systems and cattle production in Tanah Laut, South Kalimantan. In Proceedings of the IOP Conference Series: Earth and Environmental Science, Yogyakarta, Indonesia, 23–25 September 2019; Volume 387.
2. Bunmee, T.; Chaiwang, N.; Kaewkot, C.; Jaturasitha, S. Current situation and future prospects for beef production in Thailand—A review. *Asian-Australas. J. Anim. Sci.* **2018**, *31*, 968–975. [CrossRef]
3. Panjaitan, T.S.; Fordyce, G.; Quigley, S.P.; Winter, W.H.; Poppi, D.P. An integrated village management system for Bali cattle in the eastern islands of Indonesia: The 'Kelebuh' model. In Proceedings of the 13th Animal Science Congress of the Asian-Australasian Association of Animal Production Societies, Hanoi, Vietnam, 22–26 September 2008; p. 576.
4. Nulik, J.; Dahlanuddin, H.; Hau, D.K.; Pakereng, C.; Edison, R.G.; Liubana, D.; Ara, S.P.; Giles, H.E. Establishment of *Leucaena leucocephala* cv. Tarramba in eastern Indonesia. *Trop. Grassl.-Forrajes Trop.* **2013**, *1*, 111–113. [CrossRef]
5. Dahlanuddin; Yanuarianto, O.; Poppi, D.P.; McLennan, S.R.; Quigley, S.P. Liveweight gain and feed intake of weaned Bali cattle fed grass and tree legumes in West Nusa Tenggara, Indonesia. *Anim. Prod. Sci.* **2014**, *54*, 915–921. [CrossRef]
6. Dahlanuddin; Ningsih, B.S.; Poppi, D.P.; Anderson, S.T.; Quigley, S.P. Long-term growth of male and female Bali cattle fed Sesbania grandiflora. *Anim. Prod. Sci.* **2014**, *54*, 1615–1619. [CrossRef]
7. Dahlanuddin; Yuliana, B.T.; Panjaitan, T.; Halliday, M.J.; Van De Fliert, E.; Shelton, H.M. Survey of Bali bull fattening practices in central Lombok, eastern Indonesia, based on feeding of Sesbania grandiflora. *Anim. Prod. Sci.* **2014**, *54*, 1273–1277. [CrossRef]
8. Panjaitan, T.; Fauzan, M.; Dahlanuddin, H.; Halliday, M.J.; Max Shelton, H. Growth of Bali Bulls Fattened with Forage Tree Legumes in Eastern Indonesia: Leucaena leucocephala in Sumbawa. In Proceedings of the 22nd International Grassland Congress, Sydney, Australia, 15–19 September 2013; pp. 601–602.
9. Dahlanuddin; Zaenuri, L.A.; Sutaryono, Y.A.; Hermansyah; Puspadi, K.; McDonald, C.; Williams, L.J.; Corfield, J.P.; van Wensveen, M. Scaling out integrated village management systems to improve Bali cattle productivity under small scale production systems in Lombok, Indonesia. *Livest. Res. Rural Dev.* **2016**, *28*, 1–13.
10. Chang, H.S.C.; Ilham, N.; Rukmantara, A.; Wibisono, M.G.; Sisriyeni, D. A Review of Integrated Cattle Oil Palm Production in Malaysia, Papua New Guinea and Indonesia. *Australas. Agribus. Rev.* **2020**, *28*, 1–27.
11. Widi, T.S.M.; Udo, H.M.J.; Oldenbroek, K.; Budisatria, I.G.S.; Baliarti, E.; Viets, T.C.; van der Zijpp, A.J. Is cross-breeding of cattle beneficial for the environment? The case of mixed farming systems in Central Java, Indonesia. *Anim. Genet. Resour.* **2015**, *57*, 1–14. [CrossRef]
12. Widi, T.S.M.; Widyas, N.; Damai, R.G.M.F. Weaning weight of Brahman cross (BX) and Bali cattle under intensive and oil palm plantation-cattle integrated systems. In Proceedings of the IOP Conference Series: Earth and Environmental Science, Yogyakarta, Indonesia, 23–25 September 2019; Volume 387, pp. 1–5.
13. Turner, H.G.; Short, A.J. Effects of field infestations of gastrointestinal helminths and of the cattle tick (Boophilus Microplus) on growth of three breeds of cattle. *Aust. J. Agric. Res.* **1972**, *23*, 177–193. [CrossRef]
14. Turner, H.G. Genetic variation of rectal temperature in cows and its relationship to fertility. *Anim. Prod.* **1982**, *35*, 401–412. [CrossRef]
15. Frisch, J.E.; Vercoe, J.E. An analysis of growth of different cattle genotypes reared in different environments. *J. Agric. Sci.* **1984**, *103*, 137–153. [CrossRef]

16. Frisch, J.E.; O'Neill, C.J. Comparative evaluation of beef cattle breeds of African, European and Indian origins. 2. Resistance to cattle ticks and gastrointestinal nematodes. *Anim. Sci.* **1998**, *67*, 39–48. [CrossRef]
17. Office International des Epizooties. *Responsible and Prudent Use of Anthelmintic Chemicals to Help Control Anthelmintic Resistance in Grazing Livestock Species*; World Organization for Animal Health: Paris, France, 2021.
18. Burrow, H.M.; Moore, S.S.; Johnston, D.J.; Barendse, W.; Bindon, B.M. Quantitative and molecular genetic influences on properties of beef: A review. *Aust. J. Exp. Agric.* **2001**, *41*, 893–919. [CrossRef]
19. Burrow, H.M. Importance of adaptation and genotype × environment interactions in tropical beef breeding systems. *Animal* **2012**, *6*, 729–740. [CrossRef]
20. Burrow, H. Strategies for increasing beef cattle production under dryland farming systems. *Wartazoa* **2019**, *29*, 161–170. [CrossRef]
21. Copland, J.W. Bali Cattle: Origins in Indonesia. In *Proceedings of the Jembrana Disease and the Bovine Lentiviruses*; Wilcox, G.E., Soseharsono, S., Dharma, D.M.N., Copland, J.W., Eds.; ICAR: Canberra, Australia, 1996; pp. 29–33.
22. Ashari, M.; Busono, W.; Nuryadi; Nurgiartiningsih, A. Analysis of chromosome and karyotype in Bali cattle and Simmental-Bali (Simbal) crossbreed cattle. *Pak. J. Biol. Sci.* **2012**, *15*, 736–741. [CrossRef]
23. Jellinek, P.; Avenell, K.; Thahar, A.; Sitorus, P. Infertility associated with cross breeding of Bali cattle. In Proceedings of the 2nd Ruminant Seminar, Ciawi, Bogor, Indonesia, 28–30 May 1980; pp. 28–30.
24. Steinfeld, H.; Gerber, P.; Wassenaar, T.; Castel, T.; Rosales, M.; de Haan, C. *Livestock's Long Shadow: Environmental Issues and Options*; FAO: Rome, Italy, 2006.
25. Leroy, G.; Baumung, R.; Boettcher, P.; Scherf, B.; Hoffmann, I. Review: Sustainability of crossbreeding in developing countries; definitely not like crossing a meadow . . . . *Animal* **2016**, *10*, 262–273. [CrossRef]
26. Mohamad, K.; Olsson, M.; van Tol, H.T.A.; Mikko, S.; Vlamings, B.H.; Andersson, G.; Rodríguez-Martínez, H.; Purwantara, B.; Paling, R.W.; Colenbrander, B.; et al. On the origin of Indonesian cattle. *PLoS ONE* **2009**, *4*, e5490. [CrossRef]
27. Sutarno; Setyawan, A.D. Review: Genetic diversity of local and exotic cattle and their crossbreeding impact on the quality of Indonesian cattle. *Biodiversitas* **2015**, *16*, 327–354. [CrossRef]
28. Payne, W.J.A.; Hodges, J. *Tropical Cattle—Origins, Breeds and Breeding Policies*; Blackwell Science Ltd.: Oxford, UK, 1997; ISBN 0-632-04048-3.
29. Talib, C. Sapi Bali di Daerah Sumber Bibit dan Peluang Pengembangannya. *Wartazoa* **2002**, *12*, 100–107.
30. Martojo, H. Indigenous bali cattle is most suitable for sustainable small farming in Indonesia. *Reprod. Domest. Anim.* **2012**, *47*, 10–14. [CrossRef]
31. Aziz, N.; Maksudi, M.; Prakoso, Y.A. Correlation between hematological profile and theileriosis in Bali cattle from Muara Bulian, Jambi, Indonesia. *Vet. World* **2019**, *12*, 1358–1361. [CrossRef]
32. Direktorat Kesehatan Hewan. *Pedoman Pengendalian dan Penanggulangan Penyakit Jembrana*; Direktorat Jenderal Peternakan dan Kesehatan Hewan Kementrian Pertanian Republik Indonesia: Jakarta, Indonesia, 2015.
33. Astuti, M. Potensi dan Keragaman Sumberdaya Genetik Sapi Peranakan Ongole (PO). *Wartazoa* **2004**, *14*, 30–39.
34. Hartati, H.; Utsunomiya, Y.T.; Sonstegard, T.S.; Garcia, J.F.; Jakaria, J.; Muladno, M. Evidence of *Bos javanicus* x *Bos indicus* hybridization and major QTLs for birth weight in Indonesian Peranakan Ongole cattle. *BMC Genet.* **2015**, *16*, 75. [CrossRef]
35. Romjali, E. Program pembibitan sapi potong lokal di Indonesia. *Wartazoa* **2018**, *28*, 199–209. [CrossRef]
36. Seftiana, A.; Kurnianto, E.; Sutopo, S. Evaluasi keunggulan genetik sapi peranakan ongole betina dengan dua metode yang berbeda di satuan kerja Sumberejo Kendal. *J. Ilmu Teknol. Peternak. Indones.* **2019**, *5*, 1–10.
37. Sudiharta; Sudrajad, P. Keragaan bobot lahir pedet sapi lokal (Peranakan Ongole/PO) kebumen dan potensinya sebagai sumber. In *Proceedings of the Seminar Nasional: Menggagas Kebangkitan Komoditas Unggulan Lokal Pertanian dan Kelautan*; Fakultas Pertanian, Universitas Trunojoyo: Madura, Indonesia, 2013.
38. Prihandini, P.W.; Hakim, L.; Nurgiartiningsih, V.A. Seleksi pejantan berdasarkan nilai pemuliaan pada sapi Peranakan Ongole (PO) di Loka Penelitian Sapi Potong Grati–Pasuruan. *TERNAK Trop.* **2011**, *12*, 99–109.
39. Safitri, L.; Hamdani, M.D.I.; Husni, A. Estimasi nilai pemuliaan bobot sapih sapi peranakan ongole (PO) di Desa Wawasan Kecamatan Tanjungsari Kabupaten Lampung Selatan. *J. Res. Innov. Anim.* **2019**, *3*, 28–33.
40. Kusuma, S.B.; Ngadiyono, N.; Sumadi, S. Estimasi dinamika populasi dan penampilan reproduksi sapi peranakan ongole di Kabupaten Kebumen Provinsi Jawa Tengah. *Bul. Peternak.* **2017**, *41*, 230–242. [CrossRef]
41. Christi, T.R.R.F. Penampilan reproduksi sapi peranakan ongole dara. *J. Ilmu Peternak.* **2017**, *1*, 7–14.
42. Nasuha, N.; Sumadi, S.; Dyah, M. Perbandingan tampilan produktivitas sapi peranakan ongole dengan limousin-peranakan ongole di Kabupaten Tuban, Jawa Timur. In Proceedings of the Prosiding Seminar Nasional Teknologi Peternakan dan Veteriner, Bogor, Indonesia, 26–27 October 2020; pp. 323–329.
43. Putra, W.P.B.; Gunawan, M.; Kaiin, E.M.; Said, S. Kinerja reproduksi sapi peranakan ongole (*Bos indicus*) di BPPIBT-SP Ciamis, Jawa Barat. In Proceedings of the Prosiding Seminar Teknologi Agribisnis Peternakan (STAP) Fakultas Peternakan Universitas Jenderal Soedirman, Purwokerto, Indonesia, 7 July 2018; Volume 6, pp. 327–334.
44. Fauziah, L.W.; Busono, W.; Ciptadi, G. Performans reproduksi sapi peranakan ongole dan peranakan limousin pada paritas berbeda di kecamatan paciran Kabupaten Lamongan. *TERNAK Trop.* **2016**, *16*, 49–54. [CrossRef]
45. Fatah Wiyatna, M.; Gurnadi, E.; Mudikdjo, D.K. Produktivitas Sapi Peranakan Ongole pada Peternakan Rakyat di Kabupaten Sumedang. *J. Ilmu Ternak* **2012**, *12*, 22–25.

46. Aditia, E.L.; Priyanto, R.; Baihaqi, M.; Putra, B.W.; Ismail, M. Performa Produksi Sapi Bali dan Peranakan Ongole yang Digemukan dengan Pakan Berbasis Sorghum. *J. Ilmu Produksi Teknol. Has. Peternak.* **2013**, *1*, 155–159.
47. Priyanto, R.; Fuah, A.M.; Aditia, E.L.; Baihaqi, M.; Ismail, M. Peningkatan produksi dan kualitas daging sapi lokal melalui penggemukan berbasis serealia pada taraf energi yang berbeda. *J. Ilmu Pertan. Indones.* **2015**, *20*, 108–114. [CrossRef]
48. Payne, W.J.A.; Rollinson, D.H.L. Madura cattle. *Z. Tierz. Züchtungbiol.* **1976**, *93*, 89–100. [CrossRef]
49. Namikawa, T. Geographic distribution of bovine Hemoglobin-beta (Hbb) alleles and the phylogenetic analysis of the cattle in Eastern Asia. *Z. Tierz. Züchtungsbiol.* **1981**, *98*, 151–159. [CrossRef]
50. Widi, T.S.M.; Udo, H.M.J.; Oldenbroek, K.; Budisatria, I.G.S.; Baliarti, E.; van der Zijpp, A.J. Unique cultural values of Madura cattle: Is cross-breeding a threat? *Anim. Genet. Resour.* **2014**, *54*, 141–152. [CrossRef]
51. Widi, T.S.M. *Mapping the Impact of Crossbreeding in Smallholder Cattle Systems in Indonesia*; Wageningen University and Research Centre: Wageningen, The Netherlands, 2015.
52. Hartatik, T.; Mahardika, D.A.; Widi, T.S.M.; Baliarti, E. Karakteristik dan Kinerja Induk Sapi Silangan Limousin-Madura dan Madura di Kabupaten Sumenep dan Pamekasan. *Bul. Peternak.* **2009**, *33*, 143–147. [CrossRef]
53. Suwiti, N.K.; Rantam, F.A.; Besung, I.N.K. Deteksi Protein Bovine Major Histocompatibility Complex Klas I dan Klas II pada Sapi Madura. *J. Vet.* **2008**, *9*, 147–151.
54. Nurgiartiningsih, V.M.A.; Furqon, A.; Rochadi, I.; Rochman, A.; Muslim, A.; Waqid, M. Evaluation of Birth Weight and Body Measurements of Madura Cattle based on Year of Birth and Breeding System in Madura Breeding Centre, Indonesia. In Proceedings of the IOP Conference Series: Earth and Environmental Science, Changsha, China, 18–20 September 2020; Volume 478.
55. Tribudi, Y.A.; Prihandini, P.W. Repeatability estimates for birth, weaning and yearling weight in Madura cattle. *Int. Res. J. Adv. Eng. Sci.* **2019**, *4*, 207–208.
56. Sulistyoningtyas, I.; Nurgiartiningsih, V.A.; Ciptadi, G. Evaluasi performa bobot badan dan statistik vital sapi madura berdasarkan tahun kelahiran. *J. Ilm. Peternak. Terpadu* **2017**, *5*, 40–43.
57. Hartatik, T.; Widi, T.S.M.; Ismaya, D.T.; Baliarti, E. The exploration of genetic characteristics of Madura cattle. In Proceedings of the 5th International Seminar on Tropical Animal Production (ISTAP), Yogyakarta, Indonesia, 19–22 October 2010; pp. 578–584.
58. Widi, T.S.M.; Hartatik, T. The characteristics and performances of sonok compared to karapan cows as important consideration for conservation of madura cattle. In *Tropentag 2009. International Research on Food Security, Natural Resource Management and Rural Development*; DITSL GmbH: Witzenhausen, Germany, 2009.
59. Nurlaila, S.; Kurniadi, B.; Zali, M.; Nining, H. Status reproduksi dan potensi sapi Sonok di Kabupaten Pamekasan. *J. Ilm. Peternak. Terpadu* **2018**, *6*, 147–154. [CrossRef]
60. Hariansyah, A.R.R.; Raharjo, A.; Zainuri, A.; Parwoto, Y.; Prasetiyo, D.; Prastowo, S.; Widyas, N. Genetic parameters on Bali cattle progeny test population. In Proceedings of the IOP Conference Series: Earth and Environmental Science, Banda Aceh, Indonesia, 26–27 September 2018; Volume 142, p. 012003.
61. Widyas, N.; Prastowo, S.; Haryanto, R.; Nugroho, T.; Widi, T.S.M.S.M. Madura cattle stratification as a signature of traditional selection and diverse production systems. In Proceedings of the IOP Conference Series: Earth and Environmental Science, Yogyakarta, Indonesia, 23–25 September 2019; Volume 387, pp. 6–11.
62. Nielsen, H.M.; Amer, P.R.; Byrne, T.J. Approaches to formulating practical breeding objectives for animal production systems. *Acta Agric. Scand. A Anim. Sci.* **2014**, *64*, 2–12. [CrossRef]
63. Kinghorn, B.P.; Banks, R.G.; Simm, G. Genetic improvement of beef vattle. In *The Genetics of Cattle*; Garrick, D.J., Ruvinsky, A., Eds.; CAB International: Wallingford, UK, 2015; pp. 451–473. ISBN 978-1-78064-221-5.
64. Gregory, K.E.; Cundiff, L.V.; Koch, R.M. Breed effects and heterosis in advanced generations of composite populations for preweaning traits of beef cattle. *J. Anim. Sci.* **1991**, *69*, 947–960. [CrossRef]
65. Widi, T.S.M.; Budisatria, I.G.S.; Baliarti, E.; Udo, H.M.J. Designing genetic impact assessment for crossbreeding with exotic beef breeds in mixed farming systems. *Outlook Agric.* **2021**, *50*, 34–45. [CrossRef]
66. Frisch, J.E.; O'Neill, C.J. Comparative evaluation of beef cattle breeds of African, European and Indian origins. 1. Live weights and heterosis at birth, weaning and 18 months. *Anim. Sci.* **1998**, *67*, 27–38. [CrossRef]
67. Gazzola, C.; O'Neill, C.J.; Frisch, J.E. Comparative evaluation of the meat quality of beef cattle breeds of Indian, African and European origins. *Anim. Sci.* **1999**, *69*, 135–142. [CrossRef]
68. Pribadi, L.W.; Maylinda, S.; Nasich, M.; Suyadi, S. Prepubertal growth rate of Bali cattle and its crosses with Simmental breed at lowland and highland environment. *IOSR J. Agric. Vet. Sci.* **2014**, *7*, 52–59. [CrossRef]
69. Pribadi, L.W.; Maylinda, S.; Nasich, M.; Suyadi, S. Reproductive Efficiency of Bali Cattle and It's Crosses with Simmental Breed in the Lowland and Highland Areas of West Nusa Tenggara Province, Indonesia. *Livest. Res. Rural Dev.* **2015**, *27*, 1–10.
70. Dominanto, G.P.; Nasich, M.; Wahyuningsih, S. Evaluating Performance of Crossbreed Calves in West Papua Indonesia. *Res. Zool.* **2016**, *6*, 1–7. [CrossRef]
71. Prastowo, S.; Widi, T.S.M.; Widyas, N. Preliminary analysis on hybrid vigor in Indonesian indigenous and crossbred cattle population using data from published studies. In Proceedings of the IOP Conference Series: Materials Science and Engineering, Busan, Korea, 25–27 August 2017; Volume 193, p. 012028.

72. BPS Provinsi Bali Populasi Ternak Sapi Potong Menurut Kabupaten/Kota di Bali. Available online: https://bali.bps.go.id/statictable/2014/11/06/129/jumlah-sapi-dan-kerbau-menurut-kabupaten-kota-dan-jenis-kelamin-kondisi-1-mei-2013-hasil-sensus-pertanian-2013.html (accessed on 11 May 2022).
73. McCool, C.J. Spermatogenesis in Bali cattle (*Bos sondaicus*) and hybrids with *Bos indicus* and *Bos taurus*. *Res. Vet. Sci.* **1990**, *48*, 288–294. [CrossRef]
74. Inchaisri, C.; Jorritsma, R.; Vos, P.L.A.M.; van der Weijden, G.C.; Hogeveen, H. Economic consequences of reproductive performance in dairy cattle. *Theriogenology* **2010**, *74*, 835–846. [CrossRef]
75. Kocu, N.; Priyanto, R.; Salundik, S.; Jakaria, J. Produktivitas Sapi Bali Betina dan Hasil Persilangannya dengan Limousin dan Simmental yang di Pelihara Berbasis Pakan Hijauan di Kabupaten Keerom Papua. *J. Ilmu Produksi Teknol. Has. Peternak.* **2019**, *7*, 29–34.
76. Rahayu, S. The Reproductive Performance of Bali Cattle and It's Genetic Variation. *Berk. Penelit. Hayati* **2014**, *20*, 28–35. [CrossRef]
77. Burrow, H.M.; Mrode, R.; Mwai, A.O.; Coffey, M.P.; Hayes, B.J. Challenges and Opportunities in Applying Genomic Selection to Ruminants Owned by Smallholder Farmers. *Agriculture* **2021**, *11*, 1172. [CrossRef]
78. Ponzoni, R.W.; Newman, S. Developing breeding objectives for australian beef cattle production. *Anim. Prod.* **1989**, *49*, 35–47. [CrossRef]
79. Amer, P.R.; Crump, R.; Simm, G. A terminal sire selection index for UK beef cattle. *Anim. Sci.* **1998**, *67*, 445–454. [CrossRef]
80. Barwick, S.A.; Henzell, A.L. Development successes and issues for the future in deriving and applying selection indexes for beef breeding. *Aust. J. Exp. Agric.* **2005**, *45*, 923–933. [CrossRef]
81. Prayaga, K.C.; Corbet, N.J.; Johnston, D.J.; Wolcott, M.L.; Fordyce, G.; Burrow, H.M. Genetics of adaptive traits in heifers and their relationship to growth, pubertal and carcass traits in two tropical beef cattle genotypes. *Anim. Prod. Sci.* **2009**, *49*, 413–425. [CrossRef]
82. Bourdon, R.M. *Understanding Animal Breeding*, 2nd ed.; Pearson Education Limited: Edinburgh Gate, UK, 2014.
83. Muchadeyi, F.C.; Ibeagha-Awemu, E.M.; Javaremi, A.N.; Gutierrez Reynoso, G.A.; Mwacharo, J.M.; Rothschild, M.F.; Sölkner, J. Editorial: Why Livestock Genomics for Developing Countries Offers Opportunities for Success. *Front. Genet.* **2020**, *11*, 626. [CrossRef]
84. McEwin, R.A.; Hebart, M.L.; Oakey, H.; Pitchford, W.S. Within-breed selection is sufficient to improve terminal crossbred beef marbling: A review of reciprocal recurrent genomic selection. *Anim. Prod. Sci.* **2021**, *61*, 1751–1759. [CrossRef]
85. Warburton, C.L.; Costilla, R.; Engle, B.N.; Corbet, N.J.; Allen, J.M.; Fordyce, G.; McGowan, M.R.; Burns, B.M.; Hayes, B.J. Breed-adjusted genomic relationship matrices as a method to account for population stratification in multibreed populations of tropically adapted beef heifers. *Anim. Prod. Sci.* **2021**, *61*, 1788–1795. [CrossRef]
86. Meuwissen, T.H.E.; Hayes, B.J.; Goddard, M.E. Prediction of total genetic value using genome-wide dense marker maps. *Genetics* **2001**, *157*, 1819–1829. [CrossRef] [PubMed]
87. Bolormaa, S.; Hayes, B.J.; Savin, K.; Hawken, R.; Barendse, W.; Arthur, P.F.; Herd, R.M.; Goddard, M.E. Genome-wide association studies for feedlot and growth traits in cattle. *J. Anim. Sci.* **2011**, *89*, 1684–1697. [CrossRef]
88. Meuwissen, T.; Hayes, B.; Goddard, M. Genomic selection: A paradigm shift in animal breeding. *Anim. Front.* **2016**, *6*, 6–14. [CrossRef]
89. Gunawan, A.; Jakaria. Genetic and non-genetics effect on birth, weaning, and yearling weight of Bali Cattle. *Media Peternak.* **2011**, *34*, 93–98. [CrossRef]
90. Warmadewi, D.A.; Oka, L.; Sudyadnya, P.; Sudana, L.B. Heritability Estimate and Genetic Response to Selection for Yearling Weight in Bali Cattle. *J. Anim. Sci. Udayana Univ.* **2014**, *3*, 1–9.
91. Hartati; Muladno; Priyanto, R.; Gunawan, A.; Talib, C. Heritability Estimation and Non-Genetic Factors Affecting Production Traits of Indonesian Ongole Cross. *J. Ilmu Ternak Vet.* **2015**, *20*, 168–174. [CrossRef]
92. Hayes, B.J.; Daetwyler, H.D. 1000 Bull Genomes Project to Map Simple and Complex Genetic Traits in Cattle: Applications and Outcomes. *Annu. Rev. Anim. Biosci.* **2019**, *7*, 89–102. [CrossRef] [PubMed]
93. Mrode, R.; Ojango, J.M.K.; Okeyo, A.M.; Mwacharo, J.M. Genomic selection and use of molecular tools in breeding programs for indigenous and crossbred cattle in developing countries: Current status and future prospects. *Front. Genet.* **2019**, *10*, 694. [CrossRef]
94. Ojango, J.M.K.; Mrode, R.; Rege, J.E.O.; Mujibi, D.; Strucken, E.M.; Gibson, J.; Mwai, O. Genetic evaluation of test-day milk yields from smallholder dairy production systems in Kenya using genomic relationships. *J. Dairy Sci.* **2019**, *102*, 5266–5278. [CrossRef] [PubMed]
95. Marshall, K.; Gibson, J.P.; Mwai, O.; Mwacharo, J.M.; Haile, A.; Getachew, T.; Mrode, R.; Kemp, S.J. Livestock genomics for developing countries—African examples in practice. *Front. Genet.* **2019**, *10*, 297. [CrossRef]
96. Al Kalaldeh, M.; Swaminathan, M.; Gaundare, Y.; Joshi, S.; Aliloo, H.; Strucken, E.M.; Ducrocq, V.; Gibson, J.P. Genomic evaluation of milk yield in a smallholder crossbred dairy production system in India. *Genet. Sel. Evol.* **2021**, *53*, 73. [CrossRef]

Perspective

Challenges and Opportunities in Applying Genomic Selection to Ruminants Owned by Smallholder Farmers

Heather M. Burrow [1,*], Raphael Mrode [2], Ally Okeyo Mwai [3], Mike P. Coffey [4] and Ben J. Hayes [5]

1 Faculty of Science, Agriculture, Business and Law, University of New England, Armidale 2351, Australia
2 Animal Biosciences, International Livestock Research Institute, Scotland's Rural College, Roslin Institute Building, Easter Bush, Edinburgh EH125 9RG, UK; R.Mrode@cgiar.org
3 Animal Biosciences, International Livestock Research Institute, Nairobi P.O. Box 00100, Kenya; O.MWAI@CGIAR.ORG
4 Animal Breeding and Genomics, Scotland's Rural College, Roslin Institute Building, Easter Bush, Edinburgh EH125 9RG, UK; mike.coffey@sruc.ac.uk
5 Queensland Alliance for Agriculture and Food Innovation, University of Queensland, 306 Carmody Road, St Lucia 4067, Australia; b.hayes@uq.edu.au
* Correspondence: Heather.Burrow@une.edu.au; Tel.: +61-2-6773-3512

Abstract: Genomic selection has transformed animal and plant breeding in advanced economies globally, resulting in economic, social and environmental benefits worth billions of dollars annually. Although genomic selection offers great potential in low- to middle-income countries because detailed pedigrees are not required to estimate breeding values with useful accuracy, the difficulty of effective phenotype recording, complex funding arrangements for a limited number of essential reference populations in only a handful of countries, questions around the sustainability of those livestock-resource populations, lack of on-farm, laboratory and computing infrastructure and lack of human capacity remain barriers to implementation. This paper examines those challenges and explores opportunities to mitigate or reduce the problems, with the aim of enabling smallholder livestock-keepers and their associated value chains in low- to middle-income countries to also benefit directly from genomic selection.

Keywords: genomic selection; smallholder farmers; beef and dairy cattle; sheep and goats; phenotypes; reference populations; capacity-building; value of genomic information

Citation: Burrow, H.M.; Mrode, R.; Mwai, A.O.; Coffey, M.P.; Hayes, B.J. Challenges and Opportunities in Applying Genomic Selection to Ruminants Owned by Smallholder Farmers. *Agriculture* **2021**, *11*, 1172. https://doi.org/10.3390/agriculture11111172

Academic Editor: Maria Selvaggi

Received: 12 October 2021
Accepted: 16 November 2021
Published: 20 November 2021

1. Introduction

Although major differences exist between the productivity and available resources of livestock producers in advanced and low- to middle-income countries (LMICs), several very significant challenges need to be overcome by all farmers, regardless of their location, if they are to capture the new opportunities that already exist and continue to emerge.

The world's population is expected to increase from 7 billion people in 2011 to 9 or 10 billion by 2050, with most of that growth occurring in Africa and Asia [1]. The incomes of many people in LMICs are now increasing and, with rising incomes, the demand for meat and dairy products is also growing [2]. To achieve food security by 2050, livestock enterprise and industry efficiency, as measured by total factor productivity, needs to increase by 2.0–2.5% per annum. This is the equivalent of doubling outputs from constant resource inputs through to 2050 [3]. Due to the pressures on agriculture in developed countries, a significant proportion of that increased production must occur in the regions of greatest need, i.e., in Africa and Asia. This increased demand for food is leading to greater competition for inputs such as land, water, grain and labor, driving up the cost of livestock production. Climate change is adding to this challenge [4], requiring animals that are productive under hotter and drier climates and, in the tropics and sub-tropics, requiring animals that can tolerate significant increases in ecto- and endo-parasitic burdens and vector-borne diseases. There is therefore an urgent need to greatly increase the productivity of livestock herds and flocks while using less grain and water; the animals must also simultaneously tolerate more extreme climates and disease stressors and farmers must reduce their animals' greenhouse gas (GHG) emissions. An added beneficial outcome of improving production efficiency is that emissions intensity decreased for most livestock species globally between 2000 and 2018 because of increased production efficiency [5]. The same authors also showed that improving production efficiency, particularly in countries within Asia and Africa, has much greater mitigation effects than removing livestock products from global human diets [5], thereby retaining the human health and nutritional benefits of consuming livestock products in those regions.

The opportunities to significantly improve the productivity of livestock systems are greatest for extensive or pastoral production systems in tropical and sub-tropical environments, including those in Africa and Asia. These systems employ land resources with few alternative uses, including urbanization. In addition, they capitalize on the strengths of ruminants, which utilize low-quality pastures that are not suitable for humans or monogastric livestock species. The pastoral livestock industries are also far less likely than the intensive livestock industries to face inequitable demands about their production systems from urban populations, as has occurred over recent years in the intensive livestock industries.

To double outputs from the same resource base, as required for global food security by 2050, livestock farmers in the tropics and sub-tropics will need to adopt new, cost-effective, and transformational technologies for use with animals that are already well adapted to their production environments. Traditional technologies that deliver incremental changes will assist in improving productivity, in the same way they have in the past. By way of example, one study demonstrated that through the use of long-established technologies, such as animal breeding and animal nutrition, US dairy farmers now require 21% fewer cows, 23% less feed, 65% less water and 90% less land to produce 1 billion kg of milk than they did in 1944, with a 57% reduction in methane emissions and simultaneous large reductions in waste [6]. However, these traditional technologies are no longer sufficient by themselves to deliver the major increases in productivity that are now required.

Potentially, the greatest opportunity to significantly improve the productivity of livestock industries in LMICs in tropical and sub-tropical environments to 2050 is via the use of genomic (DNA-based) information through genomic selection, using genome-wide genetic markers to estimate the genetic merit of individual animals [7]. Recent significant advances in genomic technologies that support this recommendation include:

- The very rapidly reducing costs of full genome sequencing [8];
- The momentous reductions in the time taken to sequence an entire genome, down from close to 13 years for the first human genome sequence [9] to a full sequence now being achieved in a single day [10], and with potential in the near future to achieve full genome sequencing on the same day in the field rather than at laboratory sites;
- The ability to accurately impute whole-genome sequence data from lower-density, lower-cost single nucleotide polymorphism (SNP) panels [11–14];
- The potential to use whole-genome sequence data to discover the mutations causing variations among animals and, in turn, using that knowledge of functional mutations to improve the accuracy of breeding value predictions [15];
- Resolution of the "missing heritability" problem [16], proving that genomic selection approaches account for significant proportions of the genetic variation for economically important complex traits;
- Vastly improved computational capacity that is now allowing the cost-effective storage and processing of petabyte (10^{12} bytes)-scales of data [17];
- The ability to use pooled DNA samples from groups of animals to identify the average genetic merit at low cost [18], thereby enabling the development of new, cost-effective management applications based on genomic information; and
- The increased ability to capture essential individual animal performance data (phenotypes) through the use of automated or semi-automated electronic data capture methods.

Traditional genetic improvement programs, based on measuring large numbers of pedigree-recorded animals in well-defined cohort groups for the full range of economically important productive and adaptive traits, is generally not possible for smallholder farmers in LMICs. Now, the opportunity to use genomic data, in conjunction with the use of information and communication technologies, offers significant new opportunities to increase the rates of genetic gain by characterizing indigenous and crossbred animals for use in conservation, crossbreeding and within-breed selection programs, to improve economically important traits. Other technologies, such as genome editing, coupled with emerging reproductive technologies that enable rapid multiplication and decreased dependency on cold chains for the delivery of improved genetics, will potentially transform livestock breeding even further as causal mutations are found [19].

To date, there has been limited use of genomic technologies in grazing livestock in LMICs, due to several major challenges inhibiting their use. The following sections examine those challenges and identify opportunities to mitigate or remove them for the ruminant livestock species that predominate in those regions, i.e., beef and dairy cattle, sheep, and goats. Even though this paper focuses on the application of genomic selection in LMICs, no attempt is made to evaluate the ongoing refinement of the genomic selection methodology or the increasingly sophisticated demands on the computational capacity required to drive the method, because those challenges continue to be addressed more rapidly than the alternative constraints facing the use of genomic selection in LMICs.

2. The Need for Accurate Phenotyping and Record-Keeping

In both advanced economies and LMICs, the main limitation to genomic (and traditional) selection in extensively managed livestock is the difficulty and expense of measuring animals in appropriately sized contemporary groups for the full range of economically important productive and adaptive traits. As discussed by [20], technology may in the future provide the means of measuring animals, but it cannot replace the statistical imperative that, for these measurements to be beneficial for genetic improvement programs, contemporary groups of appropriate structure and sufficient size are required. Unless the design is adequate in terms of contemporary group size and structure, the measurements will not provide useful predictions of genetic merit. This applies to traditional genetic improvement programs as well as those capturing beneficial traits through genomic selection.

As suggested in Table 1, the measurement of most phenotypes required for genetic improvement programs in smallholder herds and flocks is generally not feasible in the field. Where measurement is feasible, there is an additional requirement that accurate records be maintained at the level of individual animals. Such record-keeping is often an additional challenge for smallholder farmers, mainly because platforms that can effectively collate and make sense of such highly fragmented data are lacking. This recording has been assisted in the past by animal breeding research projects, such as those described by [21]. However, with the short-term nature and eventual closure of many of those types of projects, the data capture has generally been discontinued by the smallholder farmers. More recent research projects such as the African Dairy Genetic Gains (ADGG), BAIF India and community-based breeding programs (CBBP) in Ethiopia and Malawi are now adapting digital tools, such as mobile phones and tablets, to capture performance data for easy-to-measure traits such as milk yield, body condition score and artificial insemination records [21].

By way of example in the ADGG project, milk yield, heart girth (for predicting body weight), and body condition score are collected monthly using software based on the Open Data Kit (ODK) that is installed on tablets and mobile phones, employing the services of performance-recording agents. In addition, iCow (http://www.icow.co.ke/, accessed 15 November 2021), a technological platform owned by a private company, Green Dreams, a partner in the ADGG, has provided feedback information to farmers for herd management through text messages and web-based training. This performance data has enabled the genomic prediction and selection of first-rate young bulls for breeding in Tanzania [30]. The main challenges of the data-capture system are the high cost of employing performance-recording agents and poor internet connectivity to upload the data. The most obvious data issues relate to inconsistencies in the dates for various animal events, such as birth, calving, and milking dates. This leads to a large number of animals being rejected from any meaningful genetic analysis.

Table 1. Phenotypes that should ideally be included in livestock-breeding objectives and the feasibility of recording them in smallholder herds and flocks in low- to middle-income countries.

Phenotype and Purpose	Options to Measure Key Traits for Use in Genetic Improvement Programs in Smallholder Livestock Populations
Product Quantity and Quality	
Animal live weights and weight gains for the genetic evaluation of potential meat quantity in meat-producing animals and to provide assessments of animal nutrition and the effect of environmental stressors and/or endemic diseases on individuals and groups of meat and dairy animals	Except for farmers directly engaged in well-funded genetic improvement research programs, a lack of animal-handling infrastructure and access to scales generally means that records of individual animal weights and weight gains are not feasible in smallholder herds/flocks. This is particularly true for meat animals. Future infrastructure development may enable remote walk-over weighing or similar options for measurements, although [22] concluded that current walk-over weighing systems did not justify the investment needed in individual animal electronic identification. An alternative approach is for farmers to measure the animal's circumference and length as an indicator of body weight, but this would also require access to appropriate handling facilities to enable accurate tape placement and length and height measurements. The accuracy of these assessments is reasonable in well-designed cohorts of the same breed of animals, but it varies markedly across breeds and animal size. Hence, in situations where animal breed composition is unclear, as is often the case in smallholder herds and flocks, this is not a reliable measurement for use in genetic evaluation [23]. Predictions from images based on deep learning may become available in the near future.
Milk volume as the primary product for dairy animals and a maternal trait for meat animals	Measuring milk volume is relatively straightforward in dairy animals in smallholder herds and flocks but is not feasible in meat-producing animals, other than indirectly through the offspring's weaning weights.
Meat (e.g., tenderness, flavor, juiciness) and milk quality (e.g., protein concentration, fat content, etc.) attributes	Although sensor-based meat quality assessment systems exist, they are not readily available and given the meat evaluation and pricing systems, such assessments may not provide value for money, except where animals from smallholder flocks and herds are sold through commercial value chains to meet market quality specifications. In those cases, the processor or retailer purchasing the carcasses or milk should be able to provide the phenotypes required. However, quality-based value chains are currently scarce in most LMICs.
Live animal fat depth, eye muscle area, etc., in meat animals	Subjective assessments of fat coverage or body condition scores are feasible but the value of doing so for genetic improvement programs in the absence of other measures, such as weights, is not high.
Efficiency of feed utilization, particularly while animals are grazing on pasture	While this phenotype represents the best opportunity to improve the amount of feed required by an animal to maintain its body weight, it is still not feasible to record even in sophisticated breeding programs in high-income countries, except when animals are fed grain-based diets in pens or assessments are made on monoculture crops/pastures. Those measures are unable to account for animals' diet selections and browsing behaviors, both of which impact feed utilization at pasture.
Female reproduction (most measures of male reproduction are not generally feasible in smallholder herds and flocks)	
Age at first estrus	Smallholder farmers can be trained to recognize the signs of the first estrus in female animals that are closely monitored and the use of digital cow calendar reminder systems enable subsequent signs to be easily and accurately timed and recognized. However, as the signs vary across animals, the records are unlikely to be sufficiently accurate for genetic improvement programs. Another option could be to use solar-powered sensor networks to remotely capture livestock data, such as estrus and pregnancy status, using animal ear- or neck tags (https://www.allflex.global/product/heatimepro/, accessed 15 November 2021), but to justify the expense, this type of data would need to be captured in relatively large herds or flocks.
Date of calving/lambing, age of first calving/lambing and inter-calving/lambing period	In closely managed smallholder herds and flocks, recording the date of calving/lambing of individual dams is feasible, though not currently routinely practiced except through research programs. Knowledge of this date enables the calculation of the age of first calving/lambing and inter-calving/lambing periods.
Pregnancy rate	If smallholder farmers have cost-effective access to veterinary services, pregnancy testing may be a feasible option for some. Use of estrus detection ear tags (see "Age at first estrus" section above) may also provide a useful indicator of pregnancy status if the tags are also used during the breeding season by indicating which females do not return to estrus.
Weaning rate and offspring mortality rate	If smallholder farmers practice calf/lamb weaning and the calves/lambs are routinely individually identified to their dams, it may be feasible for annual weaning rates (and, hence, annual offspring mortality rates) to be calculated if the number of females mated in the previous breeding season is also recorded.
Adaptive traits	
Resistance to ecto- and endo-parasites	As summarized by [24], "resistance" in this context refers to both the ability of the individual host to resist infection or control the parasite lifecycle (resistance) and also where an individual host may be infected by a parasite but suffer little or no harm (tolerance). These terms are used interchangeably here. An in-depth discussion of the ability of cattle to resist a wide range of individual parasites is given in [24], but measuring an individual animal's resistance to any of those parasites in advanced and low- to medium-income economies is not generally feasible due to both the intermittent nature of parasite infestations and the difficulty of measurement.
Resistance to endemic diseases transmitted by parasites	Measurement of disease resistance under pastoral conditions is generally very difficult, even in advanced economies. The simple presence/absence of disease can be subjectively assessed by a skilled recorder when the animals are observed during routine handling procedures. However, infrequent observation of animals means that incidence of disease generally goes unrecorded, except where animal deaths occur and a diagnosis giving the cause of death is possible.
Tolerance to heat stress	Traditionally, heat stress has been recorded using repeated measurements of the rectal temperature of animals under conditions of heat stress [25,26] or through subjective coat scores of animals during the summer months [27]. Increasingly, however, tolerance to heat stress is being assessed through the use of temperature–humidity indices based on the automatic meteorological recording of ambient temperatures and humidity in local regions [28,29].

Another consideration is how such record-keeping can be made sustainable beyond the life of the associated research projects, while recognizing that the farmers who provided the records remain the owners of that data beyond the life of the projects. This is an issue that needs to be directly addressed by each of those projects. ADGG has been examining several business model options for the sustainability of the record-capture system; these include piloting mobile phone-based systems for direct data capture from farmers through monthly alerts, engaging government officials in the respective country to encourage their involvement, and exploring private company participation. Currently, direct farmer-incentivized systems are being tested in Kenya as a possible long-term solution to this challenge.

Even where measurement is feasible, it is likely that many smallholder herds and flocks are unable to generate within-herd genetic linkages through the use of multiple sires to generate contemporary groups, meaning that the contemporary group design requirements also present significant challenges. For this reason, the most feasible option for smallholder herds and flocks to participate in genetic improvement programs is likely to be through the use of specifically designed reference, resource or nucleus populations aimed at the identification of genetically superior sires for subsequent use in smallholder herds and flocks. However, such reference populations need to be managed under conditions that are as similar to the smallholder and pastoral systems as possible. Past attempts, where government research centers were used, have generally failed.

3. The Role of Reference Populations

Over many decades, the dairy cattle industries in high-income countries have conducted successful genetic improvement programs using a model where individual dairy herds contributed pedigree and performance records (and more recently, genomic information) to national and international genetic evaluation schemes. These types of schemes have generally not been feasible for other livestock species, such as beef cattle and meat- and wool-yielding sheep and goats, which have traditionally focused on visual appraisal as an indicator of performance in the absence of objective, routinely recorded performance data, such as the daily milk volumes that exist in the dairy industries.

For this reason, an alternative approach was developed for use in those species, where large livestock populations were specifically designed and established to accurately manage and record animals, particularly for difficult- or expensive-to-measure phenotypes, within well-designed contemporary groups to capture data for the traits of interest. As part of the design of these populations, great effort was expended to generate strong genetic linkages within and across contemporary groups of animals and across herds and flocks being evaluated, whether by the exchange of specific bulls and rams or by the use of specified AI sires in reference herds and flocks. To achieve the levels of accuracy required for these difficult- or expensive-to-measure traits, very large animal resource populations that have been accurately recorded for the particular trait are needed [8]. These populations are known as reference, resource, or nucleus populations and, to date, have all been established as part of large and well-funded research projects.

The first beef cattle and then sheep reference populations established in Australia were needed because, at the time of their establishment, there were no breed associations or breeding companies interested in or able to undertake genetic improvement based on objective performance data (*cf.* the traditional visual appraisal approaches common at that time) and particularly for hard- or expensive-to-record traits. Examples of such populations in beef cattle in Australia are described by [31] for growth, feed efficiency and carcass and beef quality, and by [32,33] for the full range of productive and adaptive traits in the breeding objective. More recent examples include the "Repronomics®" project [34] that builds on the populations described in [32,33], and the more recent Northern Genomics project [35]. The Northern Genomics Project works with 54 collaborating herds across northern Australia (including those farmed in some very challenging environments), with 26,000 heifers and cows now genotyped and trait-recorded [35]. The collaborators and associated veterinarians collect data on cohorts of heifers in well-defined, and in some cases very large, contemporary groups. In most cases, the herds are mixed-breed, cross-breed or tropical composites (these composites being admixtures of three or four breed types, i.e., *Bos indicus*, tropically adapted *Bos taurus*, temperate *Bos taurus*—British and temperate *Bos taurus*—European, with many of the composites having ancestry from 6 or more individual breeds, as described in detail by [31,32]). The traits include heifer puberty (based on ultrasound scans to determine if the heifers have cycled or not), weight, height, and body condition score at approximately 600 days, whether they are pregnant or not four months after calving (a re-breed trait), farmer-scored temperament, tick score and buffalo fly lesion score. All traits, except heifer puberty and being pregnant or not four months after calving, are farmer-recorded following some minimal training on field days. Breed composition and *Bos indicus* percentage, derived from SNP marker predictions, were used in the models used to derive SNP prediction equations for the traits. The project has estimated genomic heritabilities that are similar to those produced from pedigree-recorded herds and has also validated useful accuracies of genomic estimated breeding values (GEBV) across breeds and composites [35]. The project clearly demonstrates that useful GEBV can be produced from data collected in commercial herds. However, a clear difference between these herds and those in LMICs is in contemporary group size and, as indicated above, this is really the key challenge when using data from smallholder herds in LMICs.

An additional study in the USA developed specific populations to record resistance/susceptibility to bovine respiratory disease in beef and dairy cattle [36]. Similar populations designed to capture data for a range of productive attributes in meat- and wool-yielding sheep in Australia are described by [37,38]. International efforts have been expended in creating an international resource population of dairy cows for feed intake records, collected in research herds [39].

Similar populations have been established more recently for smallholder dairy farmers in countries in sub-Saharan Africa and India through externally funded, highly participatory research programs, such as the ADGG project. These programs use information and communication technologies (ICT) to digitally capture and submit data that are sufficiently large for use in genetic evaluation [40,41]. In ADGG, the dairy cattle population designated for monthly monitoring and data capture involves animals located in sites from six regions of Tanzania and Ethiopia, covering the major agro-ecological zones in those countries. Therefore, genomic predictions based on the ADGG data can be used to select genetically superior animals for use across their respective countries.

In India, the BAIF Development Research Foundation has set up an excellent smartphone-based herd recording system for use by farmers and specialized milk recorders [42]. The availability of high-quality data has resulted in GEBV with moderate accuracy (~0.45) for some breed/cross-breed groupings of Indian dairy cattle [42].

Alternative CBBP have been established specifically for indigenous breeds of sheep and goats in Latin America, Africa and Asia, primarily supported by national governments in conjunction with local organizations. The implementation of CBBP combines genetic improvement programs with infrastructure, community and market development. Examples of CBBP in local sheep and goat breeds across several LMICs are described by [43–48]. Guidelines for establishing CBBP focused on small ruminants are provided by [49].

There are currently no known resource populations for smallholder beef cattle, primarily due to a lack of the significant funding necessary for their establishment and the significant length of time required to achieve genetic improvement in those herds, which have an average generation interval of 4–6 years. An attempt was made to establish linkages with populations in South Africa through the government-funded "Beef Genomics Project" that services commercial seedstock herds [50]. However, even in those seedstock herds in South Africa, challenges remain when recording the more difficult or expensive-to-measure phenotypes [50]. However, the existence of that population may, in the future, provide opportunities for smallholder beef farmers across Africa and potentially elsewhere to link with it, to drive genetic improvement programs in their own regions. This opportunity is discussed further in subsequent sections of this paper.

Opportunities to maximize the accuracy of genomic selection using multi-breed reference populations and multi-omic data are provided by [51], while another report provides guidelines to minimize the loss of genetic diversity through the use of reference populations [52]. The issue of loss of genetic diversity is of critical concern, particularly as it relates to the indigenous livestock breeds of many LMICs.

While the existence of these resource populations is currently providing significant opportunities for smallholder dairy cattle, sheep, and goats in a small number of LMICs, the greatest challenge is their sustainability on a longer-term basis. In high-income countries, the existing resource populations are in the process of being migrated from research funding to a variety of co-investment models; this will ultimately result in a model that is funded by the beneficiaries of that genetic improvement. A similar transition will ultimately be required for the small number of existing resource populations in LMICs, but how and when that will be achieved is still not clear. Meanwhile, the vast majority of LMICs have no access to the resources needed to even establish suitable resource populations to target the very significant economic, social and environmental benefits derived from the genomic selection of livestock in advanced economies.

4. Data Analyses and Estimation of Genomic Breeding Values

The basic model of best linear unbiased prediction (BLUP) evaluations [53] is:

$$Y \sim \text{mean} + \text{contemporary group} + \text{fixed effects} + \text{animal} + e \quad (1)$$

where "animal" is a random effect ~ $N(0, A\,\sigma_A)$, A is a relationship derived from pedigree and σ_A is the additive genetic variance. The contemporary group is commonly also fitted as a random effect and e always is a random effect. Fixed effects depend on traits, for example, lactation number in dairy cattle or kill-day in beef-quality traits.

In a genomic evaluation, the second model is very similar, the only difference being $N(0, G\,\sigma_G)$, with G being the genomic relationship among animals constructed from the SNP genotypes [54]. The animal solutions from BLUP in a genomic model are usually referred to as GEBV and the model itself, GBLUP.

A third model for evaluations combines both information from animals with pedigrees and phenotypes, but no genotypes, and animals with pedigrees, phenotypes and genotypes, in a "single-step approach" [55]. In this approach, an H relationship describes the relationship among animals and replaces the A in the first model, i.e., animal ~$N(0, H\,\sigma_H)$. The H includes the elements of A for non-genotyped animals and elements of G combined with A for genotyped animals. This model has been implemented very successfully in several developed countries for dairy cattle, dairy goats, and pigs [56].

A major problem in the genetic analysis of data from smallholder systems is usually the lack of pedigree information. For instance, the genomic prediction in data from Tanzania [30] has been based on model two described above; this involved 1906 genotyped cows. However, only 226 cows of those cows had either both or only one parent known; this clearly underlines the importance of the availability of genotypic information in enabling prediction of the genetic merit in smallholder systems, as the pedigree relationships are clearly inadequately recorded.

However, under the ADGG program, the combined use of the genotypic and pedigree information that is increasingly becoming available provides hope for a better future. The use of the genomic matrix, derived from SNP information, to infer relationships among animals, and the application of model three, as described above, has enabled the estimation of genetic parameters and genomic prediction in smallholder systems [30,40].

In general, the methods currently used for genomic prediction in smallholder dairy systems include GBLUP, single-step procedures and various Bayesian methods (see [30] for a detailed review). However, most of the genomic prediction systems are based largely on females and small datasets, making it very difficult to adequately define separate reference and validation populations.

Consequently, most studies have used cross-validation approaches rather than forward validation [30]. However, some studies have applied forward validation or both validation approaches [30,57,58]. The validation accuracies are mostly of low to medium value (0.21 to 0.60) for milk yield, backfat thickness and rear eye area [58–60], but some high estimates (0.71 to 0.83) have been reported for bodyweight and other beef traits [61,62].

The complexities of recording accurate pedigrees in LMICs make implementing either the single-step model, or even the original pedigree-based model, rather unattractive and, as described above, makes the pure genomic model really attractive. For the routine production of GEBV, an alternative to the GBLUP model may be useful. A model can be fitted that estimates SNP effects directly. For example, BayesR [63] fits the model thus:

$$\text{Y}{\sim}\text{mean} + \text{contemporary group} + \text{fixed effects} + \text{Zg} + \text{e} \tag{2}$$

where g = vector of SNP effects, and $\mathrm{g} \sim N\left(0, \mathrm{I}\sigma_{\mathrm{i}}^2\right)$ with four possibilities for $\sigma_{\mathrm{i}}^2 = \left\{0,\ 0.0001 * \sigma_{\mathrm{g}}^2, 0.001{*}\sigma_{\mathrm{g}}^2, 0.01{*}\sigma_{\mathrm{g}}^2\right\}$, where σ_{g}^2 is the genetic variance of the trait. Each SNP is from one of four possible normal distributions: $N(0, 0 * \sigma_g^2)$, $N(0, 0.0001 * \sigma_g^2)$, $N(0, 0.001 * \sigma_g^2)$ and $N(0, 0.01 * \sigma_g^2)$. Four distributions are used so the marker effects can be moderate to large (e.g., in the case of DGAT1), small, very small or zero. Z is the animal x-marker genotype matrix.

It has been demonstrated that BayesR results in a higher accuracy of GEBV compared to GBLUP in multi-breed populations when high-density markers are used [64].

A major advantage of the BayesR approach for LMICs is that GEBV for new selection candidates can be run very quickly and with limited computing power. GEBV for these new candidates (i.e., young sires not in the reference population) can be calculated as GEBV = Zg_hat, which takes seconds or, at worst, minutes to compute on a laptop with a reasonable random-access memory (RAM). The g_hat, estimated from running the BayesR with Gibbs sampling, for example [63], can be run on a high-performance laptop with a much larger RAM, or a high-performance computer as large numbers of new reference animals become available, for example once or twice per year. The g_hat can then be passed onto the evaluation centers for rapid routine evaluations.

The reference populations to derive the g_hat can include data from multiple countries, as demonstrated by [65], in order to expand the reference population and, therefore, make GEBV more accurate.

In theory, the highest possible marker density should be used in genomic evaluations, particularly in multi-breed populations, as this allows the SNP with the highest linkage disequilibrium (LD) with the actual mutations to be used in the predictions, and this LD should persist against breeding. The ultimate solution would be to use whole-genome sequencing in the predictions, as this would allow the actual causative mutation to be used in the prediction equation, rather than to rely on LD with a random SNP. One problem is that it is still too expensive to sequence the whole genome of all animals in the reference set. An alternative is to impute the reference set from their low-density markers (e.g., 50 k) up to the whole-genome sequence using the 1000 bull genomes database [66].

The outcome from using these imputed genotypes would, however, be an enormous prediction equation—for example, 43 million SNP long! The practical alternative that has been adopted in industry is to use an SNP panel with a relatively small number of putative causal mutations identified from sequence data (in genome-wide association studies (GWAS), for example) plus the standard panel of high-density SNP (e.g., the bovine HD array). This is much more computationally tractable and, in many cases, gives better accuracy than full sequence data [67]. Furthermore, including genome annotation information to focus on those regions more likely to harbor causal mutations can increase the accuracy of genomic predictions using sequence data [68].

It has become increasingly clear that pooling data across regions and countries is beneficial for increasing the accuracy of genomic predictions. Hence, one critical consideration during the design phase of any breeding program is the need for consistent trait definitions across the countries planning to share data, to ensure that animals in multiple populations are recorded for the same trait(s). Alternatively, where resource populations are being developed, they need to be large enough to allow an estimation of genetic correlations with indicator traits, if consistent recording of the same trait(s) cannot be achieved across all populations. Regardless, estimating these genomic correlations and genotypes by environment interactions becomes more straightforward with genomic information, as what is required is observations of the traits/environments on common chromosome segments, rather than the sires' progeny [69,70].

5. Infrastructure and Human Capacity

Two problems of major significance to smallholder farmers in LMICS are: (i) the lack of infrastructure required to undertake the on-farm management and phenotyping of animals, laboratory testing of animal samples, data capture and storage, and lack of computing facilities, etc.; and (ii) lack of human capacity, particularly in areas of technological capability, data analysis and interpretation.

The issue of data capture and storage is starting to be addressed through the use of portable devices that do not require on-site internet connection (e.g., mobile phones, tablets). However, in the absence of research projects that can assist with infrastructure development, many of these issues remain as significant challenges to the implementation of livestock genetic improvement programs in these countries, since it is unclear how business models might develop for data-recording in most LMICs. One opportunity that is currently being explored in conjunction with ADGG is the possibility of developing a web interface that would enable data from livestock resource populations from countries and industries not currently serviced by ADGG to be uploaded to the ADGG platform, and then undertake genomic prediction using the pipeline developed by ADGG at the International Livestock Research Institute in Nairobi. If that opportunity could be achieved, that would mean other livestock industries and countries would not need to develop their own separate software or pipeline, thereby generating some efficiencies.

The recent launch of the African Animal Breeders' Network (AABN—http://animalbreeding-africa.org/, accessed 15 November 2021) is in direct response to the second issue relating to the lack of human capacity, with the aim of strengthening collaboration among academia, industry, farmers' organizations, the public sector, philanthropic organizations, and development agencies to drive the development and implementation of genetic improvement programs across the African continent. Professional development and capacity-building across all sectors of the livestock genetic improvement chain, from smallholder farmers to service providers and academics, are key pillars of AABN. Similar networks will be required in LMICs in other areas of the world, such as Asia, the Pacific, Central America and the Caribbean, to build capacity among livestock keepers and service providers in those areas.

6. The Value of National and International Collaborations

One key learning from the successes of genomic selection in the livestock industries in advanced economies is that strong and effective multi-organizational, multi-disciplinary, and, often, multi-national partnerships are key to their success. Such partnerships need to be inclusive of all sectors of the animal breeding chain, from farmers through to the service providers and researchers who provide decision-making recommendations to farmers and continue to improve the technologies being used in the processes. Such processes are likely to be even more critical for genetic improvement programs in LMICs, where generating sufficient animal data independently is unlikely to be feasible for decades to come. This need for effective partnerships is behind the establishment of the AABN, referred to in the previous section. Its usefulness can also be demonstrated using a recent example presented by [59], where genomic breeding values for a very difficult- and expensive-to-measure trait (cattle resistance to ticks) were successfully estimated in relatively small numbers of beef cattle in unrelated cattle breeds in South Africa (Nguni) and Australia (Tropical Composites comprising different admixtures of four breed types and at least six individual breeds) through the use of larger phenotyped populations of Angus, Hereford, Braford and Brangus cattle in Brazil where, in effect, the Brazilian herds became effective reference populations for cattle in South Africa and Australia. This suggests that a very viable solution for genetic improvement programs in LMICs would be to formally link resource populations and genetic evaluations in LMICs with livestock-breeding programs in more advanced economies, to enable the effective implementation of genomic selection across all countries. This type of collaboration may have the added benefit of perhaps, partially, overcoming the lack of laboratory infrastructure that is a common constraint in LMICs.

7. The Ability of Genomic Information to Mitigate These Challenges

As outlined in earlier sections of this paper, the availability of genomic information is now providing exciting new opportunities to identify genetically superior animals in smallholder herds and flocks in LMICs and, based on well-documented evidence from advanced economies, to simultaneously deliver very significant economic, social and environmental benefits to those smallholder farmers and the communities and countries where they live.

The major benefit of genomic selection derives from the ability of genomic information to replace the need for pedigree recording and, specifically, generating genetic linkages within and across herds and flocks that record the same phenotypes. Another important benefit from the use of genomic information is that fewer animals are required using genomic selection approaches to achieve accurate GEBVs relative to traditional genetic improvement programs because the chromosome segments that are shared among the breeds now provide genomic linkages across the different populations. This will be particularly important if data from smaller reference populations in LMICs can in the future be combined for analysis with data from larger reference populations in more advanced economies, as occurred in the example given by [59]. The use of genomic information also enhances decision-making in crossbreeding programs by providing accurate information on the breed composition of individual animals and, in doing so, also provides a mechanism for identifying indigenous breeds that require conservation.

In the future, there is good potential for genomic information to replace an animal's phenotype, not only through the identification of causal mutations and regions of the genome impacting on particular traits but also through the use of new "-omics" technologies, such as functional genomics, gene expression, transcriptomics, proteomics and metabolomics. This will most likely occur initially for difficult- or expensive-to-measure traits with very high economic impacts, with these new technologies delivering simpler and more cost-effective diagnostic tests for both animal management and genetic improvement purposes. In that scenario, instead of data being primarily recorded for management purposes, it may in the future be more useful to imagine data being collected specifically for genetic improvement in nucleus farms (either centrally or distributed). As such, these "phenotype farms" would have a requirement to generate genomically improved genetic material for distribution to smallholder farmers in LMICs.

8. Future Opportunities

In addition to the new or adapted uses of genomic information described above, several new opportunities will become available over the coming years to assist smallholder farmers to capture some of the well-documented and very significant economic, social and environmental benefits of genomic selection that are already achieved by livestock farmers in advanced economies. These opportunities include the increasing use and availability of digital and possibly automated data capture through, for example, spatial technologies such as high-resolution satellite imaging and unmanned aerial vehicles (drones), or by using solar-powered sensor networks to remotely capture livestock data, such as live weights, estrus or pregnancy status using animal ear- or neck tags. These technologies will allow the real-time tracking of animals and animal products, providing new phenotypes for genetic improvement programs as well as improving efficiencies and data collection across the entire supply chain. The opportunity to effectively capture and analyze "big data", including publicly available information such as geographical location and meteorological information, will also allow new levels of insight and development of decision support tools, such as apps for the use of farmers in both advanced economies and LMICs.

Potentially, the greatest opportunity for smallholder farmers to capture the benefits of genomic selection over the coming years will, however, be through expansion of the very small number of existing livestock resource populations and the development of new populations in other LMICs and other livestock industries not currently serviced by the existing genetic improvement platforms in those regions. Linking those existing and new resource populations through collaborations with livestock populations in advanced economies, as outlined in Section 6, will also generate strong benefits for LMICs.

An operational framework to establish new resource populations could be along the lines of the following:

- Resource populations could be formed at relevant new regional (e.g., national or possibly even multi-national) levels within livestock species, with those populations being managed overall with input from the smallholder farmers contributing the animals, but with the responsibility for technical areas (phenotyping, genotyping, data upload, etc.) being the remit of technicians with appropriate training;
- Initial funding would be required to cover the costs of designing the populations (to ensure local relevance) and establishing them, for phenotyping the animals for the full range of economically important traits for each of the species, and for genotyping them, although only selected animals would require the use of higher-density (and thus more expensive) SNP panels;
- Designing the resource populations should be undertaken in direct collaboration with established resource populations in other countries with similar environmental systems, to ensure compatibility of the populations for the future pooling of data for genetic evaluation purposes—in LMICs, this generally means collaborations with other resource populations that operate in tropical or sub-tropical environments;
- However, due to the assumed use of genomic information, the design of the new populations does not need to specifically generate genetic linkages across the different populations, nor is there a need to restrict the design to animals of the same breed, as demonstrated by [65];
- Capture of data from the new resource populations would be achieved electronically in the field using mobile devices and subsequently uploaded to the data platform when internet coverage is available;
- Assuming that the opportunity described in Section 5 can be realized, a web interface would be built to enable the ADGG portal to capture the data from the new resource populations, thereby avoiding the need for the new populations to develop separate software and pipelines. That portal would also provide appropriate analytical models to enable multiple-country genetic evaluations within species, with data ownership continuing to be retained by the farmers who own the livestock being evaluated. It would also become the permanent repository for data collected through those new populations, beyond the life of any research projects that initially fund the data collection, thereby overcoming another of the challenges raised earlier in this manuscript;
- Many of the countries lacking existing resource populations do have trained animal geneticists with an interest in having greater involvement in data analysis. Ongoing capacity-building of those and other interested people through the AABN (also described in Section 5) would ensure that the analytical models are directly relevant to the countries or regions where they are employed;

- The animal geneticists identified to manage the new resource populations and to undertake the genetic evaluations would be encouraged to directly collaborate with researchers in other countries, to undertake cross-country genetic evaluations that generate value for farmers across all collaborating countries;
- In countries where artificial breeding centers do not exist, such centers would be established either by governments or the private sector to collect germplasm from animals proven to be genetically superior, with the germplasm (most likely semen) being made available to smallholder farmers, thereby enabling genetic improvement of their herds/flocks;
- Over time, the beneficiaries of the genetic improvement program would be expected to contribute to the ongoing costs of maintaining the genetic improvement programs, as is now occurring in Australia to sustain the earlier research-funded populations (e.g., livestock producers paying to have their own sires evaluated through the resource populations, by contributions to the costs of the resource populations or even, in some cases, establishing populations in their own herds or flocks);
- Establishing new resource populations using these guidelines would enable smallholder farmers to directly capture the benefits of genetic improvement through the use of genetically superior breeding animals but without the need to understand the complexities or overcome the major challenges of new technologies (e.g., hardware incompatibility; complexity; language barriers; lack of electricity, computers, internet access, etc.) that have proven to be major barriers to adoption in LMICs, as described by [71].

However, the expansion of existing resource populations and the development of new populations is entirely dependent on the availability of new funding for this purpose, and where that funding will come from is not at all clear. A recent presentation [72] comparing the public acceptance of biotechnologies, such as genetic engineering and gene editing with genomic selection, highlighted this major difficulty, indicating: "*There are glaring disparities when it comes to the implementation of genomic selection in the developing world . . . it is expensive to develop large populations of genotyped, phenotyped animals. It is not a scale-neutral technology, advantaging large breeds and genetic providers over small ones. Such inequality concerns would derail a genetic engineering application, yet these concerns are rarely even discussed as it relates to genomic selection . . .* ".

Therefore, perhaps the greatest opportunity to secure the proven and very significant economic, social and environmental benefits of genomic selection for smallholder farmers in LMICs is to attempt to engage a range of government, non-government and philanthropic organizations to give priority to improving the rates of genetic gain in livestock farmed by smallholders in those countries.

Author Contributions: Conceptualization and original draft preparation, H.M.B.; writing, review and editing, R.M., A.O.M., M.P.C. and B.J.H.; All authors have read and agreed to the published version of the manuscript.

Funding: This research received no external funding.

Institutional Review Board Statement: Not applicable.

Informed Consent Statement: Not applicable.

Data Availability Statement: Not applicable.

Conflicts of Interest: The authors declare no conflict of interest.

References

1. Gerland, P.E.; Raftery, A.D.; Ševčíková, H.; Li, N.; Gu, D.; Spoorenberg, T.; Alkema, L.; Fosdick, B.K.; Chunn, J.; Lalic, N.; et al. World population stabilization unlikely this century. *Science* **2014**, *346*, 234–237. [CrossRef]
2. Delgado, C.; Rosegrant, M.; Steinfeld, H.; Ehui, S.; Courbois, C. *Livestock to 2020: The Next Food Revolution*; Food, Agriculture, and the Environment Discussion Paper 28; International Food Policy Research Institute: Washington, DC, USA, 1999. Available online: http://www.fao.org/ag/againfo/resources/documents/lvst2020/20201.pdf (accessed on 15 November 2021).

3. Mullen, J. *Productivity Growth in Australian Agriculture: Trends, Sources, Performance*; Australian Farm Institute: Sydney, Australia, 2012; pp. 5–24.
4. Hughes, L. Climate change and Australia: Trends, projections and impacts. *Aust. Ecol.* **2003**, *28*, 423–443. [CrossRef]
5. Chang, J.; Peng, S.; Yin, Y.; Ciais, P.; Havlik, P.; Herrero, M. The key role of production efficiency changes in livestock methane emission mitigation. *AGU Adv.* **2021**, *2*, e2021AV000391. [CrossRef]
6. Capper, J.L.; Cady, R.A.; Bauman, D.E. The environmental impact of dairy production 1944 compared with 2007. *J. Anim. Sci.* **2009**, *87*, 2160–2167. [CrossRef] [PubMed]
7. Meuwissen, T.; Hayes, B.J.; Goddard, M.E. Prediction of total genetic value using genome-wide dense marker maps. *Genetics* **2001**, *157*, 1819–1829. [CrossRef] [PubMed]
8. Goddard, M.E.; Hayes, B.J. Mapping genes for complex traits in domestic animals and their use in breeding programmes. *Nat. Rev. Genet.* **2009**, *10*, 381–391. [CrossRef]
9. National Human Genome Research Institute. 2021. Available online: https://www.genome.gov/dna-day/15-ways/dna-sequencing (accessed on 16 August 2021).
10. Votintseva, A.A.; Bradley, P.; Pankhurst, L.; del Ojo, E.C.; Loose, M.; Nilgiriwala, K.; Chatterjee, A.; Smith, E.G.; Sanderson, N.; Walker, T.M.; et al. Same-day diagnostic and surveillance data for tuberculosis via whole-genome sequencing of direct respiratory samples. *J. Clin. Microbiol.* **2017**, *55*, 1285–1296. [CrossRef] [PubMed]
11. Hayes, B.; Goddard, M. Artificial Selection Methods and Reagents. Patent PCT/AU2007/002006, 2007.
12. Meuwissen, T.; Goddard, M.E. Accurate prediction of genetic values for complex traits by whole-genome resequencing. *Genetics* **2010**, *185*, 623–631. [CrossRef]
13. Meuwissen, T.; Goddard, M.E. The use of family relationships and linkage disequilibrium to impute phase and missing genotypes in up to whole-genome sequence density genotypic data. *Genetics* **2010**, *185*, 1441–1449. [CrossRef]
14. Meuwissen, T.; Hayes, B.J.; Goddard, M.E. Genomic selection: A paradigm shift in animal breeding. *Anim. Front.* **2016**, *6*, 6–14. [CrossRef]
15. Heard, E.; Tishkoff, S.; Todd, J.A.; Vidal, M.; Wagner, G.P.; Wang, J.; Weigel, D.; Young, R. Ten years of genetics and genomics: What have we achieved and where are we heading? *Nat. Rev. Genet.* **2010**, *11*, 723–733. [CrossRef]
16. Yang, J.; Beben, B.; McEvoy, B.P.; Gordon, S.; Henders, A.K.; Nyholt, D.R.; Madden, P.A.; Heath, A.C.; Martin, N.G.; Montgomery, G.W.; et al. Common SNPs explain a large proportion of the heritability for human height. *Nat. Genet.* **2010**, *42*, 565–569. [CrossRef]
17. Schadt, E.E.; Linderman, M.D.; Sorenson, J.; Lee, L.; Nolan, G.P. Computational solutions to large-scale data management and analysis. *Nat. Rev. Genet.* **2010**, *11*, 647–657. [CrossRef]
18. Craig, J.E.; Hewitt, A.W.; McMellon, A.E.; Henders, A.K.; Ma, L.; Wallace, L.; Sharma, S.; Burdon, K.P.; Visscher, P.M.; Montgomery, G.W.; et al. Rapid Inexpensive Genome-Wide Association Using Pooled Whole Blood. *Genome Res.* **2009**, *19*, 2075–2080. Available online: http://www.genome.org/cgi/doi/10.1101/gr.094680.109 (accessed on 15 November 2021). [CrossRef] [PubMed]
19. Hickey, J.M.; Whitelaw, B.C.; Gorjanc, G.J. Promotion of alleles by genome editing in livestock breeding programmes. *Anim. Breed. Genet.* **2016**, *133*, 83–84. [CrossRef] [PubMed]
20. Burrow, H.M.; Henshall, J.M. Relationships between Adaptive and Productive Traits in Cattle, Goats and Sheep in Tropical Environments. In Proceedings of the 10th World Congress on Genetics Applied to Livestock Production, Vancouver, BC, Canada; 2014. Available online: https://asas.org/wcgalp-proceedings/genetic-improvement-programs-selection-for-harsh-environments-and-management-of-animal-genetic-resources (accessed on 15 November 2021).
21. Mrode, R.; Dzivenu, C.E.; Marshall, K.; Chagunda, M.G.G.; Muasa, B.S.; Ojango, J.; Okeyo, A.M. Phenomics and its potential impact on livestock development in low-income countries: Innovative applications of emerging related digital technology. *Anim. Front.* **2020**, *10*, 6–11. [CrossRef] [PubMed]
22. Brown, D.J.; Savage, D.B.; Hinch, G.N. Repeatability and frequency of in-paddock sheep walk-over weights: Implications for individual animal management. *Anim. Prod. Sci.* **2013**, *54*, 207–213. [CrossRef]
23. Goopy, J.P.; Pelster, D.E.; Onyango, A.; Marshall, K.; Lukuyu, M. Simple and robust algorithms to estimate liveweight in African smallholder cattle. *Anim. Prod. Sci.* **2017**, *58*, 1758–1765. [CrossRef]
24. Burrow, H.M. Genetic aspects of cattle adaptation in the tropics. In *The Genetics of Cattle*, 2nd ed.; Garrick, D.J., Ruvinsky, A., Eds.; CABI: Wallingford, UK, 2014; pp. 571–597.
25. Turner, H.G. Genetic variation of rectal temperature in cows and its relationship to fertility. *Anim. Prod.* **1982**, *35*, 401–412. [CrossRef]
26. Turner, H.G. Variation in rectal temperature in a tropical environment and its relation to growth rate. *Anim. Prod.* **1984**, *38*, 417–427. [CrossRef]
27. Turner, H.G.; Schleger, A.V. The significance of coat type in cattle. *Aust. J. Agric. Res.* **1960**, *11*, 645–663. [CrossRef]
28. Hayes, B.J.; Carrick, M.; Bowman, P.; Goddard, M.E. Genotype x environment interaction for milk production of daughters of Australian dairy sires from test-day records. *J. Dairy Sci.* **2003**, *86*, 3736–3744. [CrossRef]
29. Ekine-Dzivenu, C.C.; Mrode, R.; Oyieng, E.; Komwihangilo, D.; Lyatuu, E.; Msuta, G.; Ojango, J.M.K.; Okeyo, A.M. Evaluating the impact of heat stress as measured by temperature-humidity index (THI) on test-day milk yield of small holder dairy cattle in a sub-Sahara African climate. *Livest. Sci.* **2020**, *242*, 104314. [CrossRef]

30. Mrode, R.; Ojango, J.M.K.; Okeyo, A.M.; Mwacharo, J.M. Genomic selection and use of molecular tools in breeding programs for indigenous and crossbred cattle in developing countries: Current status and future prospects. *Front. Genet.* **2019**, *9*, 1–13. [CrossRef] [PubMed]
31. Upton, W.; Burrow, H.M.; Dundon, A.; Robinson, D.L.; Farrell, E. CRC breeding program design, measurements and database: Methods that underpin CRC research results. *Aust. J. Exp. Agric.* **2001**, *41*, 943–952. [CrossRef]
32. Burrow, H.M.; Johnston, D.J.; Barwick, S.A.; Holroyd, R.G.; Barendse, W.; Thompson, J.M.; Griffith, G.R.; Sullivan, M. Relationships between Carcass and Beef Quality and Components of Herd Profitability in Northern Australia. *Proc. Ass. Advance. Anim. Breed. Genet.* **2003**, *15*, 359–362. Available online: http://livestocklibrary.com.au/handle/1234/5349 (accessed on 15 November 2021).
33. Johnston, D.J.; Tier, B.; Graser, H.-U. Beef cattle breeding in Australia with genomics: Opportunities and needs. *Anim. Prod. Sci.* **2012**, *52*, 100–106. [CrossRef]
34. Johnston, D.J.; Grant, T.P.; Schatz, T.J.; Burns, B.M.; Fordyce, G.; Lyons, R.E. The *Repronomics* Project–Enabling Genetic Improvement in Reproduction in Northern Australia. *Proc. Assoc. Advance. Anim. Breed. Genet.* **2017**, *22*, 385–388. Available online: http://www.aaabg.org/aaabghome/AAABG22papers/87Johnston22385.pdf (accessed on 15 November 2021).
35. Hayes, B.J.; Fordyce, G.; Landmark, S. Genomic Predictions for Fertility Traits in Tropical Beef Cattle from a Multi-Breed, Crossbred and Composite Reference Population. *Proc. Assoc. Advance. Anim. Breed. Genet.* **2019**, *23*, 283–285. Available online: http://www.aaabg.org/aaabghome/AAABG23papers/69Hayes23282.pdf (accessed on 15 November 2021).
36. Van Eenennaam, A.; Neibergs, H.; Seabury, C.; Taylor, J.; Wang, Z.; Scraggs, E.; Schnabel, R.D.; Decker, J.; Wojtowicz, A.; Aly, S.; et al. Results of the BRD CAP project progress towards identifying genetic markers associated with BRD susceptibility. *Anim. Health Res. Rev.* **2014**, *15*, 157–160. [CrossRef] [PubMed]
37. Van der Werf, J.H.J.; Kinghorn, B.P.; Banks, R.G. Design and role of an information nucleus in sheep breeding programs. *Anim. Prod. Sci.* **2010**, *50*, 998–1003. [CrossRef]
38. Swan, A.A.; Johnston, D.J.; Brown, D.J.; Tier, B.; Graser, H.-U. Integration of genomic information into beef cattle and sheep genetic evaluations in Australia. *Anim. Prod. Sci.* **2011**, *52*, 126–132. [CrossRef]
39. Veerkamp, R.F.; De Haas, Y.; Pryce, J.E.; Coffey, M.P.; Spurlock, D.M.; Vandehaar, M.J. *Guidelines to Measure Individual Feed Intake of Dairy Cows for Genomic and Genetic Evaluations*; ICAR Technical Series 19; International Committee for Animal Recording: Rome, Italy, 2015; p. 191. Available online: https://www.google.com/search?client=firefox-b-e&q=Guidelines+to+measure+individual+feed+intake+of+dairy+cows+for+genomic+and+genetic+evaluations (accessed on 15 November 2021).
40. Ojango, J.M.K.; Mrode, R.; Rege, J.E.O.; Mujibi, D.; Strucken, E.M.; Gibson, J.; Mwai, O. Genetic evaluation of test-day milk yields from smallholder dairy production systems in Kenya using genomic relationships. *J. Dairy Sci.* **2019**, *102*, 5266–5278. [CrossRef]
41. Marshall, K.; Gibson, J.P.; Mwai, O.; Mwacharo, J.M.; Hailem, A.; Getachew, T.; Mrode, R.; Kemp, S.J. Livestock genomics for developing countries–African examples in practice. *Front. Genet.* **2019**, *10*, 297. [CrossRef]
42. Al Kalaldeh, M.; Swaminathan, M.; Gaundare, Y.; Joshi, J.; Aliloo, H.; Strucken, E.M.; Ducrocq, V.; Gibson, J.P. Genomic evaluation of milk yield in a smallholder crossbred dairy production system in India. *Genet. Sel. Evol.* **2021**, *53*, 73. [CrossRef]
43. Wurzinger, M.; Sőlkner, J.; Iñiguez, L. Important aspects and limitations in considering community-based breeding programs for low-input smallholder livestock systems. *Small Rum. Res.* **2011**, *98*, 170–175. [CrossRef]
44. Mueller, J.P.; Rischkowsky, B.; Haile, A.; Philipsson, J.; Mwai, O.; Besbes, B.; Valle Zárate, A.; Tibbo, M.; Mirkena, T.; Duguma, G.; et al. Community-based livestock breeding programmes: Essentials and examples. *J. Anim. Breed. Genet.* **2015**, *132*, 155–168. [CrossRef]
45. Bhuiyan, M.S.A.; Bhuiyan, A.K.F.H.; Lee, J.H.; Lee, S.H. Community based livestock breeding programs in Bangladesh: Present status and challenges. *J. Anim. Breed. Genom.* **2017**, *1*, 77–84. [CrossRef]
46. Karnauh, A.B.; Dunga, G.; Rewe, T. Community based breeding program for improve goat production in Liberia. *MOJ Curr. Res. Rev.* **2018**, *1*, 216–221. [CrossRef]
47. Haile, A.; Gizaw, S.; Getachew, T.; Mueller, J.P.; Amer, P.; Rekik, M.; Rischkowsky, B. Community-based breeding programmes are a viable solution for Ethiopian small ruminant genetic improvement but require public and private investments. *J. Anim. Breed. Genet.* **2019**, *136*, 319–328. [CrossRef]
48. Haile, A.; Getachew, T.; Mirkena, T.; Duguma, G.; Gizaw, S.; Wurzinger, M.; Soelkner, J.; Mwai, O.; Dessie, T.; Abebe, A.; et al. Community-based sheep breeding programs generated substantial genetic gains and socioeconomic benefits. *Animal* **2020**, *14*, 1362–1370. [CrossRef]
49. Haile, A.; Wurzinger, M.; Mueller, J.; Mirkena, T.; Duguma, G.; Rekik, M.; Mwacharo, J.M.; Okeyo Mwai, A.; Sölkner, J.; Rischkowsky, B.A. *Guidelines for Setting Up Community-Based Small Ruminants Breeding Programs*; ICARDA: Beirut, Lebanon, 2018; 32p. Available online: https://hdl.handle.net/10568/103248 (accessed on 15 November 2021).
50. Marle-Köster, E.; Visser, C. Genetic improvement in South African livestock: Can genomics bridge the gap between the developed and developing sectors? *Front. Genet.* **2018**. [CrossRef]
51. Goddard, M.E.; MacLeod, I.M.; Kemper, K.E.; Xiang, R.; van den Berg, I.; Khansefid, M.; Daetwyler, H.D.; Hayes, B.J. The Use of Multi-Breed Reference Populations and Multi-Omic Data to Maximize Accuracy of Genomic Prediction. In Proceedings of the World Congress on Genetics Applied to Livestock Production, Auckland, New Zealand, 10–15 February 2018. Available online: http://www.wcgalp.org/proceedings/2018/use-multi-breed-reference-populations-and-multi-omic-data-maximize-accuracy-genomic (accessed on 15 November 2021).

52. Eynard, S.E.; Croiseau, P.; Laloë, D.; Fritz, S.; Calus, M.P.L.; Restoux, G. Which individuals to choose to update the reference population? Minimizing the loss of genetic diversity in animal genomic selection programs. *Genes Genomes Genet.* **2018**, *8*, 113–121. [CrossRef]
53. Henderson, C.R. Best Linear Unbiased Estimation and Prediction under a selection model. *Biometrics* **1975**, *31*, 423–447. [CrossRef] [PubMed]
54. VanRaden, P.M. Efficient methods to compute genomic predictions. *J. Dairy Sci.* **2008**, *91*, 4414–4423. [CrossRef]
55. Aguilar, I.; Misztal, I.; Johnson, D.L.; Legarra, A.; Tsuruta, S.; Lawlor, T.J. Hot topic: A unified approach to utilize phenotypic, full pedigree, and genomic information for genetic evaluation of Holstein final score. *J. Dairy Sci.* **2010**, *93*, 743–752. [CrossRef] [PubMed]
56. Fragomeni, B.O.; Lourenco, D.A.; Tsuruta, S.; Bradford, H.L.; Gray, K.A.; Huang, Y.; Misztal, I. Using single-step genomic best linear unbiased predictor to enhance the mitigation of seasonal losses due to heat stress in pigs. *J. Anim. Sci.* **2016**, *94*, 5004–5013. [CrossRef] [PubMed]
57. Silva, R.M.O.; Fragomeni, B.O.; Lourenco, D.A.L.; Magalhães, A.F.B.; Irano, N.; Carvalheiro, R.; Canesin, R.C.; Mercadante, M.E.Z.; Boligon, A.A.; Baldi, F.S.; et al. Accuracies of genomic prediction of feed efficiency traits using different prediction and validation methods in an experimental Nelore cattle population. *J. Anim. Sci.* **2016**, *94*, 3613–3623. [CrossRef]
58. Boison, S.A.; Utsunomiya, A.T.H.; Santos, D.J.A.; Neves, H.H.R.; Carvalheiro, R.; Mészáros, G.; Utsunomiya, Y.T.; do Carmo, A.S.; Verneque, R.S.; Machado, M.A.; et al. Accuracy of genomic predictions in Gyr (*Bos indicus*) dairy cattle. *J. Dairy Sci.* **2017**, *100*, 5479–5490. [CrossRef]
59. Fernandes Júnior, G.A.; Rosa, G.J.M.; Valente, B.D.; Carvalheiro, R.; Baldi, F.; Garcia, D.A.; Gordo, D.G.; Espigolan, R.; Takada, L.; Tonussi, R.L.; et al. Genomic prediction of breeding values for carcass traits in Nellore cattle. *Genet. Sel. Evol.* **2016**, *48*, 7. [CrossRef]
60. Brown, A.; Ojango, J.; Gibson, J.; Coffey, M.; Okeyo, M.; Mrode, R. Genomic selection in a crossbred cattle population using data from the Dairy Genetics East Africa project. *J. Dairy Sci.* **2016**, *99*, 7308–7312. [CrossRef] [PubMed]
61. Neves, H.H.; Carvalheiro, R.; Brien, R.A.M.O.; Utsunomiya, Y.T.; do Carmo, A.S.; Schenkel, F.S.; Sölkner, J.; McEwan, J.C.; Van Tassell, C.P.; Cole, J.B.; et al. Accuracy of genomic predictions in Bos indicus (Nellore) cattle. *Genet. Sel. Evol.* **2014**, *46*, 17. [CrossRef] [PubMed]
62. Mrode, R.; Ojango, J.; Ekine-Dzivenu, C.; Aliloo, A.; Gibson, J.; Okeyo, M.A. Genomic prediction of crossbred dairy cattle in Tanzania: A route to productivity gains in smallholder dairy systems. *J. Dairy Sci.* **2021**, in press. [CrossRef]
63. Erbe, M.; Hayes, B.J.; Matukumalli, L.K.; Goswami, S.; Bowman, P.J.; Reich, C.M.; Mason, B.A.; Goddard, M.E. Improving accuracy of genomic predictions within and between dairy cattle breeds with imputed high-density single nucleotide polymorphism panels. *J. Dairy Sci.* **2012**, *95*, 4114–4129. [CrossRef]
64. Hayes, B.J.; Corbet, N.J.; Allen, J.M.; Laing, A.R.; Fordyce, G.; Lyons, R.; McGowan, M.R.; Burns, B.M. Towards multi-breed genomic evaluations for female fertility of tropical beef cattle. *J. Anim. Sci.* **2019**, *97*, 55–62. [CrossRef] [PubMed]
65. Cardoso, F.F.; Matika, O.; Djikeng, A.; Mapholi, N.; Burrow, H.M.; Yokoo, M.J.I.; Campos, G.S.; Gulias-Gomes, C.C.; Riggio, V.; Pong-Wong, R.; et al. Multiple country and breed genomic prediction of tick resistance in cattle. *Front. Immunol.* **2021**, *12*, 2189. [CrossRef]
66. Hayes, B.J.; Daetwyler, H.D. 1000 Bull Genomes Project to map simple and complex genetic traits in cattle: Applications and outcomes. *Ann. Rev. Anim. Biosci.* **2019**, *7*, 89–102. [CrossRef]
67. Van den Berg, I.; Boichard, D.; Lund, M.S. Sequence variants selected from a multi-breed GWAS can improve the reliability of genomic predictions in dairy cattle. *Genet. Sel. Evol.* **2016**, *48*, 83. [CrossRef]
68. Xiang, R.; MacLeod, I.M.; Daetwyler, H.D.; de Jong, G.; O'Connor, E.; Schrooten, C.; Chamberlain, A.J.; Goddard, M.E. Genome-wide fine-mapping identifies pleiotropic and functional variants that predict many traits across global cattle populations. *Nat. Commun.* **2021**, *12*, 860. [CrossRef]
69. Visscher, P.M.; Hemani, G.; Vinkhuyzen, A.A.; Chen, G.B.; Lee, S.H.; Wray, N.R.; Goddard, M.E.; Yang, J. Statistical power to detect genetic (co) variance of complex traits using SNP data in unrelated samples. *PLoS Genet.* **2014**, *10*, e1004269. [CrossRef]
70. Hayes, B.J.; Daetwyler, H.D.; Goddard, M.E. Models for Genome x Environment interaction: Examples in livestock. *Crop Sci.* **2016**, *56*, 2251. [CrossRef]
71. Richardson, J.W. Challenges of adopting the use of technology in less developed countries: The case of Cambodia. *Comp. Educ. Rev.* **2011**, *55*, 8–29. [CrossRef]
72. Mueller, M.L.; Van Eenennaam, A. Synergistic power of genomic selection, assisted reproductive technologies and gene editing to drive genetic improvement of cattle. *CABI Agric. Biosci.* (submitted)

MDPI
St. Alban-Anlage 66
4052 Basel
Switzerland
Tel. +41 61 683 77 34
Fax +41 61 302 89 18
www.mdpi.com

Agriculture Editorial Office
E-mail: agriculture@mdpi.com
www.mdpi.com/journal/agriculture

www.ingramcontent.com/pod-product-compliance
Lightning Source LLC
LaVergne TN
LVHW071245170726
843515LV00006BA/1331
9783036571447